Vibrational Spectroscopy for Tissue Analysis

Series in Medical Physics and Biomedical Engineering

Series Editors: John G Webster, Alisa Walz-Flannigan, Slavik Tabakov, and Kwan-Hoong Ng

Other recent books in the series:

Webb's Physics of Medical Imaging, Second Edition
M A Flower (Ed)

Correction Techniques in Emission Tomography
Mohammad Dawood, Xiaoyi Jiang, and Klaus Schäfers (Eds)

Physiology, Biophysics, and Biomedical Engineering
Andrew Wood (Ed)

Proton Therapy Physics
Harald Paganetti (Ed)

Correction Techniques in Emission Tomography
Mohammad Dawood, Xiaoyi Jiang, Klaus Schäfers (Eds)

Practical Biomedical Signal Analysis Using MATLAB®
K J Blinowska and J Żygierewicz (Ed)

Physics for Diagnostic Radiology, Third Edition
P P Dendy and B Heaton (Eds)

Nuclear Medicine Physics
J J Pedroso de Lima (Ed)

Handbook of Photonics for Biomedical Science
Valery V Tuchin (Ed)

Handbook of Anatomical Models for Radiation Dosimetry
Xie George Xu and Keith F Eckerman (Eds)

Fundamentals of MRI: An Interactive Learning Approach
Elizabeth Berry and Andrew J Bulpitt

Handbook of Optical Sensing of Glucose in Biological Fluids and Tissues
Valery V Tuchin (Ed)

Intelligent and Adaptive Systems in Medicine
Oliver C L Haas and Keith J Burnham

An Introduction to Radiation Protection in Medicine
Jamie V Trapp and Tomas Kron (Eds)

A Practical Approach to Medical Image Processing
Elizabeth Berry

Series in Medical Physics and Biomedical Engineering

Vibrational Spectroscopy for Tissue Analysis

Ihtesham ur Rehman
The Kroto Research Institute
University of Sheffield
UK

Zanyar Movasaghi
Queen Mary University of London
UK

Shazza Rehman
Airedale N H S Trust
West Yorkshire, UK

CRC Press
Taylor & Francis Group
Boca Raton London New York

CRC Press is an imprint of the
Taylor & Francis Group, an **informa** business

A TAYLOR & FRANCIS BOOK

CRC Press
Taylor & Francis Group
6000 Broken Sound Parkway NW, Suite 300
Boca Raton, FL 33487-2742

First issued in paperback 2019

© 2013 by Taylor & Francis Group, LLC
CRC Press is an imprint of Taylor & Francis Group, an Informa business

No claim to original U.S. Government works

ISBN-13: 978-1-4398-3608-8 (hbk)
ISBN-13: 978-0-367-86516-0 (pbk)

Library of Congress Cataloging-in-Publication Data

Rehman, Ihtesham ur.
    Vibrational spectroscopy for tissue analysis / Ihtesham ur Rehman, Zanyar Movasaghi, Shazza Rehman.
       p. ; cm. -- (Series in medical physics and biomedical engineering ; 25)
    Includes bibliographical references and index.
    ISBN 978-1-4398-3608-8 (hardback : alk. paper)
    I. Movasaghi, Zanyar. II. Rehman, Shazza. III. Title. IV. Series: Series in medical physics and biomedical engineering.
    [DNLM: 1. Spectroscopy, Fourier Transform Infrared. 2. Spectrum Analysis, Raman. 3. Diagnostic Imaging--methods. 4. Neoplasms--diagnosis. QC 454.R36]

616.07'57--dc23   2012014642

Visit the Taylor & Francis Web site at
http://www.taylorandfrancis.com

and the CRC Press Web site at
http://www.crcpress.com

Series Preface

The Series in Medical Physics and Biomedical Engineering describes the applications of physical sciences, engineering, and mathematics in medicine and clinical research.

The series seeks (but is not restricted to) publications in the following topics:

- Artificial organs
- Assistive technology
- Bioinformatics
- Bioinstrumentation
- Biomaterials
- Biomechanics
- Biomedical engineering
- Clinical engineering
- Imaging
- Implants
- Medical computing and mathematics
- Medical/surgical devices
- Patient monitoring
- Physiological measurement
- Prosthetics
- Radiation protection, health physics, and dosimetry
- Regulatory issues
- Rehabilitation engineering
- Sports medicine
- Systems physiology
- Telemedicine
- Tissue engineering
- Treatment

The Series in Medical Physics and Biomedical Engineering is an international series that meets the need for up-to-date texts in this rapidly developing field. Books in the series range in level from introductory graduate textbooks and practical handbooks to more advanced expositions of current research. This series is the official book series of the International Organization for Medical Physics.

The International Organization for Medical Physics

The International Organization for Medical Physics (IOMP), founded in 1963, is a scientific, educational, and professional organisation of 76 national adhering organisations, more than 16,500 individual members, several corporate members, and 4 international regional organisations.

IOMP is administered by a council, which includes delegates from each of the adhering national organisations. Regular meetings of the council are held electronically as well as every three years at the World Congress on Medical Physics and Biomedical Engineering. The president and other officers form the executive committee, and there are also committees covering the main areas of activity, including education and training, scientific, professional relations, and publications.

Objectives

- To contribute to the advancement of medical physics in all its aspects
- To organise international cooperation in medical physics, especially in developing countries
- To encourage and advise on the formation of national organisations of medical physics in those countries which lack such organisations

Activities

The official journals of the IOMP are *Physics in Medicine and Biology* and *Medical Physics and Physiological Measurement*. The IOMP publishes a bulletin, *Medical Physics World*, twice a year, which is distributed to all members.

A World Congress on Medical Physics and Biomedical Engineering is held every three years in cooperation with the International Federation for Medical and Biological Engineering (IFMBE) through the International Union for Physics and Engineering Sciences in Medicine (IUPESM). A regionally-based international conference on medical physics is held between world congresses. IOMP also sponsors international conferences, workshops, and courses. IOMP representatives contribute to various international committees and working groups.

The IOMP has several programmes to assist medical physicists in developing countries. The joint IOMP Library Programme supports 69 active libraries in 42 developing countries, and the Used Equipment Programme coordinates equipment donations. The Travel Assistance Programme provides a limited number of grants to enable physicists to attend the world congresses. The IOMP website is being developed to include a scientific database of international standards in medical physics and a virtual education and resource centre.

Information on the activities of the IOMP can be found on its website at www.iomp.org.

To those who strive to seek knowledge and guidance, and

constantly strive to make the world a better place

Contents

Preface

The interest and excitement in the field of spectroscopy of biological tissues are increasing rapidly as both the clinical and nonclinical researchers are recognising that the vibrational spectroscopic techniques, infrared (IR), and Raman spectroscopic techniques have the potential to become noninvasive tissue diagnostic tools. Researchers have contributed to the literature on the diagnostic importance of different spectroscopic and imaging techniques in the area of cancer detection and other biological molecule analysis. However, there is a gap, as it appears that the details of the characteristic peak frequencies and their definitions that can be attributed to specific functional groups present in the biological tissues have not been fully understood.

In addition, there is no single source to date that addresses both the IR and Raman spectroscopic investigations of biological tissues, as researchers have to rely on a number of research sources and the majority of the time the interpretations of the spectral data differ significantly. In this book, a significant amount of spectroscopic investigation reported on biological tissues has been reviewed in detail and it reveals that there are striking similarities in defining different peak frequencies. As a result, creating a unique database containing a detailed study on the works, different chemical bands, and their assignments of spectral bands could provide significant assistance to research groups focusing on spectroscopy. This, in turn, can lead to considerable improvements in the quality and quantity of the spectroscopic research done in the field of medicine and tissue engineering.

Textbooks from a single author or even authors from the same discipline can too strongly concentrate on their own area of research, making it narrowly focused on one aspect. On the other hand, books from multiple authors can easily have varied interpretations and writing styles, which can very often clash and may provide misleading information to readers. In this book, authors from three different disciplines—a spectroscopist, an oncologist, and a dental surgeon—with a wealth of experience in using spectroscopy for the analysis of biological tissues, combine to provide a broad overview of vibrational spectroscopy and its utilisation in the analysis of natural tissues. This provides a balanced perspective on an evolving discipline by three researchers from completely different backgrounds.

Aims and Objective of This Book

This book aims to present a wide and detailed collection of interpretations of IR and Raman spectral frequencies. It is envisaged that this book will be of significant assistance to research groups working on spectroscopy of biological tissues. In addition, it is also designed as an authoritative reference book for both the clinical and nonclinical researchers employing vibrational spectroscopy as a research tool.

The main objective of this book is to introduce IR and Raman spectroscopy to scientists who are either using these spectroscopic techniques to address clinical problems or are thinking of exploring spectroscopy to analyse clinical tissues and understand their chemical composition. A significant emphasis in this book is placed on the analysis of cancerous tissues. This is deliberate because, to date, biological pathways of cancer progression as a disease are well established, but the chemical pathways are still a matter of debate. We strongly believe that providing interpretation and understanding of the spectral peaks of the biological molecules in one text will help the scientist understand that IR and Raman spectroscopy techniques can offer a powerful tool both in early diagnosis of the disease and monitoring of the progression of the disease. In addition, it will help to precisely predict the properties of chemical structures that are present in complex biological tissues.

There is a significant gap in understanding the translational power of spectroscopic techniques in medicine, and this book is an effort to bridge that gap. It will bridge the gap between the knowledge obtained through spectroscopic means and its application in the field of medicine.

In the initial chapter, vibrational spectroscopy as a powerful light scattering/absorption technique is discussed and its theory and principles are described. Furthermore, it is explained that the techniques complement each other rather than contradict each other. In the case of IR spectroscopy, it is the change in the dipole moment of the molecule that is determined; in the case of Raman, the change in polarisation of the molecules is analysed. Therefore, both techniques can complement the data that is obtained.

Once the principles of the techniques are established, the nature of the techniques is explored, specifically highlighting the potential of these techniques to analyse cancerous tissues at the cellular level.

To do that, it is important to have detailed knowledge and information on cell biochemistry; this is described in detail in Chapter 2, which provides an overall view of the most important topics of biochemistry in relation to cytology. Using the information presented in this chapter and the results obtained from Raman investigations, it would be possible for researchers to determine the functional groups and chemical bonds of the cells that undergo alterations in the process of cancer development. This chapter tries to prepare an almost detailed database of some of the most important chemical bonds and functional groups of the cell. Such information can be of significant assistance in defining the spectral peaks, and can be considered as a step toward recognition of the cytological basis of the disease. Particular attention has been paid to cellular molecules such as proteins, nucleic acids, lipids, and carbohydrates, since they are the major structures responsible for biochemical and cytological changes in the process of cancer development.

To understand the chemical structural changes due to cancer progression, it is equally important to understand the biology of cancer. This is described in Chapter 3, which introduces the nomenclature of cancer to readers reflecting its microscopic origin as well as its potential to harm host cells and tissues. Cancer develops by a multistep process, and the uncontrolled growth of cancer cells results from accumulated abnormalities affecting many of the cell regulatory mechanisms. This relationship is reflected in several aspects of cell behaviour that distinguishes cancer cells from their normal counterparts. Cancer cells typically display abnormalities in the mechanisms that regulate normal cell proliferation, differentiation, and survival. These stages can be qualitatively and quantitatively analysed by spectroscopic means; therefore, it is important for readers to understand the multistep cell growth process.

Once the fundamental topics of spectroscopy and the cancer cell growth process are covered in detail, the nature of applied spectroscopic research is highlighted by reviewing the data that has been reported in the literature. Of necessity, a comprehensive literature review covering the use of IR and Raman spectroscopic techniques to analyse cancer and normal tissues is presented in Chapter 4. This chapter gives background and provides comprehensive coverage of spectroscopic studies on natural tissues.

Breast cancer is one of the most common diseases in the modern world and has attracted enormous interest by many different research groups. Spectroscopy is a potentially effective diagnostic method and can increase the success rate of available treatment modalities by assisting in early detection and diagnosis. Chapter 5 presents the application of two spectroscopic techniques in the diagnosis of different breast malignancies. Some statistical analysis methods that can increase the reliability of the results are also discussed.

Spectroscopy is rapidly emerging as a powerful tool for analysing chemical and structural properties of the natural tissues and most types of biological samples and tissues that have been analysed by researchers. However, analysis of cell lines has not been fully carried out, which is interesting for research groups and the work is expanding rapidly. Currently, there are fewer studies on these types of samples than there are on tissue samples. This is an expanding area of research, and, in Chapter 6, spectroscopic studies on cancerous cell line samples (normally used in oncological studies) are described.

Chapter 7 focuses on applications of spectroscopy on bone. Both Raman and FTIR techniques are discussed and spectral findings are presented in detail. Also, the spectral differences between natural bone and synthetic apatite are illustrated using both spectroscopic techniques.

There are some key points that are crucial in defining the corresponding functional groups of every spectral peak. These points have been mentioned in different articles and can play an outstanding role in the characterisation of peak frequencies and are described in Chapter 8. The objective of this chapter is to collect information from literature and present it in the form of a spectral database on which research groups can trustfully rely. This is not based on personal opinions or interpretations, but agreed upon by different researchers working throughout the world. A tabulated spectral database is presented for clarity and ease of information. This database is under constant review, modification, and completion and new results will be added to make it even more comprehensive. We hope that this database can be used by researchers all over the world.

Spectroscopy can potentially play a crucial role in the armoury of lab-based diagnostic facilities. Chapter 9 presents a vision for future clinical applications of this technique and possible ways that it can help clinicians and scientists diagnose and treat different pathologic conditions.

To summarise, despite the tremendous leap in the field of spectroscopy, where new applications are emerging as fast as the technology is being developed, there are still areas of research that are crying for further exploration. It depends upon the imagination of clinical scientists who are ready to reach out to spectroscopists and vice versa working in partnership with one goal: improving the health of humans by providing solutions to different and ever-challenging disease processes.

Ihtesham ur Rehman
Zanyar Movasaghi
Shazza Rehman

About the Authors

Zanyar Movasaghi graduated as a doctor of dental surgery (DDS) from Tabriz University of Medical Sciences in Tabriz, Iran in 2001 and worked as a dental surgeon and dentist in different hospitals, clinics, and dental practices until 2005. Between 2005 and 2009 he completed a PhD in the School of Engineering and Materials Science at Queen Mary University of London under the supervision of Dr. Ihtesham Rehman. The title of his research project was "Spectroscopic Investigation of Cancerous Tissues." Since 2009, he has worked as a dental officer treating special care patients in Barts and The London NHS Trust, London. He is currently enrolled in a master of clinical dentistry course in fixed and removable prosthodontics at King's College, London. His research interests include spectroscopy of biological tissues, treatment planning and developing services for special care patients, implantology, and prosthodontics (in particular, implant-supported removable dentures). Dr. Movasaghi has also investigated reasons for tooth extraction in the Kurdish community and used guided tissue regeneration (GTR) as a clinical procedure to restore the attachment apparatus of periodontally diseased teeth. He is a member of the General Dental Council (GDC), United Kingdom, the Infrared and Raman Discussion Group (IRDG) in England, the Iran Medical Council (dentist membership), and the Iranian Dental Association. In addition, he is interested in poetry and writes poems in Persian.

Ihtesham ur Rehman PhD, MSc, MRSC Cchem, is a reader in biomedical materials at the University of Sheffield, in the United Kingdom. He has a doctorate in biomaterials from Queen Mary University of London, where he later taught for more than eighteen years. Prior to moving to the University of Sheffield, he was reader in biomedical materials at the Interdisciplinary Research Centre (IRC) in Biomedical Materials, Queen Mary University of London, and held responsibilities for postgraduate teaching programmes as Director of Biomaterials and Director of Dental Materials.

He is also the founder and the executive director of the Interdisciplinary Research Centre in Biomedical Materials (IRCBM), Comsats Institute of Technology, Lahore, Pakistan. The IRCBM is the only centre of its type conducting research on biomaterials in Pakistan.

Dr. Rehman's research interests are multidisciplinary and include: (i) chemical structural evaluations of cells (cancer cells and stem cells) and proteins by spectroscopic means, (ii) surface modification of polymers for the design and development of an auxetic biodegradable drug-eluted stent-graft in the palliation of oesophageal cancer, and (iii) new nanodental/biomaterials compositions with improved chemical, mechanical, and biological properties.

He has authored more than 100 scientific articles and his work has resulted in a number of patents. He has a diverse research career that includes more than 22 years of university experience conducting academic research related to spectroscopy and biomaterials. He has led strong teams in research and is internationally respected for his work on spectroscopy and biomaterials.

Dr. Rehman's current research is focused on the development of diagnostic techniques for cancer detection, diagnosis, and monitoring the progression of the disease. A particularly exciting avenue, which he seeks to develop further, is the structural characterisation of biomaterials and natural tissues using spectroscopic methods for the identification of tissue types and cell genotyping for in situ monitoring.

Shazza Rehman PhD, MBBS, FRCP, is a medical oncologist consultant currently working at the Airedale National Health Service Foundation Trust in West Yorkshire. She obtained her medical degree from the University of Punjab, Lahore, Pakistan and subsequently moved to the United Kingdom in 1994 for postgraduate training. She became a member of the Royal College of Physicians, London in 1999 and was elected as Fellow of the Royal College of Physicians in 2012. She completed her PhD in molecular pathology of breast cancer at the Royal Free Hospital, University College, London in 2004.

Dr. Rehman commenced oncology speciality training in 2003 at Christie Hospital, Manchester, United Kingdom, and moved to London in 2005, where she completed her training at St. George's, Maidstone, and Guy's and St Thomas's hospitals in 2008. Currently, she is working as a medical oncologist consultant treating breast and colorectal cancers at the Bradford Royal Infirmary and Airedale Hospital in West Yorkshire.

Her other activities include acting as a training programme director for second-year doctors and lead cancer clinician for the Airedale Hospital NHS Foundation Trust. She actively participates in a national cancer peer review programme and is also a question writer for the Fellow of Royal College of Physicians part 3, oncology speciality. In addition, she is the local principal investigator on many clinical trials at Airedale Hospital and actively participates in departmental, trust audit, and teaching programmes.

1

Introduction to Spectroscopy

By the Sun and its brightness
Quran, 91:1

History

The first spectroscopic experiments were recorded by Sir Isaac Newton in the seventeenth century in which he used, for the first time, a glass prism for spectral dispersion and introduced the term *spectrum* in his classic volume *Opticks* (1704). The early nineteenth century saw major achievements in spectroscopy with the invention of optical grating (1500 lines in the Sun's spectrum) by Joseph Fraunhofer (Figure 1.1). He developed the first grating spectroscope and by using this spectroscope, German physicist Kirchhoff detected sodium in the Sun's spectrum (Figure 1.2). Further developments in spectroscopic studies were achieved by many scientists in the late nineteenth and twentieth century. The following achievements have contributed tremendously to the understanding of spectroscopy.

J. J. Balmer discovered the Balmer series of lines in the hydrogen spectrum in 1885. Discovery of concave grating by Henry A. Rowland in 1887 revolutionised experimental spectroscopy. Discovery of electrons and the atomic nucleus by English physicists J. J. Thompson (1897) and Ernest Rutherford (1911), the first principles of quantum theory by Max Planck (1900), discovery of quantum mechanics by Werner Heisenberg (1932) and Erwin Schrodinger (1933), and the first real quantitative spectral analysis by G. Scheibe, W. Gerlach, and E. Schweitzer (1925) [1–3].

Spectroscopy can be defined as an analysis technique to study the interaction of electromagnetic radiation with atoms and molecules. The scattering of a light after interaction with matter grades into its component energies (colours). The resultant dissection of energies can be used to analysis that matter's physical properties. Matter is composed of atoms; in a stable state the atoms of molecules are in an electronic, vibrational, and rotational state. When electromagnetic radiation interacts with these stable atoms or molecules, they may undergo transitions in an energy state, that is, stable state

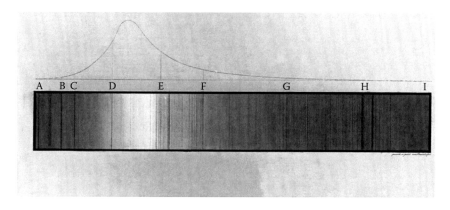

FIGURE 1.1 (See colour insert.)
Solar spectrum drawn and coloured by Fraunhofer with the dark lines named after him, 1817. (Courtesy Deutsches Museum, Munich, Germany. With permission.)(1–3).

FIGURE 1.2
First grating spectroscope; Fraunhofer, 1821. (Courtesy Deutsches Museum, Munich, Germany. With permission.)(1–3).

atoms and molecules possess minimum energy state E_0, while after radiation interaction they possess a higher energy state E^* (excited state).

$$\Delta E = E^* - E_0$$

where
  ΔE = energy difference
  E^* = excited state
  E_0 = initial state

This energy is supplied by one quantum of radiation and is absorbed by atoms or molecules. If energy supplied by one quantum radiation is smaller than excitation energy, no excitation state occurs. Two or more photons with lower energy than that required for excitation cannot combine their energies to produce the excitation energy ΔE [4].

Regions of the Spectrum

The electromagnetic spectrum (Figure 1.3) includes a wide range of electromagnetic radiations from the radiofrequency region to the γ-ray region. The approximate wavelength, frequency, energy, and their nature of interaction with atoms and molecules are given in Table 1.1 [5].

Vibrational spectroscopy is a powerful light-scattering/absorption technique used to investigate the internal structure of molecules and crystals. As the technique is specific to the chemical bonds and molecular structures, it is commonly used in chemistry and has been considerably covered by scientists and research groups from different disciplines [6,7]. However, the technique is rapidly emerging not only as a potential tool for biological studies, but also as a technique of choice for studying the chemical structural properties of biological molecules. This is due to the fact that the spectra are highly detailed, allowing subtle differences in biochemistry to be identified. Theoretically, any physiological change or pathological process

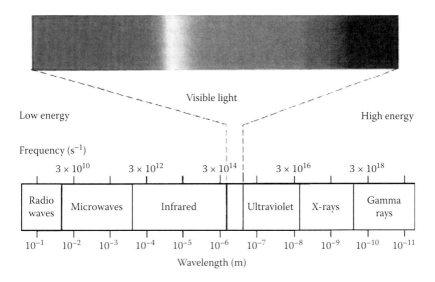

FIGURE 1.3 (See colour insert.)
Electromagnetic spectrum.

TABLE 1.1

Electromagnetic Spectrum Regions and Their Properties

Spectral Region	Frequency (Hz)	Wavelength	Energy (Joules/mole)	Nature of Interaction
Radiofrequency region	$3 \times 10^6 - 3 \times 10^{10}$	10 m – 1 cm	$10^{-3} - 10^{-1}$	Reversal spin of nucleus or electron
Microwave region	$3 \times 10^{10} - 3 \times 10^{12}$	1 cm – 100 μm	10	Rotational quantum number of molecule
Infrared region	$3 \times 10^{12} - 3 \times 10^{14}$	100 μm – 1 μm	10^3	Molecular vibration
Visible and ultraviolet regions	$3 \times 10^{14} - 3 \times 10^{16}$	1 μm – 10 nm	10^5	Change of electron distribution
X-ray region	$3 \times 10^{16} - 3 \times 10^{18}$	10 nm – 100 pm	10^7	Change of electron distribution
γ-ray region	$3 \times 10^{18} - 3 \times 10^{20}$	100 pm – 1 pm	10^9	Rearrangement of nuclear particles

that results in changes to the native biochemistry would therefore lead to changes in the Fourier transform infrared (FTIR) and Raman spectra [8]. This book investigates some of the most recent applications of vibrational spectroscopy (i.e., Raman and FTIR spectroscopy) on different biological samples, such as tissues, cells, and cell lines. The techniques can potentially determine the dissimilarities of different types of cells at the molecular level. The techniques can identify the functional groups and chemical bonds that are present in the biological tissues or/and cells. Therefore, it is possible not only to evaluate the structure of the proteins, lipids, carbohydrates, and nucleic acids that are present in a biological molecule, but also the changes that are taking place in their chemical structure due to the disease process, thus making it possible to monitor the progression of the disease process and allowing prediction of the chemical pathway of the progression.

Principles of Raman Spectroscopy

Raman spectroscopy is a vibrational spectroscopic technique that is used to optically probe the molecular changes associated with diseased tissues [9,10]. The technique is based on different types of scattering of monochromatic light, usually from a laser in the visible, near-infrared, or near-ultraviolet range. Light from the illuminated spot is collected with a lens and sent through a monochromator. When the energy of an incident photon is unaltered after collision with a molecule, the scattered photon has the same frequency as the incident photon. This is called *elastic* or *Rayleigh scattering*. When energy is transferred from the molecule to the photon or vice versa, the scattered photon has less or more than the energy of the incident photon (the energy is shifted up or down). This is *inelastic* or *Raman scattering* and was first described in 1928 by Raman, who received the Nobel prize two years later for his work in this field [11,12,13]. Photons that undergo elastic Rayleigh scattering with wavelengths close to the laser line are filtered out while the rest of the collected light is dispersed onto a detector. A very small portion (1 in 10^{10}) [14] or (1 in 10^8) [15] of the light, however, is inelastically scattered at a different wavelength to the incident light.

In theory, the Raman effect occurs when light hits a molecule and interacts with the electron cloud and the bonds of that molecule. Subsequently, a photon excites the molecule from the ground state to a different energy level. When the molecule relaxes, it returns to a different vibrational state and emits a photon. Due to the difference between the energy of the original state and the energy of the new state, the emitted photon has a shifted and different frequency from the excitation wavelength. If the final vibrational state of the molecule is more energetic than the initial state, then the emitted photon is shifted to a lower frequency in order for the total energy of the system to remain balanced. This shift in frequency is referred to as a *Stokes shift* [6,7,16,17]. In other words, the scattered photons can be *red-shifted* by providing energy to the bond vibration (Stokes Raman scattering) [15] If the final vibrational state is less energetic than the initial state, then the emitted photon will be shifted to a higher frequency, which is designated as an *anti-Stokes shift* [6,7,16,17]. Anti-Stokes Raman scattering is referred to as *blue-shifting* [15]. Raman scattering is an example of inelastic scattering because of the energy transfer between the photons and the molecules during their interaction. The difference in energy between the incident and scattered photons corresponds to the energy of the molecular vibration [15]. If the proton has a higher frequency and therefore lower energy than the incident light, this is known as *Stokes-Raman* and is due to the change in the vibrational mode of the molecule. Figure 1.4 depicts different Raman scattering processes. The Raman signal can be enhanced by up to 15 orders of magnitude in this way [18].

Raman spectra are plots of scattered intensity as a function of the energy difference between the incident and scattered photons, and are obtained by pointing a monochromatic laser beam at a sample. When light strikes a molecule, most of the light is scattered at the same frequency as the incident light (elastic scattering). As mentioned, only a small fraction is scattered at a different wavelength (inelastic or Raman scattering) due to light energy changing the vibrational state of the molecule. The loss (or gain) in the photon energies corresponds to the difference in the final and initial vibrational energy levels of the molecules participating in the interaction. The resultant spectra are characterised by shifts in wave numbers (inverse of wavelength in cm^{-1}) from the incident frequency. The frequency difference between incident- and Raman-scattered light is termed the *Raman shift*, which is unique for individual molecules and is measured by the machine's detector

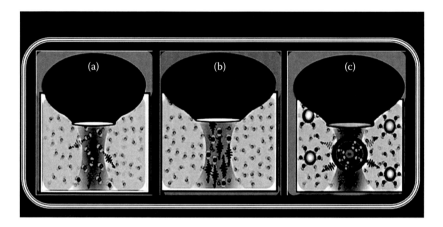

FIGURE 1.4
Different Raman scattering processes. (a) Shows spontaneous Raman scattering produced by scattered laser beam causing vibrations of chemical bonds; (b) A coherent Raman signal on the anti-Stokes side of the spectrum is generated when two or more coherent laser beams with different frequencies overlap spatially and temporally onto the specimen; (c) The Coherent anti-Stokes Raman spectroscopy (CARS) signal is produced by the nonlinear mixing of the ultrashort excitation pulses and is several order of magnitude stronger than the spontaneous Raman signal. Molecules adsorbed on metallic nanostructures will experience a strong electromagnetic field in addition to chemical enhancement, due to the presence of plasmon resonances near the surface of metals.

and represented as 1/cm. Raman peaks are spectrally narrow, and in many cases can be associated with the vibration of a particular chemical bond (or a single functional group) in the molecule [16,17,19,20,21]. The vibrations are molecular bond specific allowing a *biochemical fingerprint* to be constructed of the material [8].

Confocal Raman microspectroscopy (Figure 1.5) is the amalgamation of traditional Raman spectroscopy and confocal microscopy, and allows the examination of Raman spectra from small volumes. Briefly, a temperature-stabilised diode laser operating at 785 nm is expanded and introduced via a holographic notch filter (HNF) into an inverted microscope and passed to the sample via an objective. The backscattered Raman light is collected by the same objective and passed through the HNF. The Raman signal is then reflected by the dichroic mirror and imaged onto a confocal aperture. Finally, the beam is imaged onto the spectrograph [22].

It is important to note that depending on whether the bond length or angle is changing, vibrations are subdivided into two classes:

- Stretching (symmetric and asymmetric)
- Bending (scissoring, rocking, wagging and twisting) (Figure 1.6)

In other words, there are six different vibrational modes for chemical bands. In symmetric and asymmetric stretching types, the lengths of

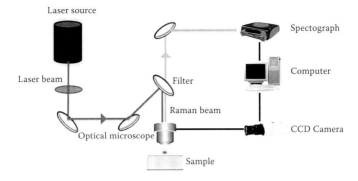

FIGURE 1.5
Raman spectroscopy, a schematic diagram representing positioning of the necessary parts of a Raman spectrometer.

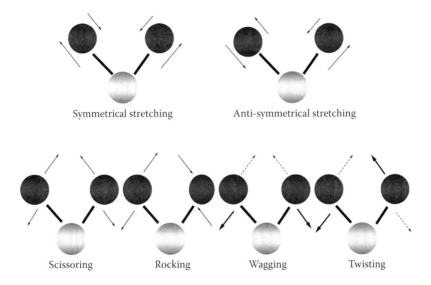

FIGURE 1.6
Different types of vibration of the chemical bonds.

the bands become unstable. In another four types, the length remains stable and the angulations of the bands changes.

Infrared (IR) Spectroscopy

Infrared spectroscopy has emerged as a powerful tool for chemical analysis because of its ability to provide detailed information on the spatial distribution of chemical composition at the molecular level [23,24]. In applications requiring qualitative and quantitative analysis, the potential of IR spectroscopy to identify chemical components via fingerprinting analysis of their vibrational spectrum is unsurpassed [23].

Infrared spectroscopy mainly deals with the infrared region of the electromagnetic spectrum and most commonly focuses on absorption spectroscopy (when the frequency of the IR is the same as the vibrational frequency of a bond, absorption occurs). When a material is exposed to infrared radiation, absorbed radiation usually excites molecules into a higher vibrational state. The wavelengths that are absorbed by the sample are characteristic of its molecular structure [25]. However, similar to other spectroscopic techniques, this technique can also identify molecular structure and investigate sample composition, and the spectral bands are indicative of molecular components and structures [26].

The technique works on the fact that chemical bonds or groups of bonds vibrate at characteristic frequencies [27]. A molecule that is exposed to infrared rays absorbs infrared energy at frequencies that are characteristic of that molecule. Infrared spectroscopy is based on the fact that molecules have specific vibrational modes that can be activated provided that they are hit by photons with specific energy levels. However, in order for a molecule or a chemical bond to be IR active, it must be in a dipole condition. This technique works almost exclusively on samples with covalent bonds. This is different from the Raman effect, which mainly deals with the polarisability of chemical bonds [28]. Therefore, in simple terms, IR spectroscopy detects changes in the dipole moment of the molecules, whereas Raman spectroscopy analyses change in the polarisation of molecules. For a molecule to be infrared active it must have a dipole moment.

FTIR spectroscopy is a type of vibrational spectroscopy technique and is the most useful tool available to a scientist when it comes to solving a problem that involves discovering the molecular structure, molecular behaviour, or the identification of unknown organic chemical substances and mixtures [29,30]. When the material under investigation is put into an FTIR spectrometer, it will absorb the radiation emitted (typically infrared radiation) and the successful absorption will display the uniqueness or *fingerprint* of the material under investigation. Discovering the frequency of a particular material can be done without the use of an FTIR spectrometer [31].

Electromagnetic radiation comes in two distinct forms—in the form of particles or in the form of waves. It is well understood that infrared radiation is, in actual fact, electromagnetic in origin [31]. The energy (*E*) of any electromagnetic radiation can be expressed by this equation:

$$E = h\nu$$

where h is Planck's constant and ν is the frequency of radiation. This equation is known as Planck's law. An infrared spectrum plots the absorption of a material (or to be more accurate, the basal molecules) onto an infrared spectrum [30].

The infrared spectrum is merely a plot of a function of radiant absorbance (or transmittance) versus frequency (cm^{-1}/μ). The most significant portion of the infrared spectrum is the part where the material absorbs radiation (where in the spectrum it *peaks*) and the amount of radiation absorbed. The absorption peaks give rise to the individuality of the absorption spectra of dissimilar materials (and different molecules) [31].

FTIR has the ability to deal with nonaqueous samples, but is unable to produce an accurate peak for aqueous samples due to the strong absorption bands of water. FTIR needs more sample preparation compared to Raman spectroscopy and is unable to perform confocal imaging. FTIR spectroscopy operates based on dipole moment changes during molecular vibration. In terms of resolution, FTIR spectroscopy provides a lateral resolution of 10–20 mm [29].

FTIR and Dipole Moment Change

The frequencies of infrared radiation are in the same order of magnitude as that of the oscillations of atoms of any molecule not at absolute zero temperature. This is particularly fortunate since this assures us that the infrared spectrum is accurate, to some degree. The oscillations of the molecules reflect a change in the dipole moment during the oscillations, given that the frequencies of the radiation correspond exactly with the frequencies of the molecular oscillations [31].

It is not necessary for a molecule to possess a permanent dipole moment. Carbon tetrachloride, CCl_4, has zero dipole but when it absorbs infrared radiation and vibrations occur on the C–Cl bond, a change in the dipole moment occurs [30].

Diatomic molecules, made up of the same atoms, are very unlikely to absorb any amount of infrared radiation (Figure 1.7). This is due to the fact

FIGURE 1.7
Diatomic molecules with the same basal atoms.

that no change in dipole moment occurs during their vibrations; the same principle applies to monatomic molecules or ions. However, diatomic molecules made up of basal atoms that are different do display some infrared absorption due to the occurrence of dipole moment changes [31].

A general principle that can be applied when it comes to infrared radiation absorption is that covalent bonding between unlike atoms gives rise to infrared absorption. Molecules and materials that are partly covalent and partly ionic, such as materials that are partly organic and partly inorganic, will show absorption because of that portion of their bonding that is covalent in nature. Another general principle that can be applied is that all organic substances absorb infrared radiation [31].

The infrared spectrum is recorded by passing a beam of infrared light through the sample and recording the changes at the energy level of the photons as a result of interactions with the sample. This can be done with a monochromatic beam, which changes in wavelength over time. However, using a Fourier transform (FT) instrument makes it possible to measure all wavelengths at once. The aim is to measure the quality and quantity of transmittance or absorbance of each different wavelength by a sample that can produce transmittance or absorbance spectrum.

FTIR and Raman Spectrometry

Raman and FTIR are two of the available spectroscopic techniques (Figure 1.8). Although there are some other spectroscopic methods, including fluorescence spectroscopy [30,31,32], nuclear magnetic resonance (NMR) [33], magnetic resonance (MR) spectroscopy [34,35,36], and ultraviolet-visible (UV-Vis) spectroscopy [37], Raman and FTIR are the techniques that are most often used. This is mainly because of the advantages of these techniques, such as having a simple analysis procedure and being noninvasive in nature.

FTIR and Raman are potential tools for noninvasive optical tissue diagnosis. These techniques have become a significant analytical method for biomedical applications. The measurements are nondestructive to tissue and only very small amounts of material (micrograms to nanograms) are required. In some studies, sensitivity and specificity values of more than 90% for distinguishing cancer from normal tissues have been reported [38–41]. Furthermore, such analysis yields structural information concerning the conformational order in complex natural tissues. The methods are employed to find more conservative ways of analysis to measure characteristics within tumour tissue and cells, which would allow accurate and precise assignment of functional groups, bonding types, and molecular conformations. Spectral bands in vibrational spectra are molecule specific and provide direct information about biochemical composition. These bands are relatively narrow,

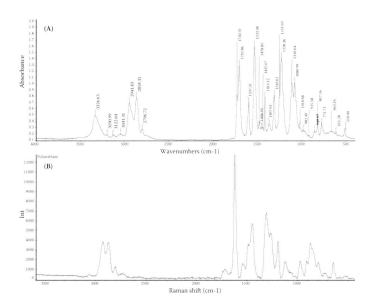

FIGURE 1.8
FTIR (A) and Raman (B) spectra of polyurethane.

easy to resolve, and sensitive to molecular structure, conformation, and environment.

Raman spectroscopy and FTIR are relevant techniques, with their respective spectra complementary to one another. In spite of this fact, there are some differences between these two techniques. Figure 1.8 shows spectra for Raman and FTIR, and the main differences between the two methods can be seen in Table 1.2. Probably the most important difference is the type of samples that can be investigated by each of these methods. FTIR mainly deals with nonaqueous samples, while Raman is as effective with aqueous samples as it is with nonaqueous ones. This is because of the problem with FTIR spectroscopy, which is due to strong absorption bands of water. In Raman, however, fluorescence and the strong effect of glass (mostly containers) are the most important problems being faced. In addition, while Raman can collect spectra in the range of 4000–50 cm⁻¹, FTIR focuses on a comparatively narrower frequency range, located in the area of 4000–700 cm⁻¹ (Figure 1.8).

In comparison to FTIR, Raman needs little or no sample preparation and can do confocal imaging. FTIR, on the other hand, requires sample preparation and does not do confocal imaging. Furthermore, the physical effect of infrared is created by absorption, and mainly influences the dipole and ionic bands such as O–H, N–H, and C=O. Raman effect originates from scattering (emission of scattered light) and changing of the polarisability of covalent bands like C=C, C–S, S–S, and aromatics. In other words, FTIR spectroscopy

TABLE 1.2

Comparison between FTIR and Raman Spectroscopy

	Infrared	Raman
Physical effect	Absorption. Analysing the change in the dipole moment of the molecules (strong: ionic bonding such as; O–H, N–H, C=O).	Scattering. (Monitoring the emission of scattered light). Observes change in the polarisation of molecules (strong: covalent bonding such as; C=C, C–S, S–S, aromatics).
Sample preparation	Sample preparation required in majority of cases (part from using photo-acoustic sampling technique, which eliminates the need of sample preparation).	Little or no sample preparation
Problems	Water (has as strong signal)	Fluorescence can be an issue and some glasses do possess extremely high fluorescence. In certain cases quartz glass is required to do analysis.
Materials	Mainly organic compounds. Certain inorganic molecules can be analysed. Synthetic and natural materials can also be analysed.	Nearly a complete range of samples, organic, inorganic, polymers, and biological (both in wet and dry conditions) can be analysed.
Resolution:	1–20 μm (varies with beam splitters)	0.05–8 μm (varies with lasers)
Frequency range	4000–400 cm^{-1} (Mid-IR) Far and Near-IR also possible, making analysable range to 30,000 to 50 cm^{-1}	4000–50 cm^{-1}
Samples	Non-aqueous	Aqueous

is due to changes in dipole moment during molecular vibration, whereas Raman spectroscopy involves a change in polarisability [14].

Another difference existing between the techniques is their resolution. While Raman has lateral resolution of 1–2 μm and confocal resolution of 2.5 μm, FTIR does not provide the possibility of confocal resolution, and its lateral resolution is 10–20 μm [29]. It also must be mentioned that FTIR is mainly for organic compounds, but Raman can deal with nearly unlimited types of samples.

Mahadevan-Jansen and Richards-Kortum [11] describe the advantages of FTIR as producing spectra with a superior signal-to-noise ratio and an improved resolution as compared to some of the conventional IR spectrometers, but note that it may increase spectra collection times. Chowdary et al. [42] report Raman spectroscopy as being superior to FTIR techniques based on having higher spatial resolution, less sample preparation, reduced water absorption bands, and less harmful near-infrared radiation.

References

1. Schmidt, W. 2005. *Optical Spectroscopy in Chemistry and Life Sciences*. Wiley-VCH Verlag Gmbh& Co. KGaA, Weinheim, 1–12.
2. Sawyer, R. A. 1963. *Experimental Spectroscopy*. New York: Dover Publishers.
3. Wilde, E. 1985. *Geschichte der optik*, Sandig Reprint Verlag, H. R. Wohlwend, Vaduz/Lichenstein.
4. Yadav, M. S. 2003. *A Text Book of Spectroscopy*. New Delhi: Anmol Publications, 1–16.
5. Banwell, C. N., and Mccash, E. 1994. *Fundamentals of Molecular Spectroscopy*. 4th ed. United Kingdom: McGraw-Hill International, 1–30.
6. Diem, M., Griffiths, P. R., and Chalmers, J. M. 2008. *Vibrational Spectroscopy for Medical Diagnosis*. London: John Wiley & Sons, 187–202.
7. Lin-Vien, D., Colthuop, N. B., Fateley, W. G., and Grasselli, J. G. 1991. *The Handbook of Infrared and Raman Characteristic Frequencies of Organic Molecules*. New York: Academic Press.
8. Harris, A. T., Lungari, A., Needham, C. J., Smith, S. L., Lones, M. A., Fisher, S. E., Yang, X. B., Cooper, N., Kirkham, J., Smith, D. A, Martin-Hirsch, D. P., and High, A. S. 2009. Potential for Raman spectroscopy to provide cancer screening using a peripheral blood sample. *Head & Neck Oncology* 1, 34.
9. Das, K., Stone, N., Kendall, C., Fowler, C., and Christie-Brown, J. 2006. Raman spectroscopy of parathyroid tissue pathology. *Lasers Med. Sci.* 21, 192–197.
10. Nicholas Stone, N., and Matousek, P. 2008. Advanced transmission Raman spectroscopy: A promising tool for breast disease diagnosis. *Cancer Res.* 68, 4424–4430.
11. Mahadevan-Jansen, A., and Richards-Kortum, R. 1997. Raman spectroscopy for cancer detection. 19th International Conference IEEE EMBS, October 30–November 2, 1997, Chicago, IL. 2722–2728.
12. Hanlon, E. B., Manoharan, R., Koo, T. W., Shafer, K. E., Motz, J. T., Fitzmaurice, M., Kramer, J. R., Itzkan, I., Dasari, R. R., and Feld, M. S. 2000. Prospects for in vivo Raman spectroscopy. *Phys. Med. Biol.* 45, 1–59.
13. Dukor, R. K. 2002. Vibrational spectroscopy in the detection of cancer. *Biomedical Applications*, 3335–3359.
14. Conroy, J., Ryder, A. G., Leger, M. N., Hennessey, K., and Madden, M. G. 2005. Qualitative and quantitative analysis of chlorinated solvents using Raman spectroscopy and machine learning. *Proc. SPIE, Int. Soc. Opt. Eng.* 5826, 131–142.
15. Wachsmann-Hogiu, S., Weeks, T., and Huser, T. 2009. Chemical analysis in vivo and in vitro by Raman spectroscopy: From single cells to humans. *Curr. Opin. Biotechnol.* 20, 63–73.
16. Choo-Smith, L. P., Edwards, H. G. M., Endtz, H. P., Kros, J. M., Heule, F., Barr, H., Robinson, J. S. Jr., Bruining, H. A., and Pupells, G. J. 2002. Medical applications of Raman spectroscopy: From proof of principle to clinical implementation. *Biopolymers* 67, 1–9.
17. Huang, Z., McWilliams, A., Lui, M., McLean, D. I., Lam, S., and Zeng, H. 2003. Near-infrared Raman spectroscopy for optical diagnosis of lung cancer. *Int. J. Cancer.* 107, 1047–1052.
18. Jess, P. R., Smith, D. D., Mazilu, M., Dholakia, K., Riches, A. C., and Herrington, C. S. 2007. Early detection of cervical neoplasia by Raman spectroscopy. *Int. J. Cancer.* 121, 2723–2728.

19. Shafer-Peltier, K. E., Haka, A. S., Fitzmaurice, M., Crowe, J., Dasari, R. R., and Feld, M. S. 2002. Raman microspectroscopic model of human breast tissue: Implications for breast cancer diagnosis *in vivo. J. Raman Spectrosc.* 33, 552–563.
20. Gazi, E., Dwyer, J., Gardner, P., Ghanbari-Siakhani, A., Wde, A. P., Lockyer, N. P., Vickerman, J. C., Clarke, N. W., Shanks, J. H., Scott, L. J., Hart, C. A., and Brown, M. 2003. Applications of Fourier transform infrared microspectroscopy in studies of benign prostate and prostate cancer: A pilot study. *J. Pathol.* 201, 99–108.
21. Alfano, R. R., Tang, G. C., Pradhan, A., and Wahl, S. J. 1987. Optical spectroscopic diagnosis of cancer and normal breast tissues. *J. Opt. Soc. Am. B.* 6, 1015–1023.
22. Wachsmann-Hogiu, S., Weeks, T., and Huser, T. 2009. Chemical analysis in vivo and in vitro by Raman spectroscopy from single cells to humans. *Curr. Opin. Biotechol.* 20, 63–73.
23. Bogomolny, E., Argov, S., Mordechai, S., and Huleihe, M. 2008. Monitoring of viral cancer progression using FTIR microscopy: A comparative study of intact cells and tissues. *Biochim. Biophys. Acta.* 1780, 1038–1046.
24. Walsh, M. J., Fellous, T. G., Hammiche, A., Lin, W. R, Fullwood, N. J., Grude, O., Bahrami, F., Nicholson, J. M., Cotte, M., Susini, J., Pollock, H. M., Brittan, M., Martin-Hirsch, P. L., Alison, M. R., and Martin, F. L. 2008. Fourier transform infrared microspectroscopy identifies symmetric PO_2 modifications as a marker of the putative stem cell region of human intestinal crypts. *Stem Cells* 26, 108–118.
25. Baker, M. J., Gazi, E., Brown, M. D., Shanks, J. H., Gardner, P., and Clarke, N. W. 2008. FTIR-based spectroscopic analysis in the identification of clinically aggressive prostate cancer. *Br. J. Cancer.* 99, 1859–1866.
26. Sahu, R. K., Mordechai, S., and Manor, E. 2008. Nucleic acid absorbance in Mid IR and its effect on diagnostic variates during cell devision: A case study with lymphoblastic cells. *Biopolymers* 89, 993–1001.
27. Dimitrova, M., Ivanova, D., Karamancheva, I., Milev, A., and Dobrev, I. 2009. Application of FTIR-Spectroscopy for diagnosis of breast cancer tumours, *J. University of Chemical Technology and Metallurgy* 44, 297–300.
28. Kondepati, V. R., Heise, H. M., Oszinda, T., Mueller, R., Keese, M., and Backhaus, J. 2008. Detection of structural disorders in colorectal cancer DNA with Fourier-transform infrared spectroscopy. *Vibrational Spectrosc.* 46, 150–157.
29. Austrian Centre for Electron Microscopy and Nanoanalysis. 2010. Infrared and Raman microspectrometry, Graz, Austrian Centre for Electron Microscopy and Nanoanalysis, http://www.felmi-zfe.tugraz.at/ (accessed on December 10, 2011).
30. Panjehpour, M., Julius, C. E., Phan, M. N., and Vo-Dinh, T. 2002. Laser-induced fluorescence spectroscopy for in vivo diagnosis of non-melanoma skin cancers. *Lasers. Surg. Med.* 31, 367–373.
31. Li, X., Lin, J., Ding, J., Wang, S., Liu, Q., and Qing, S. 2004. Raman spectroscopy and fluorescence for the detection of liver cancer and abnormal liver tissue. Proceedings of the Annual International Conference IEEE EMBS, September 1–5, 2004, San Francisco, CA, 212–215.
32. Grossman, N., Ilovitz, E., Chaims, O., Salman, A., Jagannathan, R., Mark, S., Cohen, B., Gopas, J., and Mordechai, S. 2001. Fluorescence spectroscopy for detection of malignancy: H-ras overexpressing fibroblasts as a model. *J. Biochem. Biophys. Methods.* 50(1), 53–63.
33. Ronen, S. M., Stier, A., and Degani, H. 1990. NMR studies of the lipid metabolism of T47D human breast cancer spheroids. *FEBS Letter* 266, 147–149.

34. Shukla-Dave, A., Hricak, H., and Moskowitz, C. 2007. Detection of prostate cancer with MR spectroscopic imaging: An expanded paradigm incorporating polyamines. *Radiology* 245, 499–506.
35. Smith, I. C. P., and Baert, R. 2008. *MR Spectroscopy and the Early Detection of Cancer in Human Subjects.* Netherlands: Springer.
36. Steffen-Smith, E. A., Wolters, P. L., Albert, P. S., Baker, E. H., Shimoda, K. C., Barnett, A. S., and Warren, K. E. 2008. Detection and characterization of neuro-toxicity in cancer patients using proton MR spectroscopy. *Childs. Nerv. Syst.* 24, 807–813.
37. Skala, M. C., Palmer, G. M., and Vrotsos, K. M. 2007. Comparison of a physical model and principal component analysis for the diagnosis of epithelial neopla-sias in vivo using diffuse reflectance spectroscopy. *Optics Express.* 15, 7863–7875.
38. Frank, C. J., McCreecy, R. L., and Redd, D. C. B. 1995. Raman spectroscopy of normal and diseased human breast tissues. *Anal. Chem.* 67, 777–783.
39. Laska, J., and Widlarz, J. 2005. Spectroscopic and structural characterization of low molecular weight fractions of polyaniline. *Polymer* 46, 1485–1495.
40. Stone, N., Kendell, C., Shepherd, N., Crow, P., and Barr, H. 2002. Near-infrared Raman spectroscopy for the classification of epithelial pre-cancers and cancers. *J. Raman Spectrosc.* 33, 564–573.
41. Stone, N., Kendall, C., Smith, J., Crow, P., and Barr, H. 2004. Raman spectroscopy for identification of epithelial cancers. *Faraday Discussion* 126, 141–157.
42. Chowdary, M. V. P., Kumar, K. K., Kurien, J., Mathew, S., and Krishna, M. C. Discrimination of normal, benign, and malignant breast tissues by Raman spectroscopy. *Biopolymers* 83, 556–569.

2

Spectroscopy and Molecular Changes in Cancer Cells

Introduction

The general features of neoplastic cells exhibit specific changes in nucleic acid, protein, lipid, and carbohydrate quantities and/or conformations [1]. Proteins, lipids, and nucleic acids are the marker molecules, which may be indicative of neoplasia, and changes in these molecules might be expected. Not surprisingly, both the Fourier transform infrared (FTIR) and Raman spectroscopies, due to their fingerprint character, can be used to describe and recognise the changes that take place in cancer cells. Several studies elucidate the spectral features of cancer cells and their various components such as collagen. The Raman spectrum of a sample corresponds to the characteristic molecular groups in the sample, and the technique can give detailed information on changes in structure and composition of the cellular molecules [2] and can provide important diagnostic information. These features can potentially be used as diagnostic parameters to identify malignant tumours. This chapter provides a summary of the changes that different researchers have found at the molecular level of cancer cells using FTIR and Raman spectroscopy.

Understanding Cancer

To understand chemical structural properties of cancer and associated changes with the progression of disease, it is important to understand the biology and chemistry of the cancer. The term *neoplasia*, which literally means *new growth*, is used clinically to describe pathologic tissue masses, which grow independent of and faster than normal tissues. Most of the body cells repair and reproduce themselves similarly in spite of their different functioning and appearance. As long as this process takes place in an orderly manner, everything is under control. However, when it gets out of control, a lump, called a tumour, develops. A neoplasm (tumour) may be classified as in situ, benign, or malignant, based on its potential to harm the host body. Malignant tumours are also called *cancers* [3,4]. Benign tumours are not considered cancer. They do not threaten life because they do not spread to other organs, and once they are removed, they rarely reappear. In situ tumours look like cancer in their morphology. In general, they develop in the epithelium. These small tumours remain in the epithelial layer. A malignant tumour consists of cancer cells that sometimes spread away from the original (primary) cancer to other organs through the lymphatic system and bloodstream causing damage to organs and tissues near the tumour. Worse is that when they get to another part of the body, they may divide forming a new tumour, called a *secondary* or a *metastasis tumour* [5]. A biopsy is used by doctors to decide whether a tumour is benign or cancerous by examining a small sample of cells under a microscope. Uncontrolled growth is the core property of cancers, which were known to the early Egyptians [4,6]. Hippocrates is reported to have distinguished benign from malignant tissues [7].

Although cancer has multiple causes and exceptions, an acceptable clinical definition would be: a set of diseases characterised by unregulated cell growth leading to invasion of surrounding tissues and spread (metastasis) to other parts of the body. Invasion and metastasis distinguish cancers from benign growth [6].

The nomenclature of a neoplasm reflects its microscopic origin as well as the potential to harm the host. For example, cancers arising from glandular epithelial tissues are called *adenocarcinomas*, where the suffix *carcinoma* indicates the epithelial origin and the prefix *adeno* reflects the glandular origin. On the other hand, cancers arising from the mesenchymal tissue are called *sarcomas* [3,4]. Leiomyosarcoma (sarcoma of smooth muscle) and *liposarcoma* (sarcoma of fat cells) are examples of this type [6]. Leukemias and lymphomas, which account for approximately 7% of human malignancies, arise from the blood-forming cells and from cells of the immune system, respectively [8]. All cell types are susceptible, but epithelial cells are most prone to change. This means that most human malignancies arise from epithelial tissue [6,7]. In fact, Most human cancers (approximately 90%) are carcinomas. This might be explained by two factors: cell proliferation mainly occurs in epithelia, and they are more frequently exposed to various forms of physical and chemical damage that favour the development of cancer [9]. Dedifferentiation or *anaplasia* denotes the loss of normal characteristics, which may be attributed to chemical structure, and these chemical structures can be very well studied using spectroscopic techniques. These techniques can help in identifying and establishing a chemical pathway by which the cancer progresses. The biological pathway of cancer progression is very well established, but the chemical pathway is still a matter of debate. We strongly believe that vibrational spectroscopy offers an excellent potential for studying the chemical structural characteristics of biological tissues and is only possible if the spectral bands are precisely and accurately assigned to specific chemical bonds and functional groups.

Peak Assignments of Biological Molecules

Huang et al. presented very useful information for interpreting the peak intensities of the biological samples [10,11]. Table 2.1 summarizes the assignments of the peaks investigated in these studies. It seems that these definitions play a considerably important role in peak definitions of biological samples done by other research groups [12–19]. As can be seen, proteins, lipids, nucleic acids, and polysaccharides have distinct and definable peaks in Raman spectra.

Some of the Raman peaks can be interpreted more easily, as compared to FTIR. This is because of the overlapping of the assignments, which can be seen, for example, in peak intensities of 1445 cm^{-1} and 1302 cm^{-1} with assignments due to both proteins and lipids. However, it was shown that lung cancer tumours have higher amounts of proteins, decreased intensities of phospholipids, and more widened and greater nucleic acid peaks.

However, there are differences among the definitions of some peaks. For instance, it was concluded that a shoulder at 1121 cm^{-1} is due to the symmetric phosphodiester stretching band of nucleic acids, namely ribonucleic acid (RNA). However, a shoulder at 1020 cm^{-1} was mentioned as being due to deoxyribonucleic acid (DNA). The intensity ratio of 1121/1020 cm^{-1} was used to grade the tumours, with increasing quantities for more developed malignancies. Interestingly, Huleihel et al. [18] and Andrus [20] mention that same intensity ratio, with the same interpretation. Furthermore, it was concluded that possibly greater 1050–1080 cm^{-1} erosion indicated a greater degree of oxidative damage to DNA.

Schultz and Baranska [21] made a detailed study of the characteristic peak intensities of Raman spectroscopy (Table 2.2). Although this work is concentrated on plant samples, as almost all of these compounds and chemical bands also exist in cells and tissues, the results and conclusions could be successfully used in biological studies. Considering the significant amount of information about the cellular molecules that can be obtained from these tables, this article indicates the wide range of capacities of these techniques in producing useful results in biological research.

Based on the information presented in Table 2.2, Andrus carried out an interesting study on cancer grading of lymphoid tumours [20]. They also used 1121 cm^{-1} (RNA band), 1053 cm^{-1} (DNA band), 1084 cm^{-1} (DNA and RNA), 1650 cm^{-1} (protein), and 2800–3000 cm^{-1} (lipid and protein) in their investigations, and believe that the protein:nucleic acid ratio could be analysed through the 1650 cm^{-1}:1084 cm^{-1} ratio [20].

TABLE 2.1

Peak Definitions and Assignments of Lung Cancer Samples

Peak Position (cm⁻¹)	Protein Assignment	Lipid Assignment	Others
1745w		ν(C=O), phospholipids	
1655vs	ν(C=O) amide I, α-helix, collagen, elastin		
1618s (sh)–1620w	ν(C=C), tryptophan		ν(C=C), porphyrin
1602ms (sh)	δ(C=C), phenylalanine		
1585vw–1582ms (sh)	δ(C=C), phenylalanine		
1558vw	ν(CN) & δ(NH) amide II		ν(C=C), porphyrin
1552ms (sh)	ν(C=C), tryptophan		ν(C=C), porphyrin
1518			ν(C=C), carotenoid
1445vs	δ(CH$_2$) δ(CH$_3$), collagen	δ(CH$_2$) scissoring, phospholipids	
1379vw		δ(CH$_2$) symmetric	
1336mw (sh)	δ(CH$_3$), δ(CH$_2$) twisting, collagen		
1335s (sh)	CH$_3$CH$_2$ wagging, collagen		CH$_3$CH$_2$ wagging, nucleic acids
1322s	CH$_3$CH$_2$ twisting, collagen		
1302vs	δ(CH$_2$) twisting, wagging, collagen	δ(CH$_2$) twisting, wagging, phospholipids	
1265s (sh)	ν(CN), δ(NH) amide III, α-helix, collagen, tryptophan		
1223mw (sh)			ν_{as}(PO$_2^-$), nucleic acids
1208w (sh)	ν(C–C$_6$H$_5$), tryptophan, phenylalanine		
1172vw	δ(C–H), tyrosine		
1168vw		ν(C=C), δ(COH)	ν(C–C), carotenoid
1152w	ν(C–N), proteins		ν(C–C), carotenoid
1123w	ν(C–N), proteins		
1122mw (sh)		ν_s(CC), skeletal	
1078ms		ν(C–C) or ν(C–O), phospholipids	ν_s(CC), ν_{as}(PO$_2^-$) nucleic acids
1031mw (sh)	δ(C–H), phenylalanine		
1030mw (sh)	ν(C–C) skeletal, keratin		
1004ms	ν_s(C–C), symmetric ring breathing, phenylalanine		

TABLE 2.1 (*Continued*)

Peak Definitions and Assignments of Lung Cancer Samples

Peak Position (cm⁻¹)	Protein Assignment	Lipid Assignment	Others
973mw (sh)	$\rho(CH_3)$ $\delta(CCH)$, olefinic		
963w	Unassigned		
935w	$\nu(C–C)$, α-helix, proline, valine, keratin $\rho(CH_3)$ terminal		
883mw	$\rho(CH_2)$		
876w (sh)	$\nu(C–C)$, hydroxyproline		
855ms	$\nu(C–C)$, proline $\delta(CCH)$ ring breathing, tyrosine, phenylalanine, olefinic		Polysaccharide
823w	Out-of-plane ring breathing, tyrosine		
752w	Symmetric breathing, tryptophan		

Source: Z. Huang, A. McWilliams, M. Lui, D. I. McLean, S. Lam, and H. Zeng. Near-infrared Raman spectroscopy for optical diagnosis of lung cancer. *Int. J. Cancer* 107(2003): 1047–1052; Z. Huang, H. Lui, D. I. McLean, M. Korbelik, and H. Zeng. Raman spectroscopy in combination with background near-infrared autofluorescence enhances the in vivo assessment of malignant tissues. *Photochem. Photobiol.* 81(2005): 1219–1226. With permission.

Note: ν: stretching mode, ν_s: symmetric stretch, ν_{as}: asymmetric stretch, δ: bending mode, ρ: rocking, v: very, s: strong, m: medium, w: weak, sh: shoulder.

Other articles [22–24] have mentioned the characteristic bands of lipids and proteins. For example, lipids give a weak band at 1754 cm⁻¹ (C=O), a sharp band at 1656 cm⁻¹ (C=C), a very strong band at 1440 cm⁻¹ (CH₂ bend), and a sharp band of medium intensity at 1300 cm⁻¹ (CH₂ twist). Three bands are indicative of phospholipids. These include 1656 cm⁻¹, 1367 cm⁻¹ (νCH_3, symmetric), and 1268 cm⁻¹ (δ=CH). In protein molecules, the 1754 cm⁻¹ band is absent. At 1656 cm⁻¹ we observe a broad, very strong band (amide I). The CH₂ bend comes at 1450 cm⁻¹ and several bands of moderate intensity appear in the region of 1100–1375 cm⁻¹ (amide III and other groups). A sharp band of moderate intensity is observed at 1004 cm⁻¹ (phenylalanine). The 1558 cm⁻¹ (amide II, COO⁻) and 951 cm⁻¹ (ν_s CH₃) are other bands related to proteins.

In a review paper published by Mahadevan-Jansen and Richards-Kortum [25], a collection of biological studies done by Raman are presented. In the spectra of brain tissue, for instance, primary peaks due to lipids and proteins in white and grey matter are shown. Using the characteristic peak definitions of different cellular compounds, it is shown that the spectra of white matter show a greater contribution from lipids, cholesterol, and proteins as compared to the spectra of grey matter.

TABLE 2.2

Assignment for the Most Characteristic Raman and IR Bands

Chemical Groups	Representatives	FT-Raman (cm⁻¹)	Assignment	ATR-IR (cm⁻¹)	Assignment
Proteins	α-Helix	1655	Amide I*	1655	Amide I**
		>1275	Amide III***		
	Anti-parallel β-sheet	1670	Amide I	1670	Amide I
		1235	Amide III		
	Disordered structure (solvated)	1665	Amide I	1665	Amide I
		1245	Amide III		
	Disordered structure (non-hydrogen bonded)	1685	Amide I	1685	Amide I
		1235	Amide III		
				1543–1480	Amide II****
Lipids/fatty acids		3008	ν_{as}(=C—H)		
		2970	ν_{as}(CH$_3$)		
		2940	ν_{as}(CH$_2$)		
		2885	ν_{as}(CH$_3$)		
		2850	ν_{as}(CH$_2$)		
		1750	ν(C=O)	1750	ν(C=O)
		1670	ν(C=C) *trans*	1670	ν(C=C) *trans*
		1660	ν(C=C) *cis*	1660	ν(C=C) *cis*
		1444	δ(CH$_2$)	1444	δ(CH$_2$)
		1300	τ_{tp}(CH$_2$)		
		1266	δ_{ip}(=C—H) *cis*		
		1100–800	ν(C—C)		
Monosaccharides	α-Glucose	847	(C–O–C) skeletal mode		
	β-Glucose	898	(C–O–C) skeletal mode		
	β-Fructose	868	(C–O–C) skeletal mode		

Disaccharides	Sucrose	1462	$\delta(CH_2)$	1126	$\nu(C–O)$
		1126	$\nu(C–O)+\nu(C–C)$		
		847	(C–O–C) skeletal mode		
	Maltose	868	(C–O–C) skeletal mode		
		847	(C–O–C) skeletal mode		
		898	(C–O–C) skeletal mode		
	Cellobiose	885	(C–O–C) skeletal mode		
Polysaccharides	Cellulose			1430	$\delta(CH_2)$
				1336	$\delta(CH)$, ring
		1122	$\nu_{sym}(C–O–C)$	1125/1110	$\nu(CO)$, $\nu(CC)$, ring
		1094	$\nu_{asym}(C–O–C)$	1060	$\nu(CO)$, $\nu(CC)$, $\delta(OCH)$
				1035	$\nu(CO)$, $\nu(CC)$, $\nu(CCO)$
				985	OCH_3
	Hemicellulose	1739	$\nu(C=O)$	815	*
	Amylose	941	Skeletal modes	*	
		868			
		477			
	Amylopectin	941	Skeletal modes	*	
		868			
		477			
	Pectin			1745	$\nu(C=O)$
				1605	$\nu_{as}(COO^-)$
				1444	$\delta(CH)$

(Continued)

TABLE 2.2 (*Continued*)
Assignment for the Most Characteristic Raman and IR Bands

Chemical Groups	Representatives	FT-Raman (cm⁻¹)	Assignment	ATR-IR (cm⁻¹)	Assignment
				1419	ν_s(COO⁻)
				1368	δ(CH$_2$), ν(CC)
				1335	δ(CH), ring
		1745	ν(C=O)	1150	ν(C–O–C), ring
				1107	ν(CO), ν(CC), ring
		854	ν(C–O–C) skeletal mode of α-anomers	1055	ν(CO), δ(OCH)
				1033	ν(CC), δ(OCH)
				1018	ν(CO), ν(CC), δ(OCH), ring
				1008	ν(CO), ν(CC), δ(OCH), ring
				972	OCH$_3$
				963	δ(C=O)

Source: P. G. Andrus, Cancer monitoring by FTIR spectroscopy. *Technol. Cancer. Res. Treat.* 5(2006): 157–167. With permission.

Note: Amide I: ν(C=O) + ω(N–H), Amide II: δ(N–H) + ν(C–N), Amide III: δ(N–H) + ν(C–N).

Chan et al. [24], in a study based on detecting neoplastic and normal hematopoietic cells, show that spectral bands are mainly indicative of nucleotide confirmation (600–800 cm^{-1}), backbone geometry and phosphate ion interactions (800–1200 cm^{-1}), electronic structure of the nucleotides (1200–1600 cm^{-1}), and C–C and C–H modes due to proteins and lipids. In addition, amide vibrations such as the amide I band (due to C=O stretching [18,19]) and amide III band (due to C–N stretching and C–N bending [18,19]) in proteins are identifiable. In addition, they could recognise the peaks that are due to ring breathing modes in DNA bases, such as modes at 678 cm^{-1}, 785 cm^{-1}, 1337 cm^{-1}, 1485 cm^{-1}, 1510 cm^{-1}, and 1575 cm^{-1}. Because of the reduction in the height of these peaks, it was concluded that the overall DNA concentration is significantly lower in transformed cells than in normal ones.

This was further confirmed by the similarly reduced intensity of the 1093 cm^{-1} mode of the symmetric PO_2^- vibration of the DNA backbone. However, it was illustrated that peaks that are mainly due to protein vibrations are significantly stronger in transformed cells than in normal cells. A peak located at 1447 cm^{-1} was used as a marker mode for the protein concentration, and the 1093 cm^{-1} phosphate backbone vibration as a marker mode for DNA concentration. Furthermore, the 813 cm^{-1} and 1240 cm^{-1} were considered as the position of the two most distinct peaks for RNA, indicating a slightly elevated concentration of RNA in the transformed cells. Other distinct RNA modes include the ribose vibrations at 867 cm^{-1}, 915 cm^{-1}, and 974 cm^{-1}. It was concluded that transformed cells have lower concentrations of nucleic acid and higher concentrations of proteins.

Feld et al. [23] studied liposarcomas. Raman spectra were obtained from liposarcomas and normal tissues. The spectrum of normal tissues shows lipid bands. In addition to lipid bands, the spectrum of liposarcoma shows carotenoid bands at 1528 cm^{-1} and 1156 cm^{-1}. The intensity ratio of the CH_2 bending mode at 1142 cm^{-1} to the C=C stretching band at 1667 cm^{-1} was observed to decrease with the grade of the malignancy.

It has also been shown [25] that nuclear content is increased in colon cancer. This can be concluded from more intense peaks at 1662, 1576, 1458, and 1340 cm^{-1}, corresponding to nucleic acids in carcinoma samples.

In a study focused on breast pathology [26], Hanlon et al. found three Raman bands characteristic of fatty acids (1657, 1442, and 1300 cm^{-1}). The most prominent bands in the spectra were related to structural protein modes (1667, 1452, 1260, 890, and 820 cm^{-1}).

Peak intensities related to glycogen and phosphodiester groups of nucleic acids were found to be different among normal lung cancer samples [27]. Wang et al. showed that bands at 1030 cm^{-1} and 1050 cm^{-1} are found frequently in glycogen-rich tissues, and can be assigned as CH_2OH vibrations and the C–O stretching coupled with C–O bending of the C–OH carbohydrates, respectively. The band at 1080 cm^{-1} was assigned to symmetry phosphate stretching and glycogen, and 1241 cm^{-1} to the asymmetric phosphate stretching. Some other research groups have used almost exactly the same peaks for analysing cervical malignancies [28,29] and prostate cancers [30]. It was concluded that the vibrations of the PO_2^- were mainly due to phosphodiester groups on nucleic acids, and phosphate groups of phospholipids do not generally contribute to these bands. In addition, it was noted that the spectra of DNA and RNA display extremely weak peaks at 1399 cm^{-1} and 1457 cm^{-1}. Proteins and carbohydrates showed bands at 1173 cm^{-1}, due to CO stretching of C–OH groups of serine, threonine, and tyrosine. Proteins also had characteristic bands at 1396 cm^{-1} and 1449 cm^{-1}. These bands were related to symmetric and asymmetric CH_3 bending. Lipid bands were distinguished at 1469 cm^{-1} due to CH_2 bending of acyl chains. One of the interesting findings of this group was the interpretation of the 1153, 1161, and 1172 cm^{-1} spectral bands. The peaks at 1153 and 1161 cm^{-1} were assigned to stretching vibrations of hydrogen bonding C–OH groups, whereas the band at 1172 cm^{-1} was due to the stretching vibrations of nonhydrogen bonded C–OH groups. Using these peaks it was concluded that the loss of hydrogen bonding of the C–OH of amino acid residues of proteins in lung cancer could be observed spectroscopically. In addition, it was found that the malignant cells display decreased intensity of the PO_2^- (asym) band, and increased intensity of the PO_2^- (sym) band.

Yano et al. [31] focused on peak intensities of 1545 cm^{-1} (protein), 1467 cm^{-1} (cholesterol, methylene band), 1084 cm^{-1} (DNA, PO_2^- vibrations), and 1045 cm^{-1} (glycogen, OH stretching coupled with bending) in cancerous lung tissue. It was recognised that in cancerous tissues, levels of DNA and glycogen increase and cholesterol decreases. They found that the H1045:H1467 ratio is an exceptionally useful factor for discrimination of cancerous tissues from noncancerous ones. In other words, if this ratio is larger than 1.4, one can say with confidence that tissue contains squamous cell carcinoma or adenocarcinoma.

Eckel et al. [32] focused on five different regions for peak intensity analysis of human breast samples. These areas are as follows:

1. The amide A and B bands stemming from N–H stretching modes in proteins and nucleic acids (3300 and 3075 cm^{-1}, respectively).
2. The amide I band region (1600–1720 cm^{-1}) of proteins. It was concluded that 1657 cm^{-1} can be assigned to the α-helical structure and 1635 cm^{-1} to the β-sheet structure, while peaks at 1649 cm^{-1} and 1680 cm^{-1} are mainly due to unordered random coils and turns.
3. The amide II band region (1480–1600 cm^{-1}) of proteins.
4. The amide III band region (1180–1300 cm^{-1}) of proteins. Gniadecka et al. did a similar study on human basal cell carcinoma samples [33]. However, the region considered to be due to the amide I and III bands are a bit different in this study. They mentioned 1640–1680 cm^{-1} as amide I and 1220–1330 cm^{-1} as the amide III region. In addition,

840–860 cm^{-1} has been assigned to polysaccharides, 1420–1450 cm^{-1} to CH$_2$ scissoring vibrations in lipids, and 928–940 cm^{-1} to v(C–C) stretching in amino acids proline and valine.

Frank et al. [34] presented a detailed database for collagen peak intensities in normal and cancerous human breast tissue. These intensities are tabulated in Table 2.3. According to these results, collagen has 14 peak intensities, and cancer will cause some changes in the position of these peaks. Almost all of these peaks have been used in human skin samples by Cheng et al. [35].

Pichardo-Molina et al. studied serum samples using Raman spectroscopy and different statistical methods [36]. The blood samples were obtained from 11 patients who were clinically diagnosed with breast cancer and 12 healthy volunteer controls. The wavelength differences between the spectral bands of the control and patient groups were defined. It was shown that serum samples from patients with breast cancer and from the control group can be discriminated using Raman spectroscopy. It was also found that seven band ratios were significant in the diagnosis process. These specific bands might be helpful during screening for breast cancer using Raman spectroscopy of serum samples. The group presented a detailed table documenting all the spectral information observed in the two groups of samples. It is illustrated that there are considerable spectral differences between the peak intensities of control samples and those of cancerous blood serum (Table 2.4).

TABLE 2.3

Collagen Assignments in Human Breast Tissue (34)

Peak Position in Normal Tissue (cm^{-1})	Peak Position in Infiltrating Ductal Carcinoma (cm^{-1})	Collagen Assignment
–	817	C–C stretch
727	856	C–C stretch, proline
870	876	C–C stretch, hydroxyproline
–	920	C–C proline ring
972	937	C–C backbone
–	1004	Phenylalanine
1066	1043	Proline
–	1206	Hydroxyproline, tyrosine
–	1247	Amide III
1265	1267	Amide III
1303	1303	CH$_3$CH$_2$ twisting
–	1343	CH$_3$CH$_2$ wagging
1439	1450	CH$_3$CH$_2$ deformation
1654	1657	Amide I

Source: C. J. Frank, R. L. McCreecy, and D. C. B. Redd. Raman spectroscopy of normal and diseased human breast tissues. *Anal. Chem.* 67(1995): 777–783. With permission.

TABLE 2.4

Spectral Peaks Observed in Normal Blood Samples and Blood Samples of Patients with Breast Cancer

Control	Cancer	p Values	Bands Assignment
1,714.22 ± 5.2	1,719.64 ± 5.6	<0.05	
1,658.52 ± 0.6	1,660.01 ± 1.0	<0.05	Proteins, amide, α helix, phospholipids
1,609.00 ± 1.1	1,609.32 ± 1.0	<0.05	Tyr, Phen.
1,582.42 ± 1.1	1,582.10 ± 1.0	0.44	Protein, Tyr, Arg, Adenine
1,557.50 ± 3.0	1,555.10 ± 1.2	<0.05	Protein retinal –C=C–C=O
1,523.10 ± 3.1	1,523.80 ± 2.0	0.42	Beta carotene
1,464.14 ± 1.3	1,463.01 ± 1.0	<0.05	
1,446.01 ± 1.0	1,445.16 ± 1.0	<0.05	Phospholipid, C–H scissor in CH_2
1,415.00 ± 4.0	1,416.41 ± 2.2	0.27	
1,340.00 ± 1.1	1,341.04 ± 1.1	<0.05	Trp, Adenine α helix, Phospholipids
1,311.10 ± 1.1	1,312.03 ± 1.5	0.15	Adenine
1,271.01 ± 4.0	1,266.56 ± 2.1	<0.05	Phospholipid, Amide III
1,236.20 ± 2.4	1,234.23 ± 2.1	0.06	β sheet
1,205.03 ± 0.5	1,205.10 ± 0.4	0.11	
1,176.23 ± 1.0	1,177.00 ± 0.5	0.10	Trp, Phen
1,155.30 ± 0.6	1,155.53 ± 0.4	0.30	Beta carotene
1,126.10 ± 0.6	1,126.05 ± 0.4	0.54	Protein, Phospholipid C–C stretch
1,103.40 ± 1.0	1,104.01 ± 0.6	<0.05	Phen
1,082.14 ± 1.0	1,082.00 ± 1.0	0.51	Phospholipids O–P–O and C–C
1,061.04 ± 2.1	1,062.34 ± 2.0	0.06	Phen
1,032.54 ± 0.4	1,032.50 ± 0.2	0.76	Phen
1,003.13 ± 0.6	1,003.01 ± 0.3	0.82	Phen
959.10 ± 0.4	959.05 ± 0.4	<0.05	CH_2 rock
938.04 ± 1.0	937.50 ± 1.0	<0.05	Skeletal stretch α
897.03 ± 1.0	897.00 ± 1.0	0.40	
867.30 ± 1.0	876.50 ± 0.4	0.52	Trp
852.00 ± 1.0	851.66 ± 0.3	0.95	Tyr
828.00 ± 1.0	828.00 ± 0.3	0.76	CH rock in CH_2
757.00 ± 1.4	757.41 ± 0.6	0.16	Protein
741.34 ± 2.0	742.31 ± 0.0	0.16	Phospholipid
661.00 ± 1.0	659.73 ± 1.3	0.30	
714.00 ± 3.0	715.00 ± 1.3	0.42	Polysaccharides
642.02 ± 1.0	641.40 ± 03.	0.07	
621.03 ± 1.0	620.54 ± 0.3	0.36	Protein ring mode

Source: W. T. Cheng, M. T. Liu, H. N. Liu, and S. Y. Lin. 2005. Micro-Raman spectroscopy used to identify and grade human skin pilomatrixoma. *Microsc. Res. Tech.* 68(2005): 75–79. With permission.

The information presented in this chapter covers the changes that take place at the molecular level in cancer cells and can be detected using the spectroscopic techniques described. It is important to understand the biology and chemistry of cancer cells, which will help in analysing chemical structural properties of the cancer cells by spectroscopic means. A good knowledge of cell structure will allow to us to predict and follow the chemical pathway by which the disease progresses.

References

1. Stone, N., Kendell, C., Shepherd, N., Crow, P. and Barr, H. 2002. Near-infrared Raman spectroscopy for the classification of epithelial pre-cancers and cancers. *J. Raman Spectrosc.* 33, 564–573.
2. Stone, N., Kendall, C., Smith, J., Crow, P. and Barr, H. 2004. Raman spectroscopy for identification of epithelial cancers. *Faraday Discussion* 126, 141–157.
3. Franks, L. M. 1991. *Introduction to the Cellular and Molecular Biology of Cancer.* New York: Oxford University Press.
4. LaFond, R. E. 1988. *Cancer: The Outlaw Cell*, 2nd ed. Washington, DC: American Chemical Society.
5. Lodish H., Berk A., Zipursky S. L., Matsudaira P., Baltimore D., and Darnell J. E. 2000. *Molecular Cell Biology*, 4th ed. New York: W. H. Freeman & Co.
6. King, R. J. B., and Benjamin R. J. 2006. *Cancer Biology*, 3rd ed. New York: Pearson/ Prentice Hall.
7. Dukor, R. K. 2002. Vibrational spectroscopy in the detection of cancer, *Biomedical Applications*, 3335–3359.
8. Cooper, G. M., 2003. *The Cell: A Molecular Approach*, 3rd ed. Sunderland, MA: Sinauer Associates.
9. Gortan Kumar, Robbins. 1999. *Pathologic Basis of Disease.* Philadelphia: Saunders.
10. Huang, Z., McWilliams, A., Lui, M., McLean, D. I., Lam, S., and Zeng, H. 2003. Near-infrared Raman spectroscopy for optical diagnosis of lung cancer. *Int. J. Cancer* 107, 1047–1052.
11. Huang, Z., Lui, H., McLean, D. I., Korbelik, M., and Zeng, H. 2005. Raman spectroscopy in combination with background near-infrared autofluorescence enhances the in vivo assessment of malignant tissues. *Photochem. Photobiol.* 81, 1219–1226.
12. Austrian Centre for Electron Microscopy and Nanoanalysis. 2010. Infrared and Raman microspectrometry, Graz, Austrian Centre for Electron Microscopy and Nanoanalysis, http://www.felmi-zfe.tugraz.at/ (accessed on December 10, 2011).
13. Skala, M. C., Palmer, G. M., and Vrotsos, K. M. 2007. Comparison of a physical model and principal component analysis for the diagnosis of epithelial neoplasias in vivo using diffuse reflectance spectroscopy. *Optics Express* 15(12), 7863–7875.
14. Kneipp, K., Kneipp, H., Kartha, V. B., Manoharan, R., Deinum, G., Itzkan, I., Dasari, R. R., and Feld, M. S. 1998. Detection and identification of a single DNA base molecule using surface-enhanced Raman scattering (SERS). *Phys. Rev.* E57(6), 6281–6284.

15. Yang, Y., Sule-Suso, J., Sockalingum, G. D., Kegelaer, G., Manfait, M., and El Haj, A. J. 2005. Study of tumour cell invasion by Fourier transform infrared microspectroscopy. *Biopolymers* 78, 311–317.
16. Min, Y. K., Yamamoto, T., Kohoda, E., Ito, T., and Hamaguchi, H. 2005. 1064 nm near-infrared multichannel Raman spectroscopy of fresh human lung tissues, *J. Raman Spectrosc.* 36, 73–76.
17. Kaminaka, S., Yamazaki, H., Ito, T., Kohoda, E., and Hamaguchi, H. 2001. Near-infrared Raman spectroscopy of human lung tissues: Possibility of molecular-level cancer diagnosis. *J. Raman Spectrosc.* 32, 139–141.
18. Huleihel, M., Salman, A., Erukhimovich, V., Ramesh, J., Hammody, Z., and Mordechai, S. 2002. Novel optical method for study of viral carcinogenesis in vitro. *J. Biochem. Biophys. Methods.* 50, 111–121.
19. McIntosh, L. M., Jackson, M., Mantsch, H. H., Stranc, M. F., Pilavdzic, D., and Crowson, A. N. 1999. Infrared spectra of basal cell carcinomas are distinct from non-tumour-bearing skin components. *J. Invest. Dermatol.* 112, 951–956.
20. Andrus, P. G. 2006. Cancer monitoring by FTIR spectroscopy. *Technol. Cancer. Res. Treat.* 5, 157–167.
21. Schultz, H., and Baranska, M. 2007. Identification and qualification of valuable plant substances by IR and Raman spectroscopy. *Vibrational Spectrosc.* 43, 13–25.
22. Lakshmi, R. J., Kartha, V. B., Krishna, C. M., Solomon, J. G. R., Ullas, G., and Uma Devi, P. 2002. Tissue Raman spectroscopy for the study of radiation damage: Brain irradiation of mice. *Radiat. Res.* 157, 175–182.
23. Feld, M. S., Manoharan, R., Salenius. J., Orenstein-Carndona, J., Romer, T. J., Brennan III, J. F., Dasari, R. R., and Wang, Y. 1995. Detection and characterization of human tissue lesions with near-infrared Raman spectroscopy. *SPIE Advances in Fluorescence Sensing Technology* 2388, 99–104.
24. Chan, J. W., Taylor, D. S., Zwerdling, T., Lane, S. T., Ihara, K., and Huser, T. 2006. Micro-Raman spectroscopy detects individual neoplastic and normal hematopoietic cells. *Biophys. J.* 90, 648–656.
25. Mahadevan-Jansen, A., and Richards-Kortum, R. 1997. Raman spectroscopy for cancer detection, 19th International Conference IEEE EMBS Chicago, IL, October 30–November 2.
26. Hanlon, E. B., Manoharan, R., Koo, T. W., Shafer, K. E., Motz, J. T., Fitzmaurice, M., Kramer, J. R., Itzkan, I., Dasari, R. R., and Feld, M. S. 2000. Prospects for in vivo Raman spectroscopy. *Phys. Med. Biol.* 45, 1–59.
27. Wang, H. P., Wang, H.-C., and Huang, Y.-J. 1997. Microscopic FTIR studies of lung cancer cells in pleural fluid. *Sci. Total Environ.* 204, 283–287.
28. Wood, B. R., Quinn, M. A., Burden, F. R., and McNaughton, D. 1996. An investigation into FT-IR spectroscopy as a bio-diagnostic tool for cervical cancer. *Biospectroscopy* 2, 143–153.
29. Wood, B. R., Quinn, M. A., Tait, B., Ashdown, M., Hislop, T., Romeo, M., and McNaughton, D. 1998. FTIR microspectroscopic study of cell types and potential confounding variables in screening for cervical malignancies. *Biospectroscopy* 4, 75–91.
30. Gazi, E., Dwyer, J., Gardner, P., Ghanbari-Siakhani, A., Wde, A. P., Lockyer, N. P., Vickerman, J. C., Clarke, N. W., Shanks, J. H., Scott, L. J., Hart, C. A., and Brown, M. 2003. Applications of Fourier transform infrared microspectroscopy in studies of benign prostate and prostate cancer: A pilot study. *J. Pathol.* 201, 99–108.

31. Yano, K., Ohoshima, S., Grotou, Y., Kumaido, K., Moriguchi, T., and Katayama, H. 2000. Direct measurement of human lung cancerous and noncancerous tissues by Fourier transform infrared microscopy: Can an infrared microscope be used as a clinical tool? *Anal. Biochem.* 287, 218–225.

32. Eckel, R., Huo, H., Guan, H. W., Hu, X., Che, X., and Huang, W. D. 2001. Characteristic infrared spectroscopic patterns in the protein bands of human breast cancer tissue. *Vibrational Spectrosc.* 27, 165–173.

33. Gniadecka, M., Wulf, H. C., Mortensen, N. N., Nielsen, O. F., and Christensen, D. H. 1997. Diagnosis of basal cell carcinoma by Raman spectroscopy. *J. Raman. Spectrosc.* 28, 125–129.

34. Frank, C. J., McCreecy, R. L., and Redd, D. C. B. 1995. Raman spectroscopy of normal and diseased human breast tissues. *Anal. Chem.* 67, 777–783.

35. Cheng, W. T., Liu, M. T., Liu, H. N., and Lin, S. Y. 2005. Micro-Raman spectroscopy used to identify and grade human skin pilomatrixoma. *Microsc. Res. Tech.* 68, 75–79.

36. Pichardo-Molina, J. L., Frausto-Reyes, C., Barbosa-García, O., Huerta-Franco, R., González-Trujillo, J. L., Ramírez-Alvarado, C. A., Gutiérrez-Juárez, G., and Medina-Gutiérrez, C. 2007. Raman spectroscopy and multivariate analysis of serum samples from breast cancer patients. *Lasers Med. Sci.* 22, 229–236.

3

The Biology of Cancer

To understand chemical structural properties of cancer and associated changes with the progression of disease, it is important to understand the biology and chemistry of cancer. The term *neoplasia*, which literally means *new growth*, is used clinically to describe pathologic tissue masses, which grow independent of and faster than normal tissues. Most of the body cells repair and reproduce themselves similarly in spite of their different functioning and appearance. As long as this process takes place in an orderly manner, everything is under control. However, when it gets out of control, a lump, called a *tumour*, develops. A neoplasm (tumour) may be classified as in situ, benign, or malignant, based on its potential to harm the host body. Malignant tumours are also called *cancers* [1,2,3]. Benign tumours are not considered cancer. They do not threaten life because they do not spread to other organs, and once they are removed, they rarely reappear. In situ tumours look like cancer in their morphology. In general, they develop in epithelium. These small tumours remain in the epithelial layer. A malignant tumour consists of cancer cells that sometimes spread away from the original (primary) cancer to other organs through the lymphatic system and bloodstream causing damage to organs and tissues near the tumour. When they get to another part of the body, they may divide forming a new tumour called a *secondary* or a *metastasis tumour* [4]. A biopsy is used by doctors to decide whether a tumour is benign or cancerous in which a small sample of cells is examined under a microscope. Uncontrolled growth is the core property of cancers, which were known to the early Egyptians [3,5]. Hippocrates reportedly distinguished benign from malignant tissues [6].

Although cancer has multiple causes and exceptions, an acceptable clinical definition would be: a set of diseases characterized by unregulated cell growth leading to invasion of surrounding tissues and spread (*metastasis*) to other parts of the body. Invasion and metastasis distinguish cancers from benign growths [5].

The nomenclature of a neoplasm reflects its microscopic origin as well as the potential to harm the host. For example, cancers arising from glandular epithelial tissues are called *adenocarcinomas*, where the suffix *carcinoma* indicates the epithelial origin and the prefix *adeno* reflects the glandular origin. On the other hand, cancers arising from the mesenchymal tissue are called *sarcomas* [1,3]. Leiomyosarcoma (sarcoma of smooth muscle) and liposarcoma (sarcoma of fat cells) are examples of this type [5]. *Leukemias* and *lymphomas*, which account for approximately 7% of human malignancies, arise from the blood-forming cells and from cells of the immune system, respectively [7]. All cell types are susceptible but epithelial cells are most prone to change. This means that most human malignancies arise from epithelial tissue [5,6]. In fact, most human cancers (approximately 90%) are carcinomas. This might be explained by two factors: cell proliferation mainly occurs in epithelia, and they are more frequently exposed to various forms of physical and chemical damage that favor the development of cancer [8]. Dedifferentiation or anaplasia denotes the loss of normal characteristics.

Epidemiology of Cancer

Epidemiology is the study of distribution and causes of disease in human populations [5,9]. The epidemiological approach aims to provide an accurate and representative picture of the pattern of cancer in the population as a whole, and to help uncover the causes of cancer. The basis of the epidemiological approach is simple. Information is collected in a rigorously standardized manner for every individual (or a representative sample of individuals) diagnosed with or dying from cancer in a defined population, and the resulting data set is analyzed with suitable statistical techniques [9].

In developed countries, cancer is the second-biggest cause of death [10]. It has been proven that some factors can lead to an increased risk of cancer development. For instance, lack of exercise [11], hypercholesterolemia [12], increasing age [10], geographical differences, sex variations [13], ethnic and racial differences, diet, smoking [14], and a family history of cancer can have significant effects on the epidemiology and incidence of the disease [5,9].

Although there are many kinds of cancer, only a few occur frequently. Figure 3.1 [15] and Table 3.1 show the incidence and mortality rates of some of the most common malignancies in the United Kingdom and the United States. Cancer will afflict up to one in three persons before their seventy-fifth birthday in developed countries, and one in four persons will die from it [9].

More than one million cases of cancer are diagnosed annually in the United States, and more than 500,000 Americans die of cancer each year. Cancers of 11 different body sites account for more than 80% of this total cancer incidence [7]. In the United Kingdom, almost a quarter of a million cancers are diagnosed each year, and there are some 160,000 deaths [16].

Unfortunately, it has been shown that the number of cases that are diagnosed with cancer in the age group 55–84 is much higher in the United

	Cancer Origin	Mortality	Incidence
1	Lung	33,044	37,127
2	Colorectal	16,148	35,006
3	Breast	12,417	44,091
4	Prostate	10,209	31,900
5	Oesophagus	7,230	7,611
6	Pancreas	7,045	7,156
7	Stomach	5,881	8,432
8	Bladder	4,816	10,148
9	Ovary	4,434	6,906
10	NHL	4,418	9,772

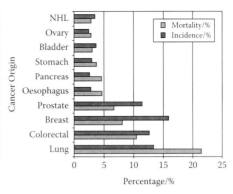

FIGURE 3.1
The 10 most common UK cancers (2004). [15]

TABLE 3.1

Most Frequent Cancers in the United States

Cancer Site	Deaths per Year	Cases per Year
Prostate	28,900 (5.2%)	220,900 (16.6%)
Breast	40,200 (7.2%)	212,600 (15.9%)
Lung	157,200 (28.2%)	171,900 (12.9%)
Colon/rectum	57,100 (10.3%)	147,500 (11.1%)
Lymphomas	24,700 (4.4%)	61,000 (4.6%)
Bladder	12,500 (2.2%)	57,400 (4.3%)
Skin (melanoma)	7,600 (1.4%)	54,200 (4.1%)
Uterus	10,900 (2.0%)	52,300 (3.9%)
Kidney	11,900 (2.1%)	31,900 (2.4%)
Pancreas	30,000 (5.4%)	30,700 (2.3%)
Leukemias	21,900 (3.9%)	30,600 (2.3%)
Subtotal	402,900 (72.4%)	1,071,000 (80.3%)

Kingdom in comparison to other European countries (Table 3.2). In addition, it has also been shown that there is a wide difference in cancer survival between middle-aged and elderly patients in Europe [17].

Cancer Development

One of the most common characteristics of the development of a neoplasm is the extended period of time between the initial application of a carcinogenic (cancer-causing) agent—physical, chemical, or biological—and

TABLE 3.2

Number of Cases for Considered Cancers, Diagnosed during the Period 1990–1994, by Sex, Age Group, and European Country (EUROCARE-3) [17]

	Number of Cases			
	Men		Women	
Country	55–64 (years)	65–84 (years)	55–64 (years)	65–84 (years)
North				
Denmark	7,249	21,304	9,186	20,858
Finland	6,350	17,716	6,559	14,849
Iceland	285	926	357	607
Norway	5,513	20,604	5,458	15,678
Sweden	10,149	43,956	12,253	32,166
United Kingdom				
England	44,590	159,987	52,139	126,556
Scotland	8,445	25,574	9,552	21,971
Wales	4,278	16,274	5,134	12,986
Centre-West				
Austria	1,049	2,771	870	2,467
France	3,261	6,899	3,130	5,885
Germany	2,434	5,764	1,899	5,819
Switzerland	1,055	13,883	1,012	10,528
The Netherlands	5,619	2,491	4,932	2,279
South				
Italy	19,667	47,802	15,360	35,822
Malta	134	425	181	350
Portugal	425	1,181	1,124	1,264
Spain	7,517	16,294	5,297	10,189
East				
Czech Republic	1,926	3,154	1,352	2,710
Estonia	2,964	3,855	2,344	4,152
Poland	4,405	6,646	3,808	6,506
Slovakia	7,462	12,922	5,623	9,507
Slovenia	3,326	4,891	2,710	4,737
All Countries	*148,103*	*435,319*	*150,280*	*347,886*

Source: A. Quaglia, R. Capocaccia, A. Micheli, et al. 2007. A wide difference in cancer survival between middle aged and elderly patients in Europe. *International Journal of Cancer* 120, no. 10 (2007): 2196–2201.

the appearance of the neoplasm. This latency phenomenon or tumour induction time occurs even when the carcinogen is administered continuously to an experimental animal [3]. This is probably because of the fact that cancer develops through a succession of stages marked by the accumulation of genetic events within the cell [18].

It is generally believed that most cancers have a monoclonal (single-cell) origin [18]. This single cell can give rise to a focus of neoplastic cells. In the case of epithelial tissues, if the neoplastic cells are confined to the epithelium, the lesion is regarded as a precancer. These precancer lesions can progress by invading the basement membrane to become full cancers [1,3]. It has been recognized that gross changes in genetic material, particularly nonrandom chromosomal abnormalities such as translocations, characterize many human cancers [9]. These abnormalities are illustrated by dividing excessively to form lumps and tumours and also by failing to differentiate normally [3]. The progression from normal cell to cancer cell is a complex multistep process, which is called *carcinogenesis*.

Carcinogenesis is a multistep process involving the cooperation of several oncogenic mutations [20–23]. Single oncogenic mutations are not sufficient to cause malignancy, as they frequently induce multiple contradictory signals inhibiting cancer cell proliferation [24]. This is exemplified in the control of cell cycle progression, where activation of signals can stimulate or inhibit cell cycle entry [25,26]. In this context, cell cycle inhibition often results from a stress response mediated by activation of the tumour suppressors (such as p. 53), which is disabled upon loss of function of these genes, leading to stimulation of uncontrolled proliferation [27–29].

Initially, genetic alterations are thought to confer a growth advantage to individual cells by either decreasing tumour suppressor gene activity or increasing oncogene activity or both. Further genetic alterations result in the development of cell clones that have the ability to invade adjacent tissues, establish metastatic deposits, and evade immune surveillance. At some point in the process, these malignant cell clones also lose the normal ability to respond to hormonal growth regulatory signatures [11]. Cell changes continue to occur even after a cancer has formed [5].

In a simple description of carcinogenesis, a tumour (or neoplasm) begins when some cell within a normal population sustains a mutation that increases its propensity to proliferate (*tumour initiation*) [7]. The altered cell and its daughters look normal, but they grow and divide too much. This condition is termed *hyperplasia*. After some time, which can be years, one of these cells suffers another mutation that further reduces control on cell growth. This cell proliferates very fast and the offspring of this cell appears abnormal in shape and orientation. This state is known as *dysplasia*. Again, after some time, a mutation occurs that alters cell behavior (*tumour progression*) [7]. The affected cells become even more abnormal in growth and appearance. If the tumour is contained within original tissue, it is called *in-situ carcinoma*. If cells break away from such a tumour, they can travel through the bloodstream or lymph system to other areas of the body and establish new tumours. They continue to grow in new locations. This process, as mentioned before, is referred to as *metastasis* [6].

The fact that cancer is predominantly a disease of the elderly, and age is the most important determinant of cancer risk for both males and females [10,16], can illustrate the multistep and successive nature of the disease and the necessity of time for its establishment. Most cancers arise after 65 years of age [16], and an increasing proportion at age 85 and over [5].

Causes of Cancer

Cancers arise mainly from deregulated proliferation. However, modulation of the normal processes that leads to cell loss (programmed cell death or apoptosis) is also sufficient, although not necessary, for tumour development [30]. Since the development of malignancy is a complex multistep process, many factors may affect the likelihood that cancer will develop, and it is overly simplistic to speak of single causes of most cancers. Nonetheless, many agents, including radiation, chemicals, and viruses, have been found to induce cancer in both experimental animals and humans. These agents are called *carcinogens* [7].

DNA mutation is the well-known explanation for the initiation of any malignancy, but the cell must also be incited into proliferation [31]. DNA mutations to facilitate such expansion must occur on one or more genes affecting cell growth, differentiation, death, and gene replication. DNA mutations may occur through environmental causes, but are most likely to occur from internal factors such as endocrine estrogen exposure and errors in DNA synthesis or replication.

It must be mentioned that in order to initiate any malignancy, such alterations should take place in several genes [32]. For instance, it has been identified that in esophageal squamous cell carcinoma (ESCC), 147 genes are up-regulated and 376 are down-regulated [33], or 167 genes have been demonstrated to be significantly overexpressed and strongly related to radioresistance of oral squamous cell carcinoma [34]. In addition, the results of some studies demonstrate the feasibility of prognosis identification using the altered genes [35–37]. A gene expression signature consisting of 70 genes, for example, has been identified as strongly predictive of good or poor prognosis in breast cancer [38].

Radiation and chemical carcinogens act by damaging DNA and inducing mutations. Some of the carcinogens that contribute to human cancers include solar ultraviolet radiation (the major cause of skin cancer), carcinogenic chemicals in tobacco smoke, and aflatoxin (a potent liver carcinogen produced by some molds that contaminate improperly stored supplies of peanuts and other grains). The carcinogens in tobacco smoke (including benzo$_{(a)}$pyrene, dimethylnitrosamine, and nickel compounds) are the major identified causes of human cancer [7].

Other carcinogens contribute to cancer development by stimulating cell proliferation, rather than by inducing mutations. Such compounds are referred to as *tumour promoters*. The phorbol esters that stimulate cell proliferation by activating protein kinase C are classic examples. Some hormones, such as estrogens, can be important tumour promoters in the development of some human cancers (such as endometrial cancer in women), as well [7].

In addition to chemicals and radiation, some viruses induce cancer both in experimental animals and in humans. The common human cancers caused by viruses include liver cancer and cervical carcinoma. Hepatitis B and C viruses, papillomaviruses, adenomaviruses, herpesviruses, retroviruses, and polioviruses are some instances of tumour viruses [7].

Mechanisms in Generation of Tumours

Tumours, as explained, result from disruption of the process that controls the normal growth and the mortality of cells. This loss of normal control mechanisms arises from the acquisition of mutations in three broad categories of cells:

1. Proto-oncogenes, the normal products of which are components of signaling pathways that regulate proliferation and which in their mutated form, become dominant oncogenes [39].
2. Tumour suppressor genes, which generally exhibit recessive behaviour, the loss of function of which leads to deregulated control of cell cycle progression and inability to undergo apoptosis.
3. DNA repair genes, mutations in which promote genetic instability.

Oncogenes

The oncogene hypothesis was first proposed by Robert J. Huebner and George J. Todaro [40]. The term *oncogene* itself is derived from the Greek word "oncos," meaning tumour [18]. Cancer results from alterations in critical regulatory genes that control cell proliferation, differentiation, and survival. Studies of tumour viruses revealed that specific genes (called *oncogenes*) are capable of inducing cell transformation. These genes were first identified in viruses capable of inducing tumours in animals and of transforming cells in vitro, and are altered versions of the normal genes that control cell growth and function at every level of growth regulation [41–43]. Although there are still a considerable number of researches being carried out on oncogenic viruses [44–47], a considerable majority (approximately 80%) of human cancers is not induced by viruses and apparently arise from other causes, such as radiation and chemical carcinogens [7,40]. Furthermore, despite the potent tumourigenic capacity of oncogenes in appropriate animal hosts, no retrovirus has yet been shown to be directly oncogenic in humans [18].

Retroviruses have RNA genomes and can replicate through a DNA intermediate in infected cells. The oncogenes carried by these viruses are strongly homologous in sequence to normal cellular genes (*proto-oncogenes*) that are highly conserved in evolution [3,18]. In other words, the normal-cell genes from which the retroviral oncogenes originated are called *proto-oncogenes*. Proto-oncogenes are found at all levels of the different signal transduction cascades that control cell growth, proliferation, and differentiation, and have roles in normal embryonic development. They are important cell regulatory genes, in many cases encoding proteins that function in the signal transduction pathways controlling normal cell proliferation (e.g., *src*, *ras*, and *raf*).

Oncogenes are mutated or abnormally expressed forms of the corresponding proto-oncogenes, whose functions are to encourage and promote the normal growth and division of cells. When proto-oncogenes mutate to become carcinogenic oncogenes, the result is excessive cell multiplication. Consequently, oncogenes are usually expressed at much higher levels than the proto-oncogenes and are sometimes transcribed in inappropriate cell types. In some cases, such abnormalities of gene expression are sufficient to convert a normally functioning proto-oncogene into an oncogene that drives cell transformation. In addition to such alterations in gene expression, oncogenes frequently encode proteins that differ in structure and function from those encoded by their normal homologs.

The process of activation of proto-oncogenes to oncogenes can include retroviral transduction or retroviral integration, point mutations, insertion mutations, gene amplification, chromosomal translocation and/or protein–protein interactions. Point mutations may arise from the action of chemicals or radiation. Mutation of the coding sequence can result in formation of hyperactive protein in normal amounts. The mechanism of gene amplification can result in overproduction of a normal protein product. DNA rearrangement (chromosome translocation) can give rise to elevated cellular concentrations of the normal protein product or to the expression of new proteins created by in-frame fusion of coding sequences from separate genes.

The DNA viruses of the adenovirus, herpesvirus, poxvirus, and papovavirus families also possess oncogenic potential. Some DNA tumour viruses are oncogenic in humans (Hepatitis B virus, Epstein-Barr virus), but most are only tumourigenic in other species. The oncogenes of DNA viruses differ from those of retroviruses and their transforming genes have not yet been shown to have proto-oncogenes homologous within the normal human genome, except for a few including the presence of *BLC2* sequences in the Epstein-Barr virus (EBV) *BHRF1* gene.

The *Myc* oncogenes are a group of well-known oncogenes. It is believed that the majority of human cancers have a deregulated *Myc* (*c- Myc, N- Myc, L- Myc*) [48,49]. Some researchers have mentioned a relationship of 70% [50]. This oncogene encodes transcription factors that bind to DNA, driving the expression of a vast number of target genes. The biological outcome is enhanced proliferation (which is counteracted by apoptosis), angiogenesis, and cancer. *Myc* is sufficient to drive cells into S-phase from a quiescent state. On the other hand, it also induces cell death by apoptosis, which is thought to reflect a cellular defense against illegitimate cell proliferation [51,52].

Oncogenes encode proteins called *oncoproteins*, which have lost important regulatory constraints on their activity, and do not need external activation signals [53]. Oncoproteins associated with the inner surface of the cellular membrane are encoded members of the largest group of oncogenes that includes the src and ras families of gene types [53]. The main classes of oncoproteins include growth factors (HSTF1, INT2, PDGFB/SIS, WNT1, WNT2, WNT3, etc.), tyrosine kinases (FMS, EGFR, KIT, MET, HER2/NEU, RET, TRK, SRC), serine-threonine kinases (AKT1, AKT2), membrane-associated guanine nucleotide binding proteins (HRAS, KRAS, NRAS), cytoplasmic regulators (CRK), cell cycle regulators (INK4A, INK4B, CYCLIN, D1, CDC25A, CDC25B), transcription factors (BCL3, E2F1, ERBA, ETS, FOS, JUN, MYB, MYC, REL, TALI, SKI), intracellular membrane factor (BCL2), RNA binding proteins (EWS), and others; and these act at different points in signaling pathways [18]. Oncogene activation is mainly the result of somatic events rather than hereditary causes transmitted by mutation in the germline.

Tumour Suppressor Genes

The existence of tumour suppressor genes was first suggested in studies of rare inherited cancer syndromes. In familial retinoblastoma, for example, a genetic change was inherited through the germline, but mutation or ablation of the other allele on the homologous chromosome was required for tumour development (Knudson's two hit hypothesis) [54,55]. In sporadic cases, independent mutational events in both homologous genes are required. As a result, sporadic retinoblastoma is less common than the familial form [56].

In normal organisms, cell proliferation is balanced by regulated cell loss through apoptosis. Many tumour suppressor genes are involved in cell proliferation, genetic stability, and cell death, and they often show mutation in cancer. Such mutations in these genes generally ablate the ability to cause apoptosis and result in increased cell numbers.

The normal function of tumour suppressor genes is to control cell proliferation. The two best understood tumour suppressor genes are the retinoblastoma (*RB1*) gene and the *TP53* (also called *P53*). Retinoblastoma provides the classical model for a recessive tumour suppressor gene in that both paternal and maternal copies of *RB1* must be inactivated for the tumour to develop. For *TP53* and some other tumour suppressor genes, mutation at one allele may be sufficient to give rise to the altered cell phenotype [56–60].

Possible Functions of Tumour Suppressor Genes

The following are some possible functions of tumour suppressor genes:

1. Direct reversal of oncogene action: In order to reverse the action of oncogenes, some tumour suppressor genes introduce more copies of the normal gene into the cell.

2. Genes that encode growth inhibitors: Most of the oncogenes act by inducing cell proliferation at the wrong time and place. A set of factors that inhibit cell growth can work against this action of tumour producing genes. Some examples of proteins with inhibitory properties are the interferons, tumour growth factor-β, and the tumour necrosis factor [62].

3. Cell–cell interaction: Several investigators have proposed that this action of some normal-cell products and drugs is correlated with presence of gap junctions, through which the cell–cell contact is achieved.

4. Tumour angiogenesis factor: Angiogenesis is the process by which a vascular network of blood vessels develops, and its importance in tumourigenesis has been recognized. Tumour angiogenesis factor (TAF), a substance synthesized and secreted by tumour cells, is essential for their survival. Some genes encode a few cell products that interfere with the action of TAFs.

5. Suppression by terminal differentiation: Cells that have undergone terminal differentiation no longer divide and thus cannot give rise to tumours. One mode of suppression, therefore, consists of dividing partially differentiated cells into the terminal state.

6. Senescence: When normal human cells are grown in culture, they stop dividing after about 50–60 doublings of the population [3]. This phenomenon is called *senescence*. However, cancer cells do not senesce, and grow indefinitely, like bacteria, if given suitable growing conditions. Some cellular products can cause the cancerous cells to senesce. This will lead to the suppression of the effect of cancer promoting genes.

The Cell Cycle

All cells reproduce by dividing in two, with each parental cell giving rise to two daughter cells on completion of each cycle of cell division. These newly formed daughter cells can themselves grow and divide, giving rise to a new cell population formed by the growth and division of a single parental cell and its progeny. In meiosis, however, a cell is permanently transformed and cannot divide again [1,63].

Cell division is usually a small segment of a larger cell cycle, and is the biological basis of life. The division cycle of most cells consists of four coordinated process: cell growth, DNA replication, distribution of the duplicated chromosomes to daughter cells, and cell division. The primary concern of cell division is the maintenance of the original cell's genome. Although cell growth is usually a continuous process, DNA is synthesized during only one phase of the cell cycle, and the replicated chromosomes are then distributed to daughter nuclei by a complex series of events preceding cell division. Progression between these stages of the cell cycle is controlled by a conserved regulatory apparatus, which not only coordinates the different events of the cell cycle, but also links the cell cycle with extracellular signals that control cell proliferation. In other words, the cell cycle machinery is itself regulated by the growth factors that control cell proliferation, allowing the division of individual cells to be coordinated with the needs of the organism as a whole. Not surprisingly, defects in cell cycle regulation are a common cause of the abnormal proliferation of cancer cells, so studies of the cell cycle and cancer have become closely interconnected [64].

Phases of the Cell Cycle

A typical eukaryotic cell cycle for human cells in culture takes place approximately every 24 hours. The cell cycle is divided into two basic parts: *mitosis* and *interphase*. Mitosis (nuclear division) is the most dramatic stage of the cell cycle, corresponding to the separation of daughter chromosomes and usually ending with cell division (*cytokinesis*). However, mitosis and cytokinesis last only about an hour, so approximately 95% of the cell cycle is spent in interphase or the period between mitosis. During interphase, the chromosomes are decondensed and distributed throughout the nucleus and appear as a network of long, thin threads called *chromatin*, so the nucleus appears morphologically uniform. At the molecular level, however, is where both cell growth and DNA replication occur in an orderly manner in preparation for cell division.

The cell grows at a steady rate throughout interphase. At some point before prophase begins, the chromosomes begin to replicate themselves to form pairs of identical chromosomes, and most dividing cells double in size between one mitosis and the next. In contrast, DNA is synthesized during only a portion of interphase. The timing of DNA synthesis thus divides the cycle of eukaryotic cells into four discrete phases. The *M phase* of the cycle corresponds to mitosis, which is usually followed by cytokinesis. This phase is followed by the G_1 *phase* (gap 1), which corresponds to the interval (gap) between mitosis and initiation of DNA replication. A rapidly dividing human cell, which divides every 24 hours, spends 9 hours in the G_1 phase [4]. The metabolic rate of the cell will be high. However, a cell may pause in the G1 phase before entering the S phase and enter a state of dormancy called the G0 phase. Most mammalian cells do this. G1 consists of four subphases:

1. Competence (G1a)
2. Entry (G1b)
3. Progression (G1c)
4. Assembly (G1d)

During G_1, the cell is metabolically active and continuously grows but does not replicate its DNA. G_1 is followed by the *S phase* (synthesis), during which DNA replication takes place. At the beginning of the S phase, each chromosome is composed of one coiled DNA double helix molecule, which is called a *chromatid*. At the end of this stage, each chromosome has two identical DNA double helix molecules, and therefore is composed of two sister chromatids. During the S phase, the centrosome is also duplicated. These two events are unconnected, but require many of the same factors to progress. The end result is the existence of duplicated genetic material in the cell, which will eventually be divided into two. Damage to DNA often takes place during this phase, and DNA repair is initiated following the completion of replication. The completion of DNA synthesis is followed by the *G2 phase* (gap 2), during which cell growth continues and proteins are synthesized in preparation for mitosis. Although chromosomes have been replicated, they cannot yet be distinguished individually because they are still in the form of loosely packed chromatin fibers. This phase is the third, final, and usually the shortest subphase during interphase within the cell cycle. At the end of this gap phase is a control checkpoint (G2 checkpoint) to determine if the cell can proceed to enter the M phase and divide. The G2 checkpoint prevents cells from entering mitosis with DNA that was damaged since the last division, providing an opportunity for DNA repair and stopping the proliferation of damaged cells. Because the G2 checkpoint helps to maintain genomic stability, it is an important focus in understanding the molecular causes of cancer.

Prophase is a stage of mitosis in which chromatin condenses into a highly ordered structure called a *chromosome*. Since the genetic material has been duplicated, there are two identical copies of each chromosome in the cell (sister chromosomes). When chromosomes are paired up and attached, each individual chromosome in the pair is called a *chromatid*, while the whole unit is called a *chromosome*. When the chromatids separate, they are no longer called chromatids, but are called chromosomes again [65].

Metaphase is a stage of mitosis in the eukaryotic cell cycle in which condensed chromosomes, carrying genetic information, align in the middle of the cell before being separated into each of the two daughter cells. The analysis of metaphase chromosomes is one of the main tools of cancer cytogenetics. In *anaphase* chromosomes separate, and each chromatid moves to opposite poles of the cell. *Telophase* is a stage reversing the effects of prophase and prometaphase events. For instance, a new nuclear envelope, using fragments of the parent cell's nuclear membrane, forms around each set of separated sister chromosomes. Finally, cytokinesis takes place. This stage is the process whereby the cytoplasm of a single cell is divided to spawn two daughter cells.

The duration of these cell cycle phases varies considerably in different kinds of cells. For a typical rapidly proliferating human cell with a total cycle time of 24 hours, the G_1 phase might last about 11 hours, the S phase about 8 hours, the G_2 about 4 hours, and the M phase about 1 hour. Some cells in adult animals cease division altogether (e.g., nerve cells) and many other cells divide only occasionally, as needed to replace cells that have been lost because of injury or cell death. Cells of the latter type include skin fibroblasts, as well as the cells of some internal organs, such as the liver. As mentioned, these cells exit G_1 to enter a quiescent stage of the cycle called G_0, where they remain metabolically active but no longer proliferate unless called on to do so by appropriate extracellular signals.

Regulation of the Cell Cycle

If the cell cycle is considered to be like an automobile in motion under normal conditions, proto-oncogenes are like the accelerator and tumour suppressor genes are like brakes. The progression of cells through the division cycle is regulated by extracellular signals from the environment, as well as by internal signals that monitor and coordinate the various processes that take place during different cell cycle phases. An example of cell cycle regulation by extracellular signals is provided by the effect of growth factors on animal cell proliferation.

Progression through the cell cycle is governed by a family of cyclin-dependant kinases (CDKs) and their regulatory subunits, the cyclins [66]. As cells enter the cell cycle from G_0, D- and E-cyclins are synthesized sequentially and both are rate limiting for S phase entry. In addition, as a result of the activity of CDKs, exit from G_0 occurs with entry into the S phase.

A major cell cycle regularity point in many types of cells occurs late in G_1 and controls progression from G_1 to S. This regulatory point was first defined by studies of budding yeast (*Saccharomyces cerevisiae*), where it is known as *start*. Once cells have passed start, they are committed to entering the S phase and undergoing one cell division cycle.

The proliferation of most animal cells is similarly regulated in the G_1 phase of the cell cycle. In particular, a decision point late in G_1, called the *restriction point* in animal cells, functions analogously to start in yeasts. In contrast to yeasts, however, the passage of animal cells through the cell cycle is regulated primarily by the extracellular growth factors that signal cell proliferation, rather than by the availability of nutrients. In the presence of the appropriate growth factors, cells pass the restriction point and enter the S phase. Once it has passed through the restriction point, the cell is committed to proceed through the S phase and the rest of the cell cycle, even in the absence of further growth factor stimulation. On the other hand, if appropriate growth factors are not available in G_1, progression through the cell cycle stops at the restriction point. Such arrested cells then enter a quiescent stage of the cell cycle called G_0, in which they can remain for long periods of time without proliferating. G_0 cells are metabolically active, although they cease growth and have reduced rates of protein synthesis. As already noted, many cells in animals remain in G_0 unless called on to proliferate by appropriate growth factors or other extracellular signals. For example, skin fibroblasts are arrested in G_0 until they are stimulated to divide as required to repair damage resulting from a wound. The proliferation of these cells is triggered by the platelet-derived growth factor, which is released from blood platelets during clotting and signals the proliferation of fibroblasts in the vicinity of the injured tissue [3,63,64]. Generally speaking, extracellular signals can thus control cell proliferation by regulating progression from the G_2 to the M phase as well as from the G_1 to the S phase of the cell cycle. It is believed that the molecular events in regulating the key G1/S and G2/M phases of the cell cycle are controlled by phosphorylation/dephosphorylation and particularly by the degradation of cell-cycle regulators [67–69].

Although the proliferation of most cells is regulated primarily in G_1, some cell cycles are instead controlled principally in G_2. One example is the cell cycle of the fission yeast *Schizosaccharomyces pombe*. The cell cycle of this group is regulated primarily by control of the transition from G_2 to M, which is the principle point at which cell size and nutrient availability are monitored. In animals, the primary example of cell cycle control in G_2 is provided by *oocytes*. Vertebrate oocytes can remain arrested in G_2 for long periods of time (several decades in humans) until their progression to the M phase is triggered by hormonal stimulation.

Properties of Cancerous Cells

Much evidence shows that cancer develops by a multistep process [3], and the uncontrolled growth of cancer cells results from accumulated abnormalities affecting many of the cell regulatory mechanisms. This relationship is reflected in several aspects of cell behaviour that distinguishes cancer cells from their normal counterparts. Cancer cells typically display abnormalities in the mechanisms that regulate normal cell proliferation, differentiation, and survival [7].

The general features of neoplastic cells result in specific changes in nucleic acid, protein, lipid, carbohydrate quantities, and/or conformations. For example, neoplastic cells are known to produce more lactate than normal cells. The DNA protein interaction is also disturbed in malignant transformations resulting in repeated duplication and amplification of DNA sequences [1].

Neoplastic cells are characterized by increased nuclear material, an increased nuclear-to-cytoplasmic ratio, increased mitotic activity, abnormal chromatin distribution, and decreased differentiation. There is a progressive loss of cell maturation, and proliferation of these undifferentiated cells results in increased metabolic activity. Histologically, neoplasms are characterized by cellular crowding and disorganization. The increased metabolic activity induces rapid angiogenesis and results in the formation of leaky vessels [1].

It has been demonstrated that the total lipid and phospholipid content of cancerous tissue are more than that of the normal tissue [70]. The total cholesterol content of malignant tissues is also reported to be higher than that of normal tissue [70]. Membranes of lung cancer tissue are reported to be more fluid than the corresponding normal tissues. Furthermore, high plasma membrane fluidity of lung tumours is associated with poor prognosis [71,72]. Cell membrane fluidization is linked with multidrug resistance [73]. Not only the lipid levels, but also the protein levels are altered due to carcinogenesis [74–76]. Wong et al. studied seven different proteins in normal and cancerous cervical tissues [75]. The structure of the proteins of the cytoskeleton and extracellular matrix (ECM) are transformed by cancer [77]. Altered protein structures also change the ability of the cancer cells to contract or stretch, by influencing their deformation. As a result, the motility of cancer cells can be very different from that of normal cells [24], causing them to migrate to different sites in the human body and induce metastasis [77]. In addition, the reduced adhesiveness of cancer cells also results in morphological and cytoskeletal alterations: Many tumour cells are rounder than normal ones, in part because they are less firmly attached to either the extracellular matrix or neighboring cells [7]. Cancer cells are capable of creating their own signals (and reprogram) for growth and division [77].

In a simple description, some of the common characteristics of malignant cells are as follows [6]:

- The morphology and size of cells are different and more variable than those of normal cells.
- The nucleus of the cells is larger than in normal cells.
- Large cells with multiple nuclei can be seen.
- Normal tissues are invaded by malignant cells.

Two additional properties of cancer cells affect their interaction with other tissue components, thereby playing important roles in invasion and metastasis. First, malignant cells generally secrete proteases (e.g., collanenase in carcinomas), which digest extracellular matrix components, allowing the cancer cells to invade adjacent normal tissues. Second, cancer cells secrete growth factors that promote the formation of new blood vessels (*angiogenesis*). Angiogenesis is needed to support the growth of a tumour because blood vessels are required to supply oxygen and nutrients to the proliferating tumour cells. The formation of such new blood vessels is important not only in supporting tumour growth, but also in metastasis. These capillaries provide a ready opportunity for cancer cells to enter the circulatory system and begin the metastatic process [7].

Another general characteristic of most cancer cells is that they fail to differentiate normally. Such effective differentiation is closely coupled to abnormal proliferation, since most fully differentiated cells cease cell division. Rather than carrying out their normal differentiation program, cancer cells are usually blocked at an early stage of differentiation, consistent with their continued active proliferation [7].

Some of the properties and specifications of cancer cells are consequences of activation of oncogenes or the ablation of the tumour suppressor genes, which may lead to the fact that the growth arrest process associated with ageing (senescence) may be overridden. Senescence is usually correlated with a decrease in the length of telomeres, which are short, tandem repeats of hexanucleotide 5′-TTAGGG-3′ at the ends of each chromosome. Telomeres are synthesized by telomerase, a ribonucleo-protein DNA polymerase that is active in many tumours but is not common in normal tissue [78]. In tumour cell lines, telomerase activity reaches a maximum in the S phase and a variety of inhibitors of cell cycle progression (e.g., transforming growth factor β1) inhibit the activity of telomerase [30].

Disordered Mechanisms in Malignant Cells

Whatever the reason for cell mutations, there are at least four mechanisms that might become disordered in malignant cells and result in aberrant proliferation [9]:

1. Abnormalities in growth factor production
2. Abnormalities in growth factor receptors
3. Disturbances of postreceptor signaling
4. The reduced production of or sensitivity to growth inhibiting factors

In fact, the proliferation and differentiation of all mammalian cells is generally controlled by extracellular signals, which are referred to as *peptide regulatory factors* (PRFs). They include platelet-derived growth factor (PDGF), epidermal growth factor (EGF), nerve growth factor (NGF), transforming growth factor α (TGFα), and fibroblast growth factor (FGF). More recently, glycoproteins produced by activated T cells (lymphokines), monocytes, and macrophages (monokines) have been added to this list, including the interleukins (IL-1, IL-2, IL-3, IL-4, IL-5, IL-6, and IL-10), granulocyte macrophage colony stimulating factor (GM-CSF), granulocyte colony stimulating factor (G-CSF), the interferons (INFs α and μ), tumour necrosis factor (TNF), and lymphotoxin. These molecules may mediate both the immune and inflammatory responses as well as constitutive haematopoiesis. In addition, other factors produced by stromal cells, e.g., M-CSF (CSF-1), IL-7, leukemia inhibitory factor (LIF), and stem cell factor (SCF) form a group of molecules called *cytokines*. At least 40 cytokines are known to act on the haemopoietic system [3,7,9,18].

Cancers escape control by their environment through having (i) decreased sensitivity to inhibitory signals from adjacent cells and the extracellular matrix, or (ii) decreased requirement for growth stimulatory factors. They can either produce more ligand or become independent of the ligand [5]. As mentioned, apart from the property of invading other tissues and organs, cancer cells attain a degree of autonomy from external regulatory signals [3]. This specification renders these cells less subject to such signals than normal cells. This autonomy is reflected in a lower requirement for growth stimulatory molecules and a diminished sensitivity to inhibitory signals provided by adjacent cells and the extracellular matrix. These stimulatory and inhibitory extracellular stimuli are recognized by receptors and conveyed to the nucleus by complex multiple pathways. The altered activity is directed, obtains a growth advantage over adjacent cells, and is achieved by increasing the efficiency of the intracellular machinery directed at proliferation and by production secreted growth factors [5]. It must be mentioned that the ability of invading other tissues results from having cancer cells mixed with other types of cells in pathologic samples [3].

A difference between normal cells and cancer cells is that many cancer cells have reduced requirements for extracellular growth factors. The growth factor requirements of many tumour cells are reduced compared to their normal counterparts, contributing to the unregulated proliferation of tumour cells both in vitro and in vivo. In some cases, cancer cells produce growth factors that stimulate their own proliferation. Such abnormal production of a growth factor leads to continuous autostimulation of cell division (*autocrine growth stimulation*), and cancer cells are therefore less dependent on growth factors.

In some cases, the reduced growth factor dependence of cancer cells results from abnormalities in intracellular signaling systems, such as unregulated activity of growth factor receptors or other proteins (e.g., protein kinases) [7]. A single cell often possesses a variety of different receptors [18]. The first step in this pathway involves the association of the ligand with the appropriate receptor [9].

The uncontrolled proliferation of cancer cells in vivo is mimicked by their behavior in cell culture. A primary distinction between cancer cells and normal cells in culture is that normal cells display *density-dependent inhibition* of cell proliferation. Normal cells stop growing when they become overcrowded [3] and proliferate until they reach a finite cell density, which is determined in part by availability of growth factors added to the culture medium (usually in the form of serum). They then cease proliferating and become quiescent, arrested in the G_0 stage of the cell cycle. The proliferation of most cancer cells, however, is not sensitive to density-dependent inhibition. Rather than responding to the signals that cause normal cells to cease proliferation and enter G_0, tumour cells generally continue growing to high cell densities in culture, mimicking their uncontrolled proliferation in vivo [7]. In addition, most normal cell types in the animal are attached to a structured network secreted by the cells, called the *extracellular matrix*. These cells usually need to attach to a surface (such as a plastic or a glass) in order to grow in culture. In contrast, many cancer cells can grow in culture even if they are not attached to a surface [3].

Cancer cells are also less stringently regulated than normal cells by cell–cell and cell–matrix interactions. Most cancer cells are less adhesive than normal cells, often as a result of reduced expression of the cell surface adhesion molecules. For example, loss of E-cadherin, the principal adhesion molecule of epithelial cells, is important in the development of carcinomas (epithelial cancers). As a result of reduced expression of cell adhesion molecules, cancer cells are comparatively unrestrained by interactions with other cells and tissue components, contributing to the ability of malignant cells to invade and metastasize [7].

A striking difference in the cell–cell interactions by normal cells and those of cancer cells is illustrated by the phenomenon of *contact inhibition*. Not only the movement but also the proliferation of many normal cells is inhibited by cell–cell contact, and cancer cells are characteristically insensitive to such contact inhibition growth [3,7].

References

1. Mahadevan-Jansen, A., and Richards-Kortum, R. 1997. Raman spectroscopy for the detection of cancers: A Review, 19th International Conference IEEE EMBS, Chicago, IL, October 30–November 2.
2. Franks, L. M. 1991. *Introduction to the Cellular and Molecular Biology of Cancer*. New York: Oxford University Press.
3. LaFond, R. E., 1988. *Cancer: The Outlaw Cell*, 2nd ed. Washington, DC: American Chemical Society.
4. Lodish, H., Berk, A., Zipursky, S. L., Matsudaira, P., Baltimore, D., and Darnell, J. E., 2000. *Molecular Cell Biology*, 4th ed., New York: W. H. Freeman & Co.
5. King, R. J. B., and Benjamin R. J., 2006. *Cancer Biology*, 3rd ed. New York: Pearson/Prentice Hall.
6. Dukor, R. K., 2002. Vibrational spectroscopy in the detection of cancer. *Biomedical Applications*, 3335–3359.
7. Cooper, G. M., 2003. *The Cell: A Molecular Approach*, 3rd ed. Philadelphia: Sinauer Associates.
8. Gortan Kumar, Robbins, 1999. *Pathologic Basis of Disease*. Philadelphia: Saunders.
9. Rubin P. C. 2001. *Clinical Oncology: A Multidisciplinary Approach for Physicians and Students*, 8th ed. Philadelphia: W. B. Saunders Co.
10. Madia, F., Gattazzo, C., Fabrizio, P., and Longo, V. D. 2006. A Simple Model System for Age-Dependent DNA Damage and Cancer. *Mechanics of Ageing and Development* 128, 45–49.
11. Gago-Dominguez, M., Jiang, X., and Castelo, J. E. 2006. Lipid peroxidation and the protective effect of physical on breast cancer. *Medical Hypotheses* 68, 1138–1143.
12. Mehta, N., Hordines, J., Volpe, C., Doerr, R., and Cohen, S. A. 1997. Cellular effects of hypercholesterolemia in modulation of cancer growth and metastasis: A review of the evidences. *Surgical Oncology* 6, 179–185.
13. Miyamoto, H., Yang, Z., Chen, Y.-T., et al. 2007. Promotion of bladder cancer development and progression by androgen receptor signals. *Journal of the National Cancer Institute* 99(7), 558–568.
14. Kabat, G. C., Miller, A. B., and Rohan, T. E. 2007. Reproductive and hormonal factors and risk of lung cancer in women: A prospective cohort study. *International Journal of Cancer* 120(10), 2214–2220.
15. Cancer Research UK, 2004. UK Cancer Statistics. [online].
16. Das, R. S., and Agrawal, Y. K. 2011. Raman spectroscopy: Recent advancements, techniques and applications. *Vibrational Spectroscopy* 57, 163–176.
17. Quaglia, A., Capocaccia, R., Micheli, A., et al. 2007. A wide difference in cancer survival between middle aged and elderly patients in Europe. *International Journal of Cancer* 120(10), 2196–2201.

18. Hesketh R., 1997. *The Oncogenes and Tumour Suppressor Genes Factbook*, 2nd ed., New York: Academic Press.

19. Shibitsin, M., Campbell, L. L., Argani, P., et al. 2007. Molecular definition of breast tumour heterogeneity. *Cancer Cell* 11, 259–273.

20. Land, H., Parada, L. F., and Weinberg, R. A. 1983. Tumorigenic conversion of primary embryo fibroblasts requires at least two cooperating oncogenes. *Nature* 304, 596–602.

21. Ruley, H. E. 1983. Adenovirus early region 1A enables viral and cellular transforming genes to transform primary cells in culture. *Nature* 304, 602–606.

22. Hahn, W. C., Christopher M. C., Lundberg A. S., et al. 1999. Creation of human tumour cells with defined genetic elements. *Nature* 400, 464–468.

23. Hanahan, D., and Weinberg, R. A. 2000. The hallmarks of cancer. *Cell* 100, 57–70.

24. Xia, M., and Land, H. 2007. Tumour suppressor p53 restricts Ras stimulation of RhoA and cancer motility. *Nature Structural and Molecular Biology* 14, 215–223.

25. Lloyd, A. C., et al. 1997. Cooperating oncogenes converge to regulate cyclin/cdk complexes. *Genes Dev.* 11, 663–677.

26. Zhu, J., Woods, D., McMahon, M., and Bishop, J. M. 1998. Senescence of human fibroblasts induced by oncogenic Raf. *Genes Dev.* 12, 2997–3007.

27. Serrano, M., Lin, A. W., McCurrach, M. E., Beach, D., and Lowe, S. W. 1997. Oncogenic ras provokes premature cell senescence associated with accumulation of p53 and p16(ink4a). *Cell* 88, 593–602.

28. Palmero, I., Pantoja, C., and Serrano, M. 1998. p19ARF links the tumour suppressor p53 to Ras. *Nature* 395, 125–126.

29. Kamijo, T., et al. 1997. Tumor suppression at the mouse INK4a locus mediated by the alternative reading frame product p19ARF. *Cell* 91, 649–659.

30. Zhu X., Kumar R., Mandal M., et al. 1996. Cell cycle-dependant modulation of telomerase activity in tumour cells. *Proceedings of the National Academy of Sciences USA* 93, 6091–6095.

31. Parsons, R. 1999. Oncogenic basis of breast cancer. In D. F. Roses, ed., *Breast Cancer*. Philadelphia: Churchill Livingston, 13–24.

32. Imyanitov, E. N., and Moiseyenko, V. M. 2007. Molecular-based choice of cancer therapy: Reliabilities and expectations. *Clinica Chimica Acta* 379, 1–13.

33. Yamabukr, T., Daigo, Y., Kato, T., et al. 2006. Genome-wide gene expression profile analysis of esophageal squamous cell carcinomas. *International Journal of Oncology* 28, 1375–1384.

34. Ishigami, T., Uzawa, K., Higo, M., et al. 2007. Genes and molecular pathways related to radioresistance of oral squamous cell carcinoma cells. *International Journal of Cancer* 120(10), 2262–2270.

35. Perou, C. M., Sorlie, T., Eisen, M. B., et al. 2000. Molecular portraits of human breast tumours. *Nature* 406, 747–752.

36. Sorlie, T., Perou, C. M., Tibshirani, R., et al. 2001. Gene expression patterns of breast carcinomas distinguish tumour subclasses with clinical implications. *Proceedings of the National Academy of Science USA* 98, 10869–10874.

37. Sorlie, T., Tibshirani, R., Parker, J., et al. 2003. Repeated observation of breast tumour subtypes in independent gene expression data sets. *Proceedings of the National Academy of Science USA* 100, 8418–8423.

38. Gjerstorff, M. F., Benoit, V. M., Laenkholm, A.-V., et al. 2006. Identification of genes with altered expression in medullary breast cancer vs. ductal breast cancer and normal breast epithelia. *International Journal of Oncology* 28, 1327–1335.

39. Bishop, J. M., 1995. Cancer: The Rise of the Genetic Paradigm. *Genes Development* 9, 1309–1315.

40. Friedberg, E. C., 1986. *Cancer Biology: Readings from Scientific American*. New York: W. H. Freeman.

41. Rotenberg, S. A., and Weinstein, I. B., 1991. Protein Kinase C in neoplastic cells. In T. G. Pretlow and T. P. Pretlow, eds., *Biochemical and Molecular Aspects of Selected Cancers, Vol. 1*, 25–73. San Diego: Academic Press.

42. Bishop, J. M. 1987. The molecular genetics of cancer. *Science* 235, 305–311.

43. Slamon, D. J. 1987. Protooncogenes and human cancer. *N. Engl. J. Med.* 317, 955–957.

44. Hu, J., Cladel, N. M., Budgeon, L. R., and Christensen, N. D. 2004. Characterization of three rabbit oral papillomavirus oncogenes. *Virology* 325, 48–55.

45. Cid-Arregui, A., Juarez, V., and zur Hausen, H. 2003. A synthesis E7 gene of human papillomavirus type 16 that yields enhanced expression of the protein in mammalian cells and is useful for DNA immunization studies. *Journal of Virology* 77, 4928–4937.

46. Embers, M. B., Budgeon, L. R, Pickel, M., and Christensen, N. D. 2002. Protective immunity to rabbit oral and cutaneous papillomaviruses by immunization with short peptides of 12, the minor capsid protein. *Journal of Virology* 76, 9798–9805.

47. Peh, W. L., Middleton, K., Christensen, N., et al. 2002. Life cycle heterogeneity in animal models of human papillomavirus-associated disease. *Journal of Virology* 76, 10401–10416.

48. Nilsson, J. A., and Cleveland, J. L. 2003. Myc pathways provoking cell suicide and cancer. *Oncogenes* 22, 9007–9021.

49. Sears, R. C. 2004. The life cycle of C-Myc: From synthesis to degradation. *Cell Cycle* 3, 1133–1137.

50. Rimpi, S., and Nilsson, J. A. 2007. Metabolic enzymes regulated by the Myc oncogene are possible targets for chemotherapy or chemoprevention. *Biochemical Society Transactions* 35(2), 305–310.

51. Askew, D. S., Ashmun, R. A., Simmonis, B. C., and Cleveland, J. I. 1991. Constitutive C-Myc expression in an IL-3-dependent myeloid cell line suppresses cell cycle arrest and accelerates apoptosis. *Oncogenes* 6, 1915–1922.

52. Evan, G. I., Wyllie, A. H., Gilbert, C. S., et al. 1992. *Cell* 69, 119–128.

53. Druker B. J., Mamon H. J., and Roberts T. M. 1989. Oncogenes, growth factors, and signal transduction. *N. Engl. J. Med.* 321(20), 1383–1391.

54. Knudson, A. G. 1993. Antioncogenes and human cancer. *Proceedings of the National Academy of Sciences USA* 90, 10914–10921.

55. Knudson, A. G. 1971. Mutation and cancer: Statistical study of retinoblastoma. *Proceedings of the National Academy of Sciences USA* 68, 820–823.

56. Pharoah P. D., Day N. E., and Caldas C. 1999. Somatic mutations in the p53 gene and prognosis in breast cancer meta-analysis. *British J. of Cancer* 80, 1968–1973.

57. Horowitz J. M., Park S. H., Bogenmann E., et al. 1990. Frequent inactivation of the retinoblastoma inactivation of the retinoblastoma antioncogene is restricted to a subset of human tumour-cells. *Proceedings of the National Academy of Sciences USA* 87, 2775–2779.

58. Friend S. H., Harowitz J. M., Gerber M. R., et al. 1987. Deletions of a DNA-sequence in retinoblastoma and mesenchymal tumours: Organization of the sequence and its encoded protein. *Proceedings of the National Academy of Sciences USA* 84, 9059–9063.

59. Weichselbaum, R. R., Becket M., and Diamond A. 1988. Some retinoblastomas, osteosarcomas, and soft tissue sarcomas may share a common etiology. *Proceedings of the National Academy of Sciences USA* 65, 2106–2109.

60. Ruggeri, B., Zhang S.Y., and Camaano, J. 1992. Human pancreatic carcinomas and cell lines reveal frequent and multiple alterations in the p53 and rb-1 tumour suppressor genes. *Oncogenes* 7, 1503–1511.

61. Lane, D. P. 1992. p53, guardian of the genome. *Nature* 358, 15–16.

62. Fermisco, J., Ozzane, B., and Stiles, C. 1985. *Cancer Cells: Growth Factors and Transformation.* Cold Spring Harbor, New York 389–399.

63. John, P. C. L. 1981. *The Cell Cycle.* Cambridge, UK: Cambridge University Press.

64. Hutchison, C. 1995. *Cell Cycle Control.* New York: IRL Press.

65. Raven, P. H., Ray F. E., and Susan E. E. 2005. *Biology of Plants*, 7th ed. New York: W. H. Freeman and Company.

66. Sherr, C. J., and Roberts J. M., 1995. Inhibitors of mammalian g(1) cyclin-dependant kinase. *Genes Development* 9, 1149–1163.

67. Almendral, J. M., Sommer D., Macdonaldbravo, H., et al. 1988. Complexity of the early genetic response to growth: Factors in mouth fibroblasts. *Molecular Cell Biology* 8, 2140–2148.

68. Zipfel, P. F., Irving, S. G., Kelly, K., et al. 1989. Complexity of the primary genetic response to mitogenic activation of human T-cells. *Molecular and Cellular Biology* 9, 1041–1048.

69. Mayer, R. J., and Fujita, J. 2006. Gankyrin, the 26 S proteasome, the cell cycle and cancer. *Biochemical Society Transactions* 34, 746–748.

70. Preetha, A., Huigol, N., and Banerjee, R. 2005. Interfacial properties as biophysical markers of cervical cancer. *Biomedicine and Pharmacotherapy* 59, 491–497.

71. Sok, M., Sentjurc, M., and Schara, M. 1999. Membrane fluidity characteristics of human lung cancer. *Cancer Letter* 139, 215–220.

72. Sok, M., Sentjurc, M., Schara, M., Stare, J., and Rott, T. 2002. Cell membrane fluidity and prognosis of lung cancer. *The Annals of Thoracic Surgery* 73, 1576–1571.

73. Fu, L. W., Zhang, Y. M., Liang, Y. J., Yang, X. P., and Pan, Q. C. 2002. The multidrug resistance of tumour cells was reversed by tetrandrine in vitro and in xenografts derived from human breast adenocarcinoma. *European Journal of Cancer* 38, 418–426.

74. Coppe, J. P., Smith, A. P., and Desprez, P. 2003. Id proteins in epithelial cells. *Experimental Cell Research* 285, 131–145.

75. Wong, Y. F., Cheung, T. H., Lo, K. W. K., et al. 2004. Protein profiling of cervical cancer by protein-biochips: Proteomic scoring to discriminate cervical cancer from normal cervix. *Cancer Letters* 211, 227–234.

76. Gam, L.-H., Leow, C.-H., Man, C. N., Gooi, B.-H., and Singh, M. 2006. Analysis of differentially expressed proteins in cancerous and normal colonic tissues. *World Journal of Gastroenterology* 12(31), 4973–4980.

77. Suresh, S. 2007. Biomechanics and biophysics of cancer cells. *Acta Biomaterialia* 3, 413–438.

78. Mehle, C., and Piatyszek, M. A., 1996. Telomerase activity in human renal cell carcinoma. *Oncogenes* 13, 161–166.

4

Vibrational Spectroscopy and Cancer

Introduction

In recent years, applications of IR and Raman spectroscopy have increased a great deal. It seems that still the majority of studies being carried out using these methods are concentrated on areas related to chemistry and polymers, such as different types of coatings on membranes, analysing the process of progress in chemical reaction, etc. [1–30]. However, some other research groups have concentrated on applications of spectroscopy in other fields, such as astronomy [31], archaeology [32,33] and forensic sciences [34,35]. A newly emerged field, the application of these spectroscopic methods on biological studies, particularly on clinical studies related to malignancy and cancer detection, has attracted different research groups all over the world. These methods have been reported on a number of biological tissues including, bone, skin, colon, lung, breast, heart, cornea, liver, prostate, gastric, larynx, oral, cervix, and endometrium. This chapter reviews the studies that have been reported in the literature.

A few researchers have focused on reviewing the work of others in the past (Table 4.1). In particular, there are three review articles that deal with the applications of Raman and vibrational spectroscopy in the detection of cancer, written by A. M. Jansen and R. R. Kortum [36], Hanlon et al. [37], and R. K. Dukor [38]. They cover several topics including principles of the techniques, cancer and cell biology, the technique's sensitivity to structures of biological molecules such as proteins, nucleic acids, and lipids, IR and Raman sampling techniques, instrumentation considerations, a summary of clinical applications of Raman spectroscopy in diagnosis of cancers of different organs, and the perspectives of the future.

Skin Cancer

The article by Choo-Smith et al. [39] is on medical applications of Raman spectroscopy, from proof of principle to clinical implementation. They named some of the various biomedical issues that are addressed by the technique,

TABLE 4.1

Review Articles

Research Group	Research Method	Investigated Tissue or Sample	Effectiveness of the Technique	Reference Number
Mahadevan-Jansen and Richards-Kortum	Review	n/a	n/a	36
Hanlon et al.	Review	n/a	n/a	37
Dukor et al.	Review	n/a	n/a	38
Choo-Smith et al.	Review	n/a	n/a	39
Swinson et al.	Review	n/a	n/a	40
Shaw and Mantsch	Review	n/a	n/a	41
Barr et al.	Review	n/a	n/a	42
W. Petrich	Review	n/a	n/a	43
Zeng et al.	Different methods of spectroscopy	Lung	n/a	44
Schultz and Baranska	Review	Plant substances	n/a	45
Moreira et al.	Review	n/a	n/a	46
Krishna et al.	Review	Breast	n/a	47
Mirali Krishna et al.	Review	Cervical tissue	n/a	48

including early detection of cancers, monitoring the effect of various agents on the skin, determination of atherosclerotic plaque composition, and rapid identification of pathogenic microorganisms. They also presented a discussion on the status of the field today, as well as the problems and the issues that still need to be resolved to bring this technology to hospital settings (i.e., medical laboratories, surgical suites, or clinics). A path is proposed for the clinical implementation of the technique.

Head and Neck Cancer

Swinson et al. [40] have published a review paper on optical techniques used in diagnosis of head and neck malignancies. The article is primarily about three different methods: fluorescence spectroscopy (FS), elastic scattering spectroscopy (ESS), and Raman spectroscopy (RS). According to the information presented, these techniques have not only been shown to have a role in the detection of different kinds of malignancies, but also in performing guided biopsies for monitoring of haemoglobin tissue perfusion in free flaps and therapeutic drug levels during chemo- and photodynamic therapy [40].

R. A. Shaw and H. H. Mantsch [41] published a historical perspective on different applications of vibrational biospectroscopy. The focus of this review is biological and medical applications of near-infrared (NIR) Raman spectroscopy. It is mentioned that as these methods are finding new niches in medical realms, and are proving to be valuable (but not infallible) adjuncts to conventional spectral interpretation.

Other Biological Molecules

Schultz and Baranska published a detailed review study on identification of plant substances by infrared (IR) and Raman spectroscopy. Their article provides the characteristic bands for identifying different chemical bands such as proteins, lipids, saccharides, etc. Although this work is concentrated on plant samples, because almost all of these substances exist in cells and tissues, the results and conclusions could be successfully used in biological studies. In addition, with the information presented in this paper, a comparison between the two methods of spectroscopy, namely IR and Raman, can be conducted [45].

Moreira et al. reported a detailed review of the recent applications of Raman spectroscopy on biochemical analysis and diagnosis of several biological materials [46]. It was concluded that the technique can promote a significant improvement in the chemical identification and characterization of biological systems, clinical diagnosis, and prognosis regarding several diseases and quality of life of innumerous patients. The study also reveals that the potential for sample characterization with Raman spectroscopy associated with the possibility of analysis in situ makes this instrumental technique a very auspicious tool of biochemical analysis.

Breast Cancer

Krishna et al. reviewed a Raman spectroscopic approach to metabolic fingerprinting in breast cancer detection [47]. This review opens with a brief background on anatomical and etiological aspects of breast cancers. It then presents an overview of conventional detection approaches in breast cancer screening and diagnosis methods, followed by a concise note on the basics of optical spectroscopy and its applications in the screening/diagnosis of breast malignancy. The authors also present the recent developments in Raman spectroscopic diagnosis of breast cancers.

Cervical and Ovarian Cancers

Murali Krishna et al. published another review article on applications of optical spectroscopy in cervical cancers [48]. This review includes a brief background on the anatomy and histology of the uterine cervix and the etiology of cervical cancers, along with a brief discussion of the optical spectroscopic approach to cervical cancer diagnosis. Radiation therapy, radiation resistance, and Raman spectroscopic methodologies in cervical cancer diagnosis were also covered in this publication.

Cervical and ovarian tissues have been widely investigated by different research groups (Table 4.2). B. R. Wood et al. made an investigation on Fourier transform infrared (FTIR) spectroscopy as a biodiagnostic tool for cervical cancer. The spectra of the normal epithelial cells illustrated intense glycogen bands at 1022 cm^{-1} and 1150 cm^{-1}, and a pronounced symmetric phosphate stretch at 1078 cm^{-1}. The spectral features suggestive of dysplastic or malignant transformations were mainly pronounced symmetric and asymmetric phosphate modes and a reduction in the glycogen band intensity. This study demonstrated the potential of the automated FTIR cervical screening technology in the clinical environment [49].

Chiriboga et al. carried out research on differentiation and maturation of the epithelial cells in the human cervix, using infrared spectroscopy.

TABLE 4.2

Articles on Cervical and Ovarian Tissue

Research Group	Research Method	Investigated Tissue or Sample	Effectiveness of the Technique	Reference Number
Wood et al.	FTIR	Cervical tissue	+	49
Chiriboga et al.	FTIR	Cervix	+	50
Wood et al.	FTIR	Cervical tissue	+	51
Utzinger et al.	Raman	Cervical tissue	+	52
Sindhuphak et al.	FTIR	Cervical tissue	+	53
Mordechai et al.	FTIR	Cervical tissue	+	54
Chiriboga et al.	FTIR	Cervical tissue	+	55
Wong et al.	FTIR	Endocervix and ectocervix	+	56
Fung et al.	FTIR	Endometrium	+	57
Krishna et al.	Raman + Principal Components Analysis (PCA)	Human uterine cervix	+	58
Maheedhar et al.	Raman	Ovarian tissue	+	59
Mo et al.	Raman	Cervical tissue	+	60
Jess et al.	Raman	Cervical tissue	+	61
El-Tawil et al.	FTIR	Cervical tissue	+	62

Different layers of human cervical squamous tissue, representing different cellular maturation stages, exhibited quite dissimilar spectral patterns. Thus, it was concluded that this technique presents a powerful tool to monitor cell maturation and differentiation. In addition, a proper interpretation of the state of health of cells exfoliated from such tissues would be obtained through a detailed understanding of the spectra of the individual layers [50].

B. R. Wood et al. performed an FTIR microspectroscopic investigation of cell types and potential confounding variables in screening for cervical malignancies. The aim of the study was to determine the effectiveness of infrared spectroscopy in the diagnosis of cervical cancer and dysplasia. It was found that leukocytes, and in particular lymphocytes, have spectral features in the phosphodiester region (1300–900 cm^{-1}), suggestive of changes indicative of malignancy. The use of ethanol as a fixative and dehydrating agent resulted in retention of glycogen and thus minimized the spectral changes in the glycogen region due to sampling technique. Erythrocyte spectra exhibited a reduction in glycogen band intensity, but could be discerned by a relatively low-intensity $v_sPO_2^-$ band. Endocervical mucin spectra exhibit a reduction glycogen band and a very pronounced $v_sPO_2^-$ band, which was similar in intensity to the corresponding band in HeLa $v_sPO_2^-$ cells [51].

Utzinger et al. introduced a near-infrared Raman spectroscopic method for in vivo detection of cervical precancers. The main focus of the project was on squamous dysplasia, a precursor of cervical cancer. A pilot clinical trial was carried out at three clinical sites. Raman spectra were measured from one normal and one abnormal area of the cervix. These sites were then biopsied and submitted for routine histologic analysis. Twenty-four measures were made in vivo in 13 patients. Cervical tissue Raman spectra contained peaks in the vicinity of 1070, 1180, 1195, 1210, 1245, 1330, 1400, 1454, 1505, 1555, 1656, and 1760 cm^{-1}. The ratio of intensities at 1454 to 1656 cm^{-1} was greater for squamous dysplasia than all other tissue types, while the ratio of intensities at 1330 to 1454 cm^{-1} was lower for samples with squamous dysplasia than all other tissue types. A simple algorithm based on these two intensity ratios separated high-grade squamous dysplasia from all others, misclassifying only one sample. Spectra measured in vivo resembled those measured in vitro. It was believed that cervical epithelial cells may contribute to tissue spectra at 1330 cm^{-1}, a region associated with DNA. In contrast, epithelial cells probably do not contribute to spectra at 1454 cm^{-1}, a region associated with collagen and phospholipids [52].

R. Sindhuphak et al. screened cervical cell samples of Thai women by using FTIR spectroscopy, and compared the results to the histologic diagnosis. A total of 275 cervical cell specimens were received from patients undergoing hysterectomy. Histological examinations showed 108 normal cases and 167 abnormal cases. FTIR results versus histology showed sensitivity of 96.3% and specificity of 96.4%. False negative and false positive rates were 3.7 and 3.6%, respectively [53].

A study was conducted by Mordechai et al. on formalin fixed melanoma and cervical cancer by FTIR microspectroscopy (FTIR-MSP), to detect common biomarkers that occur in both types of cancer, distinguishing them from the respective nonmalignant tissues. The spectra were analysed for changes in levels of biomolecules such as RNA, DNA, phosphates, and carbohydrates. Whereas carbohydrate levels showed a good diagnostic potential for detection of cervical cancer, this was not the case for melanoma. However, variation of the RNA/DNA ratio as measured at 1121/1020 cm^{-1} showed similar trends between nonmalignant and malignant tissues in both types of cancer. The ratio was higher for malignant tissues in both types of cancer [54].

L. Chiriboga et al. carried out a comparative study of spectra of biopsies of cervical squamous epithelium and of exfoliated cervical cells using infrared spectroscopy. A comparison of infrared absorption spectra obtained from the different layers of squamous epithelium from the human cervix, and infrared spectra obtained from exfoliated cervical cells was done. It was shown that the technique is a sensitive tool to monitor maturation and differentiation of human cervical cells. Therefore, it was concluded that this spectroscopic method provides new insights into the composition and state of health of exfoliated cells [55].

Wong carried out research on exfoliated cells and tissues from human endocervix and ectocervix by FTIR and attenuated total reflectance (ATR)/FTIR spectroscopy. They measured the transmission infrared spectra of exfoliated endocervical mucin-producing columnar epithelial cells and the ATR infrared spectra of the single-columnar cell layer on the endocervical tissues and compared with the corresponding infrared spectra of ectocervical squamous cells and squamous epithelium. The effects of the contaminated connective tissue on the infrared spectra of the endocervical columnar epithelial tissue demonstrated that ATR/FTIR is a more desirable method than the transmission method to obtain meaningful and good-quality infrared spectra of tissue samples, especially samples consisting of thin layers of different types of tissues. Substantial differences in the infrared spectra between the columnar cells and squamous cells on the endocervical and ectocervical tissues, respectively, were evident. The strong glycogen bands in the infrared spectrum of the ectocervical squamous cells were absent in the spectrum of endocervical columnar cells. These spectral changes were similar to that observed in malignant squamous cells. Therefore, if the decrease in the intensity of the glycogen bands is used as the only criterion for the determination of the cellular abnormalities in the cervix, the presence of a large number of normal endocervical columnar cells in the cervical specimen would lead to a false result. Consequently, it was concluded that in addition to the glycogen bands, other features in the infrared spectra should be considered for evaluation of abnormalities in exfoliated cervical epithelial cells [56].

Adenocarcinoma

A pressure-tuning FTIR spectroscopic study of carcinogenesis in human endometrium was reported by M. F. K. Fung et al. The spectra of normal tissues differed from the obtained grade I and grade III adenocarcinoma. Changes in the spectra of malignant samples were observed in the symmetric and asymmetric stretching bands of the phosphodiester backbones of nucleic acids, the CH stretching region, the C–O stretching bands of the C–OH groups of carbohydrates and cellular protein residuals, and the pressure dependence of the CH_2 stretching mode. These spectral changes in the endometrium were reproducible. It was also found for the first time that the epithelium in the normal endometrium exhibits unique structural properties compared with the epithelium of other normal human tissues [57].

Krishna et al. reported on application of Raman spectroscopy for diagnosis of cancers in the human uterine cervix [58]. Raman spectra of normal and malignant tissues were recorded in the fingerprint region. It was recognised that the main spectral differences between the two groups of samples are observed in the amide I, amide III, and 853 and 938 cm^{-1}, which can be attributed to structural proteins such as collagen. In order to develop an objective discrimination method, the group carried out elaborate data analysis using principal components analysis (PCA). Standard sets for normal and malignant were prepared and tested retrospectively and prospectively. The results produced very clean clustering of normal and malignant spectra, and sensitivity and specificity of 99.5% were achieved.

Maheedhar et al. assessed the efficacy of conventional Raman spectroscopy in differentiation of normal and malignant ovarian tissues [59]. A total of 72 Spectra from eight normal and seven malignant ovarian tissues were recorded by conventional NIR Raman spectroscopy (excitation wavelength of 785 nm). Spectral data were analyzed by PCA and it was shown that malignant spectra exhibit a broader amide I band, a stronger amide III band, a minor blue shift in the delta CH_2 band, and a hump around 1480 cm^{-1} compared to a normal spectrum. The normal spectra show relatively stronger peaks around the 855 and 940 cm^{-1} region. It was demonstrated that conventional Raman spectroscopy and the statistical methodologies in discrimination of normal from malignant ovarian tissues are efficient.

The purpose of Mo et al.'s study is to explore the ability of NIR Raman spectroscopy in the high wavenumber region (2800–3700 cm^{-1}) for the in vivo detection of cervical precancer [60]. A rapid NIR Raman spectroscopy system associated with a fibre-optic Raman probe was used for the in vivo spectroscopic measurements. Multivariate statistical techniques including PCA and linear discriminant analysis (LDA) were employed to develop the diagnostic algorithm based on the spectral data from 2800–3700 cm^{-1}. Classification result based on PCA-LDA showed that high-wavenumber NIR

Raman spectroscopy can achieve the diagnostic sensitivity of 93.5% and specificity of 95.7% for precancer classification.

Jess et al. reported on application of Raman Spectroscopy for early detection of cervical neoplasia [61]. Raman microspectroscopy was applied to live and fixed cultured cells and Raman spectra were acquired from primary human keratinocytes (PHKs), PHK expressing the E7 gene of human papillomavirus (HPV) 16 (PHK E7) and CaSki cells, an HPV16-containing cervical carcinoma-derived cell line. Principal component analysis produced good discrimination between the cell types, with sensitivities of up to 100% for the comparison of fixed PHK and CaSki. The results demonstrated the ability of Raman spectroscopy to discriminate between cell types representing different stages of cervical neoplasia.

El-Tawal et al. compared smear cytology and FTIR spectroscopy in diagnosing cervical cancer [62]. A total of 800 cervical scrapings were taken by cytobrush. After preparation stages, the samples were dried over an infrared transparent matrix. Beams of infrared light were directed at the dried samples at frequencies of 4000 to 400 cm^{-1}. Data were compared with cytology (gold standard). It was shown that FTIR spectroscopy can differentiate normal from abnormal cervical cells in the samples examined. The sensitivity was 85%, specificity 91%, positive predictive value 19.5%, and negative predictive value of 99.5%.

Epithelial Cancer

Some research groups have focused on epithelial tissues (Table 4.3). N. Stone et al. for instance, applied NIR Raman spectroscopy for the classification of epithelial cancers and precancers. The aim of their research was to evaluate the use of the technique in interrogating epithelial tissue biochemistry and distinguishing between normal and abnormal tissues. Tissues were selected for clinical significance and included those that develop into carcinoma from squamous, transitional, or columnar epithelial cells. Rigorous histopathological protocols were followed. The epithelial tissues were obtained from larynx, tonsil, oesophagus, stomach, bladder, and prostate. Sensitivities and specificities of up to 100% in separating the samples were obtained [63].

TABLE 4.3

Articles on Epithelial Tissues

Research Group	Research Method	Investigated Tissue or Sample	Effectiveness of the Technique	Reference Number
Stone et al.	Raman	Epithelial tissue	+	63
Stone et al.	Raman	Epithelial tissue	+	64

The same research group performed Raman spectroscopic investigations for identification of epithelial cancer. They evaluated the potential for the technique to develop a noninvasive real-time probe for accurate and repeatable measurement of pathological samples. The study followed rigorous sample collection protocols and histopathological analysis using a board of expert pathologists. Only the data from samples with full agreement of a homogeneous pathology were used to construct a training data set of Raman spectra. Measurements of tissue specimens from the full spectrum of different pathological groups found in each tissue were made. Diagnostic predictive models were constructed and optimised. High levels of discrimination between pathology groups were demonstrated (greater than 90% sensitivity and specificity for all tissues). However, it was outlined that larger sample numbers are required for successful implementation of in vivo Raman detection of early malignancies [64].

Lung Cancer

Lung tissue has received huge attention in the field of vibrational spectroscopy (Table 4.4). H. P. Wang et al. focused on the microscopic FTIR studies of lung cancer cells in pleural fluid. The results demonstrate significant spectral differences between normal, lung cancer and tuberculous cells. The most considerable differences were in the ratio of peak intensities of 1030 and 1080 cm^{-1} bonds (originated mainly in glycogen and phosphodiester groups of nucleic acids). One of the important advantages of this method was the possibility of obtaining quick and reliable results [65].

Zeng et al. produced a review paper on different methods of optical spectroscopy and imaging for early lung cancer detection. In this article,

TABLE 4.4

Articles on Lung Tissue

Research Group	Research Method	Investigated Tissue or Sample	Effectiveness of the Technique	Reference Number
Wang et al.	FTIR	Lung	+	65
Yano et al.	FTIR	Lung	+	66
Yang et al.	FTIR	Lung	+	67
Min et al.	Raman	Lung	+	68
Kaminaka et al.	Raman	Lung	+	69
Huang et al.	Raman	Lung	+	70
Alfano et al.	Fluorescence spectroscopy	Lung	+	71
Huang et al.	Raman	Bronchial tissue	+	72
Short et al.	Raman	Lung	+	73

the optical principles behind white-light and autofluorescence bronchoscopy (AFB), as well as the role of AFB in early diagnosis of lung cancer and the overall managements of patients are discussed. Other newest developments such as Raman spectroscopy were also highlighted. According to the information presented in the article, AFB is the most successfully developed technique and has significantly improved the detection sensitivity of early lung cancer [44].

The research of K. Yano et al. was about direct measurement of human lung cancerous and non-cancerous tissues by FTIR microscopy, in order to answer the question of whether this technique can be used as a clinical tool or not. The corrected peak heights (H1045 and H1467) obtained from the bands at 1045 cm^{-1} and 1465 cm^{-1}, which are due to glycogen and cholesterol, were chosen for a quantitative evaluation of the malignancy. It was concluded that these peaks are an exceptionally useful factor for discrimination of the cancerous tissues from the non-cancerous ones. If the H1045/H1467 ratio from measured spectrum is larger than 1.4, it could be said with confidence that the tissue contains squamous cell carcinoma (SCC) or adenocarcinoma at least partially. Furthermore, they carried out the microscopic mapping of the tissues containing both cancerous and non-cancerous sections, demonstrating that the colour map reflects small changes in the spatial distribution of cancer cells in the tissues [66].

Y. Yang et al. reported on tumour cell invasion by FTIR microspectroscopy. In this study, a three-dimensional artificial membrane using collagen type I, one of the main components of basal membranes of the lung tissue, was established in order to investigate tumour cell invasion of lung cancer. The mapping images obtained with FTIR Microspectroscopy were validated with standard histological section analysis. The FTIR image produced using a single wave number at 1080 cm^{-1}, corresponding to PO_2^- groups in DNA from cells, correlated well with the histological section, which clearly revealed a cell layer and invading cells within the membrane. Furthermore, the peaks corresponding to amid A, I, and II in the spectra of the invading cells shifted compared to the non-invading cells, which may relate to the changes in conformation and/or heterogeneity in the phenotype of the cells. The data presented in this study demonstrate that FTIR Microspectroscopy can be a fast and reliable technique to assess tumour invasion in vitro [67].

Y. K. Min et al. reported on near-infrared 1064-nm multichannel Raman spectroscopy of fresh human lung tissues. Excitation at 785 nm failed to detect any Raman bands because of an extremely high fluorescence backbone. As a result, it was confirmed that 1064-nm excitation was a requisite for the Raman study of the fresh lung tissue. The observed Raman spectra of lung tissues made a clear distinction between the normal and cancerous states. It was demonstrated that 1064-nm near-infrared multichannel Raman spectroscopy is a feasible tool for in vivo, noninvasive and molecular-level clinical diagnosis of diseases including cancer [68].

The possibility of molecular-level cancer diagnosis of human lung tissues using near-infrared Raman spectroscopy was investigated by S. Kaminaka et al. They used an Nd:YAG laser of 1064 nm and could collect totally fluorescence-free Raman spectra of normal and cancerous lung tissues. It was concluded that the technique probed lung cancer unambiguously at the molecular level and could be used as a tool for cancer diagnosis [69].

Z. Huang et al. reported on diagnosis of lung cancer using near-infrared Raman spectroscopy. The objective of their study was to explore the technique for distinguishing tumour from normal bronchial tissue. A rapid-acquisition dispersive-type NIR Raman spectroscopy system was used for tissue Raman studies at 785 nm. Raman spectra differed significantly between normal and malignant tumour tissues, namely, squamous cell carcinoma (SCC) and adenocarcinoma. Tumours showed higher-percentage signals for nucleic acids, tryptophan, and phenylalanine and lower-percentage signals for phospholipids, proline, and valine, in comparison with normal tissues. Raman spectral shape differences between normal and tumour tissues were also observed particularly in the regions of 1000–1100, 1200–1400, and 1500–1700 cm^{-1}, which contain signals related to CH stretching modes of protein, lipid, and nucleic acid conformations. The ratio of Raman intensities at 1445 to 1655 cm^{-1} provided good differentiation between normal and malignant bronchial tissue ($p < 0.0001$). The results of this exploratory study indicated that NIR Raman spectroscopy provides significant potential for noninvasive diagnosis of lung cancer in vivo based on the optic evaluation of biomolecules [70].

R. R. Alfano et al., using fluorescence spectra, could distinguish normal and cancerous tissues of human lung and breast tissues. An argon ion laser beam at 488 and 457.9 nm was focused on the front surface of the tissue to a spot size of 200 μm and the spectra collected from different sample groups were dissimilar [71].

Huang et al. employed near-infrared Raman spectroscopy to study the effect of formalin fixation of normal and cancerous human bronchial tissues. The aim of the study was to find out whether the variations of the Raman spectra caused by formalin fixation would affect the potential diagnostic ability for the lung cancer detection. A rapid dispersive type of NIR Raman system with an excitation wavelength of 785 nm was used. Bronchial tissue samples were obtained from six patients with known or suspected malignancies of the lung. Raman spectra of fresh normal and tumour tissue were compared with spectra of formalin-fixed normal and tumour tissue. Changes of the ratios of Raman intensities at 1445 to 1655 cm^{-1} and 1302 to 1265 cm^{-1} versus formalin fixing times varying from 2 to 24 hours were also examined. The major Raman spectral peaks were found at 1265, 1302, 1445, and 1655 cm^{-1} in both fresh and fixed bronchial tissues. However, bronchial tissues preserved in formalin showed a progressive decrease in overall intensities of these Raman peaks. The results showed that NIR Raman spectra of human bronchial tissues were significantly affected by formalin fixing and tissue hydration. Diagnostic markers at the 980–1100 and 1500–1650 cm^{-1} regions derived from fixed tissues did not appear to be applicable for in vivo lung cancer detection. It was shown that for yielding valid Raman diagnostic information for in vivo applications, fresh tissue should be used. If only fixed tissue is available thorough rinsing of specimens in phosphate-buffered saline (PBS) before spectral measurements may help reduce the formalin fixation artefacts on tissue Raman spectra [72].

Short et al. developed a near-infrared Raman spectroscopy system to collect real-time, noninvasive, in vivo human lung spectra [73]. The 785 nm excitation and the collection of tissue emission were accomplished by using a reusable fibre-optic catheter that passed down the instrument channel of a bronchoscope. The group reported the spectra of 20 patients, indicating differences between spectra from tumour and normal tissue.

Breast Cancer

Breast tissue is another interesting topic for researchers (Table 4.5). R. Eckel et al. analysed the IR spectra of normal, hyperplasia, fibroadenoma and carcinoma tissues of human breast. They worked on characteristic spectroscopic patterns in the proteins bands of the tissue. Some of the results of their experiments are as follows: (A) In carcinomatous tissues the bands in the region of 3000-3600 cm^{-1} shifted to lower frequencies. (B) The 3300 cm^{-1}/3075 cm^{-1} absorbance ratio was significantly higher for the fibroadenoma (C) For the malignant tissues, the frequency of α-helix amide I band decreased, while the corresponding β-sheet amide I band frequency increased (D) 1657 cm^{-1}/1635 cm^{-1} and 1553 cm^{-1}/1540 cm^{-1} absorbance ratios were

TABLE 4.5

Articles on Breast Tissue

Research Group	Research Method	Investigated Tissue or Sample	Effectiveness of the Technique	Reference Number
Eckel et al.	FTIR	Breast	+	74
Kline and Treado	Raman	Breast	n/a	75
Shafer-Peltier et al.	Raman	Breast	+	76
Ronen et al.	NMR	Breast	+	77
Frank et al.	MR spectroscopy	Breast	+	78
Fabian et al.	FTIR	Breast	n/a	79
Tam et al.	Raman	Breast	+	80
Abramczyk et al.	Raman	Breast	+	81
Kast et al.	Raman	Breast	+	82
Marzullo et al.	Raman	Breast	+	83

the highest for fibroadenoma and carcinoma (E). The 1680 cm^{-1}/1657 cm^{-1} absorbance ratio decreased significantly in the order of normal, hyperplasia, fibroadenoma and carcinoma (F). The 1651 cm^{-1}/1545 cm^{-1} absorbance ratio increased slightly for fibroadenoma and carcinoma (G). The bands at 1204 cm^{-1} and 1278 cm^{-1}, assigned to the vibrational modes of the collagen, did not appear in the original spectra as the resolved peaks and were distinctly stronger for the carcinoma tissues (H). The 1657 cm^{-1}/1204 cm^{-1} and 1657 cm^{-1}/1278 cm^{-1} absorbance ratios, both yielding information on the relative content of collagen increased in the order ofnormal, hyperplasia, carcinoma, and fibroadenoma [74].

N. J. Kline and P. J. Treado reported on chemical imaging of breast tissue using Raman spectroscopy. Raman chemical imaging of lipid and protein distribution in breast was performed without the use of invasive contrast agents. Instead, tissue component discrimination was based on the unique vibrational spectra intrinsic to lipids and proteins. It was suggested that visualization of breast tissue components is an essential step in the development of a quantitative Raman 'optical biopsy' technique suitable for the non-invasive detection and classification of breast cancer [75].

K. E. Shafer-Peltier et al. reported on a Raman spectroscopic model of human breast tissue and its implications for breast cancer diagnosis in vivo. They believe that Raman spectroscopy has the potential to provide real-time, in situ diagnosis breast cancer during needle biopsy or surgery via an optical fibre probe. To understand the relationship between the Raman spectrum of a sample of breast tissue and its disease state, near-infrared Raman spectroscopic images of human breast tissue were acquired using a confocal microscope. These images were then compared with phase contrast and hematoxylin- and eosin-stained images to develop a chemical/morphological model of breast tissue Raman spectra. The model explained the spectral features of a range of normal and diseased breast tissue samples, including breast cancer, and it also could be used to relate the Raman spectrum of a breast tissue sample to diagnostic parameters used by pathologists [76].

In addition to FTIR and Raman spectroscopy, some other techniques such as nuclear magnetic resonance (NMR) have been employed for research in the field of cancer. S. M. Ronen et al. studied the metabolism of the lipids of the human breast cancer using the ^{31}P NMR technique [77].

C. J. Frank et al. compared Raman spectra of histologically normal human breast biopsy samples to those exhibiting infiltrating ductal carcinoma (IDC) or fibrocystic change. Experiments at 785 nm with charge-coupled device (CCD) detectors reduced fluorescence interference. Sample-to-sample and patient-to-patient variation for normal specimens were less than 5% for the ratios of major Raman bands. The Raman spectra changed dramatically in diseased specimens, with much weaker lipid bands being evident. The spectra of IDC samples were similar to that of human collagen. Differences between benign (fibrocystic) and malignant (IDC) lesions were smaller than those between normal and IDC specimens, but were still reproducible [78].

The main focus area of the comparative infrared spectroscopic study of H. Fabian et al. was on human breast tumours, human breast tumour cell lines, and xenografted human tumour cells. The results indicated that substantial differences exist on a macroscopic level between the tumours, tumour cell lines, and xenografted tumour cells, which are related to the presence of a significant connective tissue matrix in the tumours. On a macroscopic level tumour cell xenografts appear, in spectroscopic terms, to be relatively homogenous with a relatively weak signature characteristic of connective tissue. Differences on a microscopic level between adjacent small (30-μm^2) areas of the same xenografted tumour could be detected, which were due to local variations in collagen content. In addition to variations in collagen content, variations in the deposition of microscopic fat droplets throughout both human and xenografted tumours could be detected. The results indicated the care with which infrared spectroscopic studies of tissues must be carried out to avoid incorrect interpretation of results due to an incomplete understanding of tissue pathology [79].

Tam et al. carried out a study on sample processing techniques for breast cancer using Raman spectroscopy. Fifty breast biopsies were studied using Raman spectroscopy prior to receipt of pathology reports. This was applied to at least two of the three available tissue processing techniques using point spectroscopy, mapping, and imaging. Differences in the spectra were related to the various sample processing methods [80].

Kast et al. used Raman spectroscopy with near-infrared light excitation to study normal breast tissue and tumours from 11 mice injected with a cancer cell line [82]. Spectra were collected from 17 tumours, 18 samples of adjacent breast tissue and lymph nodes, and 17 tissue samples from the contralateral breast and its adjacent lymph nodes. Discriminant function analysis was used for classification with principal component analysis scores as input data. Discriminant function analysis and histology agreed on the diagnosis of all contralateral normal, tumour, and mastitis samples, except one tumour that was found to be more similar to normal tissue. Normal tissue adjacent to each tumour was examined as a separate data group and classification of these tissues showed that some of them were diagnostically different from normal, tumour, and mastitis tissue. It was suggested that this may reflect malignant molecular alterations prior to morphologic changes, as expected in preneoplastic processes. Overall, it was concluded that Raman spectroscopy not only distinguishes tumour from normal breast tissue, but also defects early neoplastic changes prior to definite morphologic alteration.

The main goal of the research carried out by Marzullo et al. was to investigate spectra of the borders of lesions of samples of infiltrating ductal carcinoma (IDC) and to determine the characteristics of these spectra [83]. For this purpose, a total of 93 spectra were collected from five samples of healthy tissues and from 13 samples of IDC breast tissues using FT-Raman spectroscopy. Cluster analysis was used to separate the spectra into different groups.

The results obtained from histopathological analysis were used to confirm the results of the statistical analysis. The results showed that the only significant difference between the peaks of the spectra of normal tissues and those of lesion borders is a peak at 538 cm^{-1}. This peak is related to disulfide bridges in cysteine, and it seems to be the main factor for the FT-Raman determination of the boundaries between healthy and pathological tissue.

Skin Cancer

A number of researchers have been working on skin cancer employing spectroscopic techniques (Table 4.6). The study of S. Sukuta and R. Bruch was on factor analysis of cancer FTIR fibreoptic evanescent wave (FTIR-FEW) spectra. The purpose of the research was to isolate pure biochemical compounds and spectra and to classify skin cancer tumours. Apart from fulfilling the primary goals, it was demonstrated that the combination of the FTIR-FEW

TABLE 4.6

Articles on Skin Cancer

Research Group	Research Method	Investigated Tissue or Sample	Effectiveness of the Technique	Reference Number
Sukuta and Bruch	FTIR	Skin	+	84
Cheng et al.	Raman	Skin	+	85
Kaminaka et al.	Raman	Skin and lung	+	86
Sigurdsson et al.	Raman	Skin	+	87
Gniadecka et al.	Raman	Skin	+	88
Panjehpour et al.	Fluorescence spectroscopy	Skin	+	89
Wong et al.	Pressure-tuning infrared spectroscopy	Skin	+	90
Lucassen et al.	ATR-FTIR	Skin	+	91
McIntosh et al.	FTIR	Skin	+	92
Barry et al.	FTIR and FT-Raman	Skin	+	93
Huang et al.	Raman and NI autofluorescence	Skin	+	94
Caspers et al.	Raman	Skin	+	95
Ly et al.	FTIR imaging	Skin and colon	+	96
Lieber et al.	Raman	Skin	+	97
Hammody et al.	FTIR	Melanoma and Epidermis	+	98

technique and chemical factor analysis has the potential to be a clinical diagnostic tool [84].

W. T. Cheng et al. reported on micro-Raman spectroscopy used to identify and grade human skin pilomatrixoma (PMX). The normal skin dermis, collagen type I, hydroxyapatite (HA) were used as a control. The Raman spectrum of normal skin dermis was found to be similar to that of collagen type I, confirming that the collagen was a predominant component in normal skin dermis. The most significant differences of the collected spectra of normal skin dermis and soft and hard PMX were the peaks at 1665 cm^{-1}, which assigned to the amide I band, and 1246 cm^{-1}, which assigned to the amide III band. The considerable changes in collagen content and its structural conformation, the higher content of tryptophan, and disulfide formation in PMX masses were markedly evident. In addition, the peak at 960 cm^{-1} assigned to the stretching vibration of PO_4^{3-} HA also appeared respectively in the Raman spectra of hard and soft PMX masses, suggesting the occurrence of calcification of HA in the PMX tissue. The results indicated that the micro-Raman spectroscopy may provide a highly sensitive and specific method for identifying normal skin dermis and how it differs in chemical composition from different PMX tissues [85].

S. Kaminaka et al. reported on NIR multichannel Raman spectroscopy toward real-time in vivo cancer diagnosis. The method used enabled them to measure an in vivo Raman spectrum of live human tissue (skin) in one minute using fibre probe optics. By applying the system to human lung tissue, they found that Raman spectroscopy makes a clear distinction not only between normal and cancerous tissues, but also between two different parts of lung carcinoma. The results indicated a promising future for the noninvasive real-time Raman diagnosis of cancer [86].

The research of S. Sigurdsson et al. was about detection of skin cancer by classification of Raman spectra. The classification framework was probabilistic and highly automated. Correct classification of $80.5\% \pm 5.3\%$ for malignant melanoma and $95.8\% \pm 2.7\%$ for basal cell carcinoma was reported, which are excellent and similar to that of trained dermatologists. The results were shown to be reproducible and small distinctive bands in the spectrum, corresponding to specific lipids and proteins, were also shown to hold discriminating information which they used to diagnose skin lesion [87].

Diagnosis of the most common skin cancer, basal cell carcinoma (BCC), by Raman spectroscopy was carried out by M. Gniadecka et al. Biopsies of histopathologically verified BCC and normal skin were harvested and analysed by NIR-FT Raman spectroscopy using a 1064 nm Nd:YAG laser as a radiation source. The results indicated alterations in protein and lipid structures in skin cancer samples. Spectral changes were observed in protein bands, amide I (1640–1680 cm^{-1}), amide III (1220–1300 cm^{-1}), and ν(C–C) stretching (probably in amino acids proline and valine, 928–940 cm^{-1}), and in bands characteristic of lipids, CH_2 scissoring vibration (1420–1450 cm^{-1}), and $-(CH_2)_n-$ in-phase twist vibration around 1300 cm^{-1}. Moreover, possible changes in polysaccharide structure were found in the region 840–860 cm^{-1}.

Analysis of band intensities in the regions of 1220–1360, 900–990, and 830–900 cm^{-1} allowed for a complete separation between BCC and normal skin spectra. In conclusion, Raman spectra of BCC differed considerably from those of normal skin and the technique can be viewed as a promising tool for the diagnosis of skin cancer [88].

Approximately similar results have been reported by M. Panjehpour and colleagues on in vivo diagnosis of nonmelanoma skin cancers, using laser-induced fluorescence. A nitrogen/dye laser of 410 nm was used and samples of normal, BCC, SCC, precancerous, and benign tissues could have been classified with an accuracy of even 100% [89].

Wong et al. applied infrared spectroscopy combined with high pressure (pressure-tuning infrared spectroscopy) for studying the paired sections of BCCs and normal skin from ten patients. In this study, atmospheric pressure IR spectra from BCC were dramatically different from those from the corresponding normal skin. Compared to their normal controls, BCCs displayed increased hydrogen bonding of the phosphodiester group of nucleic acids, decreased hydrogen bonding of the C–OH groups of proteins, increased intensity of the band at 972 cm^{-1}, a decreased intensity ratio between the CH$_3$ stretching and CH$_2$ stretching bands, and accumulation of unidentified carbohydrates [90].

Lucassen et al. used attenuated total reflectance Fourier transform infrared (ATR-FTIR) spectroscopy to measure hydration of the stratum corneum. It was believed that determination of the hydration state of the skin is necessary to obtain basic knowledge about the penetration and loss of water in the skin stratum corneum. In this study, direct band fitting of the water bending, combination, and OH stretch bands over the 4000–650 cm^{-1} wavenumber range were applied. Separate band fits of water, normal stratum corneum, and occluded hydrated stratum corneum spectra were obtained yielding band parameters of the individual water contributions in the bending mode at 1640 cm^{-1}, the combination band at 2125 cm^{-1}, and the OH stretches in the hydrated skin stratum corneum spectra. They concluded that band fit analysis of hydrated skin stratum corneum ATR-FTIR spectra offers the possibility for quantitative determination of individual water band parameters [91].

McIntosh et al. used infrared spectroscopy to examine basal cell carcinoma to explore distinctive characteristics of BCC versus normal skin samples and other skin neoplasms. Spectra of epidermis, tumour, follicle sheath, and dermis were acquired from unstained frozen sections, and analyzed qualitatively by t-tests and by linear discriminant analyses. Dermal spectra were significantly different from the other skin components mainly due to absorptions from collagen in dermis. Spectra of normal epidermis and basal cell carcinoma were significantly different by virtue of subtle differences in protein structure and nucleic acid content. Linear discriminant analysis characterised spectra as arising from basal cell carcinoma, epidermis, or follicle sheath with 98.7% accuracy. Use of linear discriminant analysis accurately classified spectra as arising from epidermis overlying basal cell carcinoma versus epidermis overlying nontumour-bearing skin in 98.0% of cases. Spectra of basal cell carcinoma, squamous cell carcinoma, nevi, and malignant melanoma were qualitatively similar. Distinction of basal cell carcinoma, squamous cell carcinoma, and melanocytic lesions by linear discriminant analyses, however, was 93.5% accurate. Therefore, spectral separation of abnormal versus normal tissue was achieved with high sensitivity and specificity [92].

Barry et al. recorded FT Raman and infrared spectra of the outermost layer of human skin, the stratum corneum. Assignments consistent with the FT Raman vibrations were made for the first time and compared with assignments from the FTIR spectrum. The results demonstrated that FT Raman spectroscopy holds several advantages over FTIR in studies of human skin. The molecular and conformational nature of human skin, and modifications induced by drug or chemical treatments, may be assessed by FT Raman spectroscopy [93].

Huang et al. used a combination of Raman spectroscopy with near-infrared autofluorescence for assessment of skin fibrosarcoma in mice. A murine tumour was implanted into the subcutaneous region of the lower back. Diagnostic algorithms were developed for differentiating tumours from normal tissue based on the spectral features. Thirty-two in vivo Raman, NIR fluorescence, and composite Raman and NIR fluorescence were analysed (16 normal, 16 tumours). Classification results showed diagnostic sensitivities of 81.3%, 93.8%, and 93.8%; specificities of 100%, 87.5%, and 100%; and overall diagnostic accuracies of 90.6%, 90.6%, and 96.9%, respectively for tumour identification [94].

A study on confocal Raman microspectroscopy as a noninvasive in vivo optical method to measure molecular concentration profiles in the skin was carried out by Caspers et al. It was shown that the technique can be applied to determine the water concentration in the stratum corneum as a function of distance to the skin surface, with a depth resolution of 5 μm. The resulting in vivo concentration profiles were in qualitative and quantitative agreement with published data. No other noninvasive in vivo technique exists that analyses skin molecular compositions as a function of distance to the skin surface with similar detail and spatial resolution [95].

Ly et al. applied FTIR spectral imaging on formalin-fixed paraffin-embedded biopsies from colon and skin cancerous lesions [96]. The samples were fixed in formalin and then embedded in paraffin. This technique can preserve molecular structures and is the gold standard in tissue storage. However, paraffin absorption bands are significant in the mid-infrared region and can mask some molecular vibrations of the tissue. Then direct data processing was applied on spectral images without any chemical dewaxing of the tissues. The signal of paraffin was modelled and paraffin spectra were removed from the raw images based on an outlier detection. Afterwards, pseudo-colour images were computed by K-means clustering in order to highlight histological structures of interest. Using this method, tumour areas were successfully demarcated in both types of tissues.

Lieber et al. reported on near-infrared Raman microspectroscopy for in vitro detection of skin cancer [97]. Thirty-nine skin tissue samples consisting of normal, basal cell carcinoma, squamous cell carcinoma, and melanoma from 39 patients were investigated. Raman spectra were recorded at the surface and at 20 um intervals below the surface for each sample, down to a depth of at least 100 um. Data reduction algorithms based on the nonlinear maximum representation and discrimination feature (MRDF) and discriminant algorithms using sparse multinomial logistic regression (SMLR) were developed for classification of the Raman spectra relative to histopathology. The tissue Raman spectra were classified into pathological states with a maximal overall sensitivity and specificity for disease of 100%.

Hammody et al. examined the differences in the IR spectra of melanoma tissues and the surrounding epidermis in skin biopsies [98]. Biopsies of 55 patients were analysed and it was demonstrated that the technique could differentiate melanoma from the epidermis using parameters derived from absorbance bands. Additionally, the absorbances from tyrosine and phosphate that are abnormally elevated in malignant melanoma could be used as markers.

Gastrointestinal Cancer

Spectroscopy has also been employed to study gastrointestinal tissues (apart from the colon), the details of which are tabulated in (Table 4.7). For example, N. Fujioka et al. reported on discrimination between normal and malignant human gastric tissues by FTIR spectroscopy. Their aim was to determine whether malignant and normal human gastric tissues can be distinguished by the technique. As a result, 22 out of 23 gastric tissue samples and 9 out of 12 gastric normal samples were correctly segregated, yielding 88.6% accuracy. Subsequently, they concluded that FTIR spectroscopy can be a useful tool for screening gastric cancer [99].

H. Barr et al. has published an excellent review paper about the role of optical spectroscopy for an early diagnosis of gastrointestinal malignancy and stated that fluorescence spectroscopy combined with vibrational spectroscopy offers the most realistic prospect of an early clinical condition and is currently under evaluation. In addition, optical coherence tomography can differentiate the layers of the oesophageal wall. Although complicated, Raman spectroscopy offers the greatest information with possible development of a molecular endoscope [42].

W. Petrich has published a review with the title of "Mid-Infrared and Raman Spectroscopy for medical diagnostics." The article gives examples of the potential of the method in supporting medical diagnostics, most of which have been performed in the field of internal medicine, namely, angiology, haematology, rheumatology, endocrinology, gastroenterology, and nephrology. Further potential applications in neurology, gynaecology, obstetrics

TABLE 4.7

Articles on Gastrointestinal Tissue (Other than Colon)

Research Group	Research Method	Investigated Tissue or Sample	Effectiveness of the Technique	Reference Number
Barr et al.	Review	n/a	n/a	42
Petrich	Review	n/a	n/a	43
Fujioka et al.	FTIR	Gastrointestinal tissue	+	99
Tan et al.	Raman	Gastrointestinal tissue	+	100
Weng et al.	Raman and FTIR	Gastrointestinal tissue	+	101
Mordechai et al.	FTIR	Intestine	+	102
Shetty et al.	Raman	Oesophageal tissue	+	103
Li et al.	FTIR	Gastric tissue	+	104
Xu et al.	FTIR-ATR	Stomach tissue	+	105
Suzuki et al.	Raman	Pancreatic tumour tissue	+	106
Kawabata et al.	Raman	Gastric tissue	+	107
Teh et al.	Raman	Gastric tissue	+	108
Pandya et al.	Raman	Pancreatic tissue	+	109
Hu et al.	Raman	Gastric tissue	+	110
Maziak et al.	FTIR	Oesophageal tissue	+	111
Xu et al.	ATR-FTIR	Gall bladder	+	112
Kondepati et al.	FTIR	Pancreatic tissue	+	113
Sun et al.	FTIR	Gastric tissue	+	114

and dermatology are considered. In addition, first steps towards in vivo applications are described [43].

The design of an auto-classifying system and its application in Raman spectroscopy diagnosis of gastric carcinoma was investigated by Tan et al. They developed a tentative user-friendly system to auto-classify Raman spectra of gastric carcinoma tissues. They also suggested that the software could be applied into classification of other tissues with some necessary alterations [100].

S.-F. Weng et al. studied tumours from the stomach, small intestine, colon, rectum, liver, and other parts of the digestive system, with FTIR fibre optics and FT-Raman spectroscopic techniques. The spectra of samples were recorded on a Magna750 FTIR spectrometer with a mercury cadmium telluride (MCT) detector and mid-infrared optical fibre. The measurements were carried out by touching the sample with an attenuated total reflectance probe. FT-Raman spectra of the samples were recorded on a 950 FT-Raman spectrometer. The results indicate that (i) the C=O stretching band of adipose can be observed in some normal tissues but is rarely found in malignant tissues, and (ii) the relative intensities I_{1460}/I_{1400} are high for normal tissue but low for malignant tissue. For most normal tissues, the intensities for the 1250 cm^{-1} band are stronger than for band around 1310 cm^{-1}, while the 1310 cm^{-1} band in malignant tissues is often stronger than the 1250 cm^{-1} bands [101].

Colon Cancer

Mordechai et al. used spectroscopy to study adenocarcinoma and normal colonic tissues by employing microscopic infrared study (FTIR microscopy) of thin tissue specimens and a direct comparison with normal histopathological analysis, which served as a gold reference. Several unique differences between normal and cancerous intestinal specimens were observed. The cancerous intestine showed weaker absorption strength over a wide region. They also effectively compared the results from microscopic IR absorption spectra from intestinal tissues (normal and cancerous) with other biological tissue samples [102].

Oesophageal Cancer

Shetty et al. demonstrated the potential of Raman spectroscopy for the identification and classification of the malignant changes in oesophageal carcinomas. Their study was aimed at understanding the biochemical changes that distinguish between the different stages of disease through Raman mapping studies. This technique was used to analyse 20-μm sections of tissue from 29 snap-frozen oesophageal biopsies. Contiguous haematoxylin and eosin sections were reviewed by a consultant pathologist. Changes were noted in the distribution of DNA, glycogen, lipids, and proteins. The main spectra obtained from selected regions demonstrated increased levels of glycogen in the squamous area compared with increased DNA levels in the abnormal region. It was concluded that Raman spectroscopy is a highly sensitive and specific technique for demonstration of biochemical changes in carcinogenesis, and there is a potential for in vivo application for real-time endoscopic optical diagnosis [103].

Diagnostic research was carried out by Li et al. that aimed at classifying endoscopic gastric biopsies into healthy, gastritis, and malignancy through the use of FTIR spectroscopy. A total of 103 endoscopic samples, including 19 cases of cancer, 35 cases of chronic atrophic gastritis, 29 cases of chronic superficial gastritis, and 20 healthy samples, were investigated by ATR-FTIR Significant differences were observed in FTIR spectra of these four types of gastric biopsies. It was demonstrated that the sensitivity of the method for healthy, superficial gastritis, atrophic gastritis, and gastric cancer was 90%, 90%, 66%, and 74%, respectively. It was concluded that FTIR spectroscopy can distinguish disease processes in gastric endoscopic biopsies [104].

Maziak et al. investigated the usefulness of FTIR spectroscopy in the diagnosis of oesophageal cancer [111]. Tissue samples from the diseased and normal sites of 10 patients with adenocarcinoma of the oesophagus were analysed and compared using FTIR spectroscopy and histopathology.

Specific changes were observed and recorded in the spectral features of oesophageal cancer, and spectral criteria were established for the detection of malignant tissues.

Stomach Cancer

Xu et al. studied 90 stomach tissue samples from 48 patients, including 32 normal and 58 malignant tissue samples using mid-IR fibre-optic ATR spectroscopy [105]. It was concluded that malignancy can be characterised by the absence of CH and C=O bands, a weak amide II band near 1545 cm⁻¹, a shift of the amide I band to a lower wavenumber, a decrease in the similar-to-1450 cm⁻¹ peak to less than the similar-to-1400 cm⁻¹ peak. Furthermore, subtraction of the spectra indicated that the amide I and amide II bands of normal and malignant tissues have larger differences in peak positions and relative intensities. Based on the results, the group successfully realized the detection of the tumour tissues of the digestive tract in vivo and in situ.

Pancreatic Cancer

Suzuki et al. studied living pancreatic cancer tissues grown subcutaneously in nude mice by in vivo microscope Raman spectroscopy [106]. It was aimed at comparing the spectra of living pancreatic cancer tissue to that of dead pancreatic cancer tissue. It was found that they are different from each other, and the main differences are seen in the peak positions of 937, 1251, 1447, 1671, 966, and 1045 cm⁻¹. These results strongly suggest that the spectral changes mainly reflect the protein conformational changes in the tumour tissue with death of the host animal. They also demonstrate the importance of in vivo, real-time studies of biomedical tissues using Raman spectroscopy.

Gastric Cancer

Kawabata et al. investigated whether Raman spectroscopy can be used to diagnose gastric cancer [107]. For this purpose, a total of 251 fresh biopsy specimens of gastric carcinoma and nonneoplastic mucosa were obtained from 49 gastric cancer patients at endoscopy. The fresh specimens were analysed with a near-infrared Raman spectroscopic system with an excitation wavelength of 1064 nm, without any pretreatment. Principal component analysis (PCA) was performed to distinguish gastric cancer and nonneoplastic tissue, and discriminant analysis was used to evaluate the accuracy of the gastric cancer diagnosis. It was reported that the main difference between the two groups of samples was observed in the area of 1644 cm^{-1}. The accuracy of diagnosis using this single peak was 70%, consistent with the PCA result. The overall sensitivity was 66%, and the specificity was 73%.

The purpose of a research carried out by Teh et al. was to explore near-infrared (NIR) Raman spectroscopy for identifying precancer (dysplasia) from normal gastric mucosa tissues [108]. A total of 76 gastric tissue samples obtained from 44 patients who underwent endoscopy investigation or gastrectomy operation were used in this study. The histopathological examinations showed that 55 tissue specimens were normal and 21 showed dysplasia. Both the empirical approach and multivariate statistical techniques, including PCA and LDA, together with the leave-one-sample-out cross-validation method, were employed to develop effective diagnostic algorithms for classification of Raman spectra. Raman spectra showed significant differences between normal and dysplastic tissue, particularly in the spectral ranges of 850–900, 1200–1290, and 1500–1800 cm^{-1} which contained signals related to hydroxyproline, amide III and amide I of proteins, and C=C stretching of lipids, respectively. The empirical diagnostic algorithm based on the ratio of the Raman peak intensity at 875 cm^{-1} to the peak intensity at 1450 cm^{-1} gave diagnostic sensitivity of 85.7% and specificity of 80.0%, whereas the diagnostic algorithms based on PCA-LDA yielded diagnostic sensitivity of 95.2% and specificity 90.9% for separating dysplasia from normal gastric tissue. It was also demonstrated that the ratio of peak intensities at 875 to 1450 cm^{-1} provided good differentiation between normal and dysplastic gastric tissue.

Pandya et al. evaluated the ability of Raman spectroscopy to differentiate normal pancreatic tissue from malignant tumours in a mouse model [109]. They collected 920 spectra, 475 from 31 normal pancreatic tissue and 445 from 29 tumour nodules using a 785-nm near-infrared laser excitation. Using principal component analysis, subtle chemical differences in normal and malignant tissue were successfully highlighted. Using histopathology as the gold standard, Raman analysis gave sensitivities between 91% and 96% and specificities between 88% and 96%. In this study, pancreatic tumours were characterized by increased collagen content and decreased DNA, RNA, and lipid components compared with normal pancreatic tissue.

Hu et al. carried out a study on classification of normal and malignant human gastric mucosa tissue with confocal Raman microspectroscopy [110]. They analysed thirty-two samples from human gastric mucosa tissue, including 13 normal and 19 malignant tissue samples. Spectra were obtained by this technique without any sample preparation. Comparing preprocessed spectra of malignant gastric mucosa tissues with those of counterpart normal ones, there were obvious spectral changes, including intensity increase at similar to 1156 cm^{-1} and intensity decrease at similar to 1587 cm^{-1}. The peak ratio of 1156/1587 could successfully classify the normal and malignant gastric mucosa tissue samples.

Xu et al. assessed the sensitivity and specificity of ATR-FTIR in the rapid detection of gallbladder cancer [112]. The edges of surgically resected specimens were examined rapidly. It was shown that there are a series of differences between the spectra of normal and malignant tissues, and subtraction technique is helpful in identifying gallbladder cancer. Compared with pathological diagnosis results, one benign specimen was classified as malignant and 6/6 malignant specimens and 39 cases of benign specimens were correctly classified. The sensitivity was 100% (6/6), specificity 97% (39/40), and accuracy was 97% (45/46).

Kondepati et al. reported on the detection of structural disorders in pancreatic tumour DNA with FTIR spectroscopy [113]. The DNA was isolated from the tumour and the adjacent histological normal tissue from 15 pancreatic tumour patients. Structural disorders in the tumour DNA were detected in the phosphodiester-deoxyribose spectral region. Based on such data within the spectral interval of 1192–1059 cm^{-1}, classification was achieved with a sensitivity and specificity of 87% and 80%, respectively. The test with blinded samples from eight patients based on data from the same spectral region showed a sensitivity and specificity of 100 and 62.5%, respectively.

Sun et al. carried out a FTIR spectroscopic study on gastric tissue samples from gastroscopy [114]. The spectra of 184 specimens from gastroscopy of small-sized gastric endoscope samples were collected using the FTIR spectrometer. The results showed that the spectra of chronic superficial gastritis are similar to those of the normal block tissue. But there were significant differences in band location and relative intensity between the spectra of chronic superficial gastritis and those of normal gastric tissues. There were also differences between the spectra of chronic atrophic gastritis and those of malignant gastric tissues. In addition, it was demonstrated that the subtraction technique can provide more information to correctly identify the gastric samples of chronic atrophic gastritis, superficial gastritis, and gastric cancer.

Brain Tissues

Spectroscopic investigation of brain tissues has also been carried out by a number of researchers, the details of which are tabulated in Table 4.8. L. P. Choo et al. applied infrared spectra of human central nervous system tissue for diagnosis of Alzheimer's disease (AD). In addition, they presented

TABLE 4.8

Articles on Brain Tissue

Research Group	Research Method	Investigated Tissue or Sample	Effectiveness of the Technique	Reference Number
Choo et al.	FTIR	CNS	+	115
Dovbeshko et al.	FTIR	Brain	+	116
Lehnhardt et al.	MR spectroscopy	Brain	+	117
Howe and Opstad	MR spectroscopy	Brain	+	118
Yoshida et al.	FTIR	Rat brain	+	119
Lakshmi et al.	Raman	Mice brain	+	120
Krafft et al.	Raman	Brain	+	121

a means of classifying the spectroscopic data nonsubjectively using several multivariate methods. The results demonstrated that IR spectroscopy can potentially be used in the diagnosis of AD from autopsy tissue. It was shown that correct classification of white and grey matter from brains identified by standard pathological methods as heavily, moderately, and minimally involved can be achieved with success rates of greater than 90% using appropriate methods. Classification of tissue as either control or AD was achieved with a success rate of 100% [115].

FTIR spectra of RNA isolated from tumour brain (glioma) and DNA isolated from low-dose gamma-irradiated epididymis cells of rats from the Chernobyl accident zone were investigated by G. I. Dovbeshko et al. Their aim was to study the nucleic acid damage and reported the existence of damage in the primary, secondary, and tertiary structure of nucleic acid, which seem to be connected with modification of bases and sugars, and redistribution of the H-bond network. They also mentioned that a great amount of statistical data and good mathematical approaches are needed for the use of these data as diagnostic criteria [116].

F. G. Lehnhardt et al. carried out a study on ^{1}H- and ^{31}P-MR spectroscopy of primary and recurrent human brain tumours in vitro. They worked on malignancy-characteristic profiles of water-soluble lipophilic spectral components. Tissue samples of primary tumours and their first recurrences were examined. The spectra were recorded from the samples of meningioma, astrocytoma, and glioblastoma. The findings indicated that the collected data and information enables differentiation not only among tumour types but also between primary and recurrent glioblastoma, reflecting an evolving tumour metabolism [117].

In a spectroscopic study, F. A. Howe et al. investigated brain tumours using ^{1}H MR spectroscopy [118].

S. Yoshida et al. measured the Fourier transform infrared spectra of brain microsomal membranes prepared from rats fed under two dietary oil conditions with and without brightness-discrimination learning tasks: one group was fed α-linolenate-deficient oil (safflower oil) and the other group was fed the sufficient oil (perilla oil) from mothers to offspring. The infra-red spectra of microsomes under the two dietary conditions without the learning task showed no significant difference in the range 1000–3000 cm^{-1}. Only after the learning task were the infrared spectral differences noted between the microsomal membranes from both groups. Spectral differences were observed mainly in the absorption bands of fatty acid ester at around 1730 cm^{-1} (sn-2 position), those of phosphate and oligosaccharides in the range of 1050–1100 cm^{-1}, and a band at around 1145 cm^{-1}. These results suggest changes in hydration of the membrane surface and modification in the oligosaccharide environment (removal or modification) of microsomes, which may be correlated in part with dietary oil–induced changes in learning performance [119].

Lakshmi et al. conducted a Raman study on radiation damage of brain tissue in mice. A set of studies were carried out on brain tissue from mice subjected to irradiation to identify the biochemical changes in tissue as the result of radiotherapy and radiation injury. It was shown that brain irradiation produces drastic spectral changes even in tissue far removed from the irradiation site. The changes were very similar to those produced by the stress of inoculation and restraint and the administration of an anaesthetic drug. While the changes produced by stress or anesthetics last for only a short time (a few hours to 1 or 2 days), radiation-induced changes persist even after one week. The results also supported the hypothesis that various protective factors are released throughout the body when the central nervous system (CNS) is exposed to radiation [120].

Human brain tissue, in particular white matter, contains high lipid content. These brain lipids can be divided into three principal classes: neutral lipids including the steroid cholesterol, phospholipids, and sphingolipids. Major lipids in normal human brain tissue are phosphatidylcholine, phosphatidylethanolamine, phosphatidylserine, phosphatidylinositol, phosphatidic acid, sphyngomyelin, galactocerebrosides, gangliosides, sulfatides, and cholesterol. Minor lipids are cholesterol ester and triacylglycerides. Detailed research on near-infrared Raman spectra was carried out by Krafft et al. They recorded the Raman spectra of 12 major and minor brain lipids with 785-nm excitation in order to identify their spectral fingerprints for qualitative and quantitative analysis [121].

Oral Squamous Cell Carcinoma

Oral tissues have also been widely investigated by a number of researchers (Table 4.9). Y. Fukuyama et al. used FTIR microscopy to study the differences between oral squamous cell carcinoma and normal mucosa (normal gingival epithelium or normal subgingival tissue). The tissue spectra were compared

TABLE 4.9

Articles on Tissues Related to the Oral Cavity

Research Group	Research Method	Investigated Tissue or Sample	Effectiveness of the Technique	Reference Number
Fukuyama et al.	FTIR	Oral tissue	+	122
Malini et al.	Raman	Oral tissue	+	123
Wu et al.	Raman	Oral tissue	+	124
Majmuder and Ghosh	Autofluorescence	Oral tissue	+	125
Lau et al.	Raman	Nasopharynx	+	126
Lau et al.	Raman	Larynx	+	127
Wu et al.	FTIR-ATR + Raman	Oral tissue	+	128
Teh et al.	Raman	Larynx	+	129
Wang et al.	FTIR	Salivary gland	+	130

with the purified human collagen and keratin. One half of every tissue specimen was measured with FTIR and the other half was investigated histologically. The data obtained suggested that this technique is applicable to clinical diagnostics [122].

R. Malini et al. reported on discrimination of normal, inflammatory, premalignant, and malignant oral tissue using Raman spectroscopy. Spectral profiles of different samples showed pronounced differences. It was demonstrated that all the four tissue types could be discriminated and diagnosed correctly. The biochemical differences between normal and pathological conditions of oral tissue were also discussed [123].

The application of FTIR fibre-optic technique for distinguishing malignant from normal oral tissues was reported by J. G. Wu et al. According to the results, the 1745 cm^{-1} band, which is assigned to the ester group (C=O) vibration of triglycerides, is a reliable marker that is present in normal tissue but absent or weak in malignant oral tissues. In addition, other bands such as C–H stretching and the aide bands are also helpful in distinguishing the two groups of samples. Raman spectroscopic measurements were in agreement with results observed from FTIR spectra [124].

The study of S. K. Majmuder and Ghosh concerned the use of a relevance vector machine (RVM) for optical diagnosis of cancer. It reported the use of the theory of the RVM for development of a fully probabilistic algorithm for autofluorescence diagnosis of early-stage cancer of the human oral cavity. The diagnostic algorithms were developed using in vivo autofluorescence spectral data acquired from the human oral cavity with a N$_2$ laser–based portable fluorimeter. The sensitivity and specificity toward cancer were up to 91% and 96%, respectively. When implemented on the spectral data of the uninvolved oral cavity sited from the patients, it yielded a specificity of up to 91% [125].

Nasopharyngeal Carcinoma

D. P. Lau et al. reported on Raman spectroscopy for optical diagnosis in normal and cancerous tissue of the nasopharynx. The tissues obtained from biopsies were studied using a rapid-acquisition Raman spectrometer. The spectra were collected in five seconds and consistent differences were noted between normal and cancerous tissue in three bands of 1290–1320, 1420–1470, and 1530–1580 cm^{-1} [126].

In another work, Lau et al. studied Raman spectroscopy for optical diagnosis of the larynx. The objective of the research was to determine if Raman spectra could be obtained rapidly from laryngeal tissue in vitro, and compared Raman spectra from normal, benign, and cancerous laryngeal tissue. Good-quality spectra were obtained with a 5-second signal acquisition time (SAT). Spectral peak analysis showed prediction sensitivities of 89%, 69%, and 88%, and specificities of 86%, 94%, and 94% for normal tissue, carcinoma, and papilloma. Spectral differences appeared to exist between different samples and it was concluded that the ability to obtain the spectra rapidly supports potential for future in vivo studies [127].

Wu et al. reported on distinguishing malignant from normal oral tissues using FTIR fibre-optic ATR and Raman spectroscopy [128]. It was concluded that the 1745 cm^{-1} band, which is assigned to the ester group (C=O) vibration of triglycerides, is a reliable marker that is present in normal tissues but absent or a weak band in malignant oral tissues. Other bands such as C–H stretching bands and the amide bands were also helpful in distinguishing malignant tissues from normal tissues. Subtraction spectra and the two spectroscopic methods were in agreement with one another.

Teh et al. carried out a study on the diagnostic ability of NIR Raman spectroscopy in identifying malignant tumours from normal tissues in the larynx [129]. A rapid NIR Raman system was utilized. Multivariate statistical techniques were employed to develop effective diagnostic algorithms. Raman spectra in the range of 800–1800 cm^{-1} differed significantly between normal and malignant tumour tissues. Diagnostic sensitivity of 92.9% and specificity of 83.3% for separating malignant tumours from normal laryngeal tissues were reported and it was concluded that NIR Raman spectroscopy with multivariate statistical techniques has a potential for noninvasive detection of malignant tumours in the larynx.

Wang et al. applied FTIR spectroscopy to detect pleomorphic adenoma of the salivary gland in vivo [130]. In this study, FTIR spectra of 20 patients with salivary pleomorphic adenoma were analyzed and compared. It was found that there were significant differences in the spectral features of the skin covering normal salivary gland, pleomorphic adenoma, and carcinoma change of pleomorphic adenoma. The most important differences were changes in peak position, band shape, and relative intensity of the bands in the ranges of 1000–1800 cm^{-1} and 2800–3000 cm^{-1}. The results suggested that the spectra of pleomorphic adenoma are the intermediate between normal salivary gland and carcinoma change of pleomorphic adenoma.

Blood-Related Tissue and Organs

A considerable number of studies have been conducted on blood-related tissue and organs (Table 4.10). X. Li et al. applied Raman spectroscopy and fluorescence for the detection of liver cancer and abnormal liver tissue. They measured laser-induced human serum, Raman spectra of liver cancer and

TABLE 4.10

Articles on Blood-Related Tissues and Organs

Research Group	Research Method	Investigated Tissue or Sample	Effectiveness of the Technique	Reference Number
Li et al.	Raman and fluorescence spectroscopy	Liver	+	131
Chiang et al.	Raman	Heme protein	+	132
Van de Poll et al.	Raman	Atherosclerotic plaque	+	133
Duarte et al.	Raman	Cat serum	+	134
Rohleder et al.	Raman	Serum and serum infiltrate	+	135
Silveira et al.	Raman	Human coronary arteries	+	136
Andrus and Strickland	FTIR	Lymphoid tissue	+	137
Puppels et al.	Raman	Lymphocytes	+	138
Deng et al.	Raman	Human red blood cells	+	139
Li and Jin	Raman	Human serum	+	140
Mordechai et al.	FTIR	Lymphocytes (childhood leukaemia)	+	141
Andrus	FTIR	Non-Hodgkin's lymphoma	+	142
Isabelle et al.	Raman and FTIR	Oesophageal lymph nodes	+	143
Chan et al.	Raman	Leukaemia	+	144
Khanmohammadi et al.	ATR-FTIR	Human blood	+	145
Pichardo-Molina et al.	Raman	Serum	+	146
Huleihel et al.	FTIR	Blood of diarrheic patients	+	147

analysed the spectral differences between normal people and liver cancer patients. The results from more than 200 case measurements showed that the spectral diagnosis was in good agreement with the clinical results. The experiment indicated differences in the blue shift of the fluorescence peak between the normal, liver fibrosis, and liver cirrhosis [131].

Another FT-Raman spectroscopy was carried out on the carcinogenic polycyclic aromatic hydrocarbons (PAHs) in biological systems and their banding to heme proteins by H. P. Chiang et al. The Raman spectra of benzo[a]pyrene (BaP), a typical carcinogenic PAH, were acquired under different conditions and analysed. It was concluded that CH wagging and ring stretching mixed strongly with CH in-plane bending are the most significantly affected vibrations [132].

Van de Poll et al. reported on Raman spectroscopic evaluation of the effects of diet and lipid-lowering therapy on atherosclerotic plaque development in mice. Through this technique, they could quantitatively characterise the plaque without using the standard destructive histopathological methods such as sectioning. Raman spectra were obtained over the full width and entire length of the ascending aorta and aortic arch. Spectra were modelled to calculate the relative dry weights of cholesterol and calcium salts, and quantitative maps of their distribution were created. In conclusion, Raman spectroscopy could be used to quantitatively study the size and distribution of depositions of cholesterol and calcification. It also could be used for the quantitative investigation of atherosclerosis and lipid-lowering therapy in larger animals or humans in vivo [133].

J. Duarte et al. investigated on the use of near-infrared Raman spectroscopy to detect immunoglobulin G and immunoglobulin M antibodies against *Toxoplasma gondii* in serum samples of domestic cats. The aim of this work was to investigate a new method to diagnosis *Toxoplasma gondii*, instead of serological tests which usually have a high cost and are time consuming as well. In conclusion, the possibility of antibody detection by Raman spectroscopy was confirmed [134].

D. Rohleder et al. carried out a study on quantitative analysis of serum and serum ultrafiltrate by Raman spectroscopy. They explored the technique as a reagent-free tool for predicting the concentrations of different parameters in serum and serum ultrafiltrate, such as glucose, triglycerides, urea, total protein, cholesterol, high-density lipoprotein, low-density lipoprotein, and uric acid. The parameters were determined with accuracy within the clinically interesting range. After creating a multivariate algorithm for data analysis, concentrations were predicted blindly based solely on the Raman spectra. Moreover, differentiation between HDL and LDL cholesterol and the quantification of uric acid was accomplished for serum-based Raman spectroscopy for the first time [135].

The title of the article published by L. Silveira et al. is "Correlation between near-infrared Raman spectroscopy and the histopathological analysis of atherosclerosis in human coronary arteries." The objective of the study was to obtain feasible diagnostic information to detect atheromatous plaque using NIR spectroscopy (NIRS). An 830-nm Ti:sapphire laser pumped by an argon laser was used. A spectrograph dispersed light scattered from arterial tissue and a liquid nitrogen–cooled CCD detected the Raman spectra. A total of 111 arterial fragments were scanned and Raman results were compared with histopathology. An algorithm was modelled for tissue classification into three categories: nonatherosclerotic (NA), noncalcified (NC), and calcified (C) using Raman spectra. Spectra were randomly separated into training and prospective groups. It was found that for the NA tissue, the algorithm has sensitivity of 84% and 78% and specificity of 91% and 93% for training and prospective groups, respectively. For NC tissue, the algorithm has sensitivity of 88% and 90% and specificity of 88% and 83%. For the C tissue, both sensitivity and specificity were at maximum 100% [136].

FTIR Spectroscopy for Cancer Grading

P. G. L. Andrus and R. D. Strickland used FTIR spectroscopy for cancer grading. Freeze-dried tissue samples from lymphoid tumours were studied. The absorbance ratio of 1121 cm^{-1}/1020 cm^{-1} increased, along with the emergence of an absorbance pulse at 1121 cm^{-1}, with increasing clinicopathological grade of malignant lymphoma. This study proposed the above ratio as an index of the cellular RNA/DNA ratio after subtraction of the overlapping absorbances, if present, due to collagen or glycogen. Absorbance attributable to collagen increased lymphoma grade and was greater in benign inflammatory tumours than in low-grade lymphomas. It was also suggested that the ratio trend may form the basis of a universal cancer grading parameter to assist with cancer treatment decisions and may also be useful in the analysis of cellular growth perturbation induced by drugs or other therapies [137].

G. J. Puppels et al. investigated carotenoids located in human lymphocyte subpopulations (CD4+, CD8+, T-cellreceptor-$\gamma\delta$+, and CD19+), and natural killer cells (CD16+) using Raman microspectroscopy. In CD4+ lymphocytes, a high concentration of carotenoids was found in the Gall body (about 10^{-3} M). In other cell groups, except CD19+ groups, carotenoids appeared to be concentrated in the Golgi complex (about 10^{-4} M). The concentration of carotenoids in CD19+ lymphocytes was found to be below the present detection limit (about 10^{-6} to 10^{-5} M). The results provided new possibilities for investigation of the mechanisms behind the suggested protective role of carotenoids against the development of cancers [138].

J. L. Deng et al. carried out a study on the effect of alcohol on single human red blood cells (RBCs) using near-infrared laser tweezers Raman spectroscopy. A low-power diode laser at 785 nm was applied for the trapping of a living cell and the excitation of its Raman spectrum. The denaturation process of single RBCs in 20% alcohol solution was investigated by detecting the time evolution of the Raman spectra at the single-cell level. The vitality of RBCs was characterised by Raman bands at 752 cm^{-1}, which corresponds to the porphyrin breathing mode. They found that the intensity of this band decreased by 34.1% over a period of 25 minutes after the administration of alcohol. In a further study of the dependence of denaturation on alcohol concentration, it was discovered that the decrease in the intensity of the 752 cm^{-1} band became more rapid and more prominent as the alcohol concentration increased [139].

Application of Raman spectroscopy of serum for cancer detection was investigated by Li and Jin. The spectra of serum from cancerous and normal individuals were analysed. Three Raman peaks with intensities of 1005, 1156, and 1523 cm^{-1} were consistently observed from normal blood serum samples, whereas no peaks or only very weak peaks were detected from tumourous cases [140].

Mordechai et al. applied FTIR microspectroscopy for the follow-up of childhood leukaemia chemotherapy. A case study was presented where lymphocytes isolated from two children before and after the treatment were characterized using microscopic FTIR spectroscopy. Significant changes in the spectral pattern in the wavenumber region between 800–1800 cm^{-1} were found after the treatment. Preliminary analysis of the spectra revealed that the protein content decreased in the T-type acute lymphoblastic leukaemia (ALL) patient before the treatment in comparison to the age-matched controls. It was shown that the chemotherapy treatment results in decreased nucleic acids, total carbohydrates, and cholesterol contents to a remarkable extent in both B- and T-type ALL patients [141].

Andrus collected the spectra of various grades of malignant non-Hodgkin's lymphoma. Structural changes to lipids and proteins in the wavenumber region of 2800–300 cm^{-1}, seen as an increasing CH_3/CH_2 ratio and decreasing symmetric CH_2/asymmetric CH_2 ratio, were found to occur with increasing lymphoma grade. Rising ribose content (1121 cm^{-1}) was seen to correlate with a rising 996/966 cm^{-1} ratio (an index of RNA/DNA) with increasing malignancy grade as well. It was concluded that this method can potentially be applied to clinical cancer grading, or in vitro cancer treatment sensitivity testing [142].

Isabelle et al. analysed oesophageal lymph nodes with Raman and FTIR spectroscopy in combination with multivariate statistical analysis tools, to investigate some of the major biochemical and morphological changes taking place during carcinogenesis and metastasis and to develop a predictive model to correctly differentiate cancerous from benign tissue [143]. The results of this study showed that Raman and infrared spectroscopy managed to correctly differentiate between cancerous and benign oesophageal lymph nodes with a training performance greater than 94% using PCA and LDA. It was also concluded that cancerous nodes have higher nucleic acid but lower lipid and carbohydrate content compared to benign nodes, which is indicative of increased cell proliferation and loss of differentiation.

Chan et al. analysed populations of normal T and B lymphocytes from four healthy individuals and cells from three leukaemia patients using laser trapping Raman spectroscopy [144]. A combination of two multivariate statistical methods, PCA, and LDA, was used. The results indicated that, on average, 95% of the normal cells and 90% of the patient cells were accurately classified in their respective cell types.

Khanmohammadi et al. applied ATR-FTIR spectroscopy to discriminate the blood samples obtained from healthy people and those with basal cell carcinoma [145]. Soft independent modelling class analogy (SIMCA) chemometric technique was also used. It was aimed at classifying the normal case and cancer case blood samples through the use of ATR-FTIR spectroscopy as a rapid method because the sample preparation is so easy in comparison with the common pathologic methods. For this purpose, a total of 72 blood samples, including 32 cancer and 40 normal cases, were analysed in 900–1800 cm^{-1} spectral region. Results showed 97.6% technique accuracy.

Pichardo-Molina et al. studied serum samples using Raman spectroscopy and analysed using the multivariate statistical methods of PCA and LDA [146]. The blood samples were obtained from 11 patients who were clinically diagnosed with breast cancer and 12 healthy volunteer controls. The PCA allowed definition of the wavelength differences between the spectral bands of the control and patient groups. It was found that only seven band ratios were significant in the diagnostic process. These specific bands might be helpful during screening for breast cancer using Raman spectroscopy of serum samples. It was also shown that serum samples from patients with breast cancer and from the control group can be discriminated when the LDA is applied to their Raman spectra.

Huleihel et al. reported on the application of FTIR spectroscopy for the analysis of human blood samples from healthy subjects and patients suffering from diarrhea [147]. The results obtained showed similar and consistent spectral peaks in all the examined sera.

Rehman et al. used Raman and FTIR spectroscopy to analyse the same specimen used for histopathological evaluation. Malignant breast tissue specimens were analysed to demonstrate the hypothesis that chemical changes taking place in biological tissues can reliably and reproducibly be identified by spectroscopy. It was concluded that the techniques provide distinct specimens that can be used to distinguish between the nuclear grades of ductal carcinoma in situ (DCIS) and invasive ductal carcinoma (IDC) of the breast. These studies reported for the first time spectral differences between DCIS grades. It was concluded that spectroscopy can objectively distinguish between DCIS and IDC grades and is non-destructive and reproducible [148,149].

Prostate Cancer

Prostate tissue has been investigated by vibrational spectroscopic techniques (Table 4.11). E. Gazi et al. employed FTIR microspectroscopy to differentiate

TABLE 4.11

Articles on Prostate Tissue

Research Group	Research Method	Investigated Tissue or Sample	Effectiveness of the Technique	Reference Number
Gazi et al.	FTIR	Prostate	+	150
Paluszkiewicz and Kwiatek	FTIR and SRIXE	Prostate	+	151

samples of paraffin-embedded benign and cancerous prostate tissue. They also successfully differentiated the prostate cancer cell lines derived from different metastatic sites by using FTIR spectroscopy. It was found that the ratio of peak areas of 1030 and 1080 cm^{-1} (corresponding to glycogen and phosphate vibrations, respectively) suggests a potential method for the differentiation of benign from malignant cells. It should be mentioned that the tissues were analysed after mounting onto a BaF$_2$ plate and subsequent removal of wax using Citroclear followed by acetone [150].

Paluszkiewicz and Kwiatek reported on human cancer prostate tissues using FTIR microspectroscopy and synchrotron radiation induced x-ray emission (SRIXE) techniques. The tissue samples were also analysed by a histopathologist. In this research, differences between cancerous and noncancerous parts of the analysed tissues were observed for both methods [151].

Colon Cancer

Colon, as a part of the gastrointestinal canal, has also been analysed spectroscopically (Table 4.12). The FTIR microspectroscopy study of S. Argov et al. concerned inflammatory bowel disease as an intermediate stage between normal and cancer. In this work, FTIR microspectroscopy was used to evaluate inflammatory bowel disease (IBD) cases and to study the IR spectral characteristics with respect to cancer and normal tissues from formalin-fixed colonic biopsies from patients. IBD tissues could be segregated from cancer or normal ones using certain parameters such as phosphate content and RNA/DNA ratios. The results exhibited that FTIR microspectroscopy can detect biochemical changes in morphologically identical IBD and cancer tissues and suggest which cases of IBD may require further evaluation for carcinogenesis [152].

T. Richter et al. used FTIR spectroscopy in combination with positron emission tomography to identify tumour tissues. Thin tissue sections of human squamous carcinoma from hypopharynx and human colon adenocarcinoma grown in nude mice were investigated. Tumour tissues were successfully

TABLE 4.12

Articles on Colon Tissue

Research Group	Research Method	Investigated Tissue or Sample	Effectiveness of the Technique	Reference Number
Argov et al.	FTIR	Colon	+	152
Richter et al.	FTIR & PET	Colon	+	153
Rigas et al.	FTIR	Colorectal tissue	+	154
Rigas and Wong	FTIR	Colon	+	155
Ly et al.	FTIR imaging	Skin and colon	+	96
Pinzaru et al.	SERS-Raman	Colon	+	156
Widjaja et al.	Raman	Colonic tissue	+	157
Kondepati et al.	FTIR	Colorectal cancer	+	158
Li et al.	ATR-FTIR	Colorectal cancer	+	159
Sahu et al.	FTIT	Colonic tissue	+	160

identified by FTIR spectroscopy, while it was not possible to accomplish this with positron emission tomography (PET) alone. On the other hand, PET permitted noninvasive screening for suspicious tissues inside the body, which could not be achieved by FTIR [153].

Rigas et al. employed FTIR spectroscopy to study the tissue sections of human colorectal cancer. Pairs of tissue samples from colorectal cancer and histologically normal mucosa 5–10 cm away from the tumour were obtained from 11 patients who underwent partial colectomy. All cancer specimens displayed abnormal spectra compared with the corresponding normal tissues. These changes involved the phosphate and C–O stretching bands, the CH stretch region, and the pressure dependence of the CH_2 bending and C=O stretching modes. It was indicated that in colonic malignant tissue, there are changes in the degree of hydrogen bonding of (i) oxygen atoms of the backbone of nucleic acids (increased); (ii) OH groups of serine, tyrosine, and threonine residues (any or all of them) of cell proteins (decreased); and (iii) the C=O groups of the acyl chains of membrane lipids (increased). In addition, it was shown that there were changes in the structure of proteins and membrane lipids (as judged by the changes in their ratio of methyl to methylene groups) and in the packing and the conformational structure of the methylene chains of membrane lipids [154].

Rigas and Wong studied seven human colon cell lines by infrared spectroscopy including several spectral parameters under high pressure (pressure-tuning spectroscopy). The seven adenocarcinoma cell lines displayed almost all of the important spectroscopic features of colon cancer tissues: (a) increased hydrogen bonding of the phosphodiester groups of nucleic acids, (b) decreased hydrogen bonding of the C–OH groups of carbohydrates and proteins, (c) a prominent band at 972 cm^{-1}, and (d) a shift of the band normally appearing at 1082 cm^{-1} to 1086 cm^{-1}. These cell lines differed spectroscopically from the colon cancer tissues in that: (a) they displayed a band at 991 cm^{-1}, which is weak in colon tissues; and (b) the packing and degree of disorder of membrane lipids were close to those observed in normal colonic tissues. They concluded three important aspects from their study: (i) IR spectroscopy can be used in combination with pressure tuning as a useful method to address problems of tumour biology in cell culture systems, (ii) cell lines offer a useful experimental model to explore the origin of the spectroscopic changes that were observed in colon cancer tissues, and (iii) the malignant colonocyte is the likely source of all or most spectroscopic abnormalities of human colon cancer [155].

Pinzaru et al. reported for the first time on the application of surface-enhanced Raman scattering (SERS) on normal and altered epithelial layer in human colon carcinoma tissues [156]. It was demonstrated that different tissue structures of tumour and normal colon have characteristic features in SERS spectra.

Widjaja et al. combined NIR Raman spectroscopy (785-nm laser excitation) with support vector machines (SVMs) for improving multiclass classification among different histopathological groups in tissues [157]. A total of 105 colonic tissue specimens from 59 patients including 41 normal, 18 hyperplastic polyps, and 46 adenocarcinomas were used for this purpose. A total of 817 tissue Raman spectra were acquired and subjected to PCA, in which 324 Raman spectra were from normal, 184 from polyps, and 300 from adenocarcinomatous colonic tissue. Two types of SVMs (i.e., C-SVM and v-SVM) with three different kernel functions (linear, polynomial, and Gaussian radial basis function (RBF) in combination with PCA was used to develop effective diagnostic algorithms for classification of Raman spectra of different colonic tissues. The results showed that using different methods, a diagnostic accuracy of 98.5–99.9% can be achieved. It was also shown that all the polyps can be identified as normal and adenocarcinomatous.

Kondepati et al. used FTIR spectroscopy for the detection of structural disorders in colorectal cancer DNA [158]. The DNA was isolated from cancer and its adjacent histological normal tissue of 43 colorectal cancer patients. Spectral differences among grade 1, 2, and 3 cancers were observed. Minor structural disorders in cancer DNA were also detected by differences in the spectral regions assigned to nucleotide bases and the phosphodiester-deoxyribose backbone. Using linear discriminant analysis, diagnostic accuracies between 70% and 86% were achieved using different validation strategies, the lowest achieved by using a variable selection scheme independent from the validation test data.

Li et al. reported on application of ATR-FTIR spectroscopy and fibre-optic technology for detecting in vivo and in situ colorectal cancer [159]. In this study, a total of five patients with large intestine cancer were detected in vivo and in situ. Of them, three cases of colon cancer and one case of cecum cancer, were detected intra-operatively and in vivo by using an FTIR spectrometer during surgical operation, and one case of rectum cancer was explored noninvasively and in vivo before the surgical operation. It was shown that significant differences between FTIR spectra of normal and malignant colorectal tissues can be detected in vivo and in situ.

Sahu et al. carried out a study on detection of abnormal proliferation in histologically "normal" colonic biopsies using FTIR microspectroscopy [160]. The technique was used to distinguish between normal and abnormal crypts from colon biopsies that show normal histopathological features. The results indicated that the spectra of abnormal crypts show deviations in the pattern of absorbance in the mid-IR region when compared to the spectra of normal crypts. It was also possible to classify the crypts into three groups, such as crypts having a normal absorbance pattern for all biochemical components, crypts with abnormal absorbance pattern for some biochemical components, and crypts with a completely abnormal absorbance pattern. It was concluded that FTIR can be used for diagnosis of abnormal metabolism at the molecular level of histologically completely normal-looking crypts, and this could give rise to a reduction in false negative results during examination of biopsies using conventional methods.

Investigating Cell Lines

Some research groups have used both FTIR and Raman spectroscopic techniques to analyse different types of cell lines (Table 4.13). In vitro viral carcinogenesis was studied by M. Huleihel. They used FTIR microscopy to investigate spectral differences between normal and malignant fibroblasts transformed by retrovirus infection. Significant differences were observed between cancerous and normal cells. It was concluded that the contents of vital cellular metabolites were significantly lower in the transformed cells than in the normal cells. In addition, as an attempt to identify cellular

TABLE 4.13

Articles on Different Types of Cell Lines

Research Group	Research Method	Investigated Tissue or Sample	Effectiveness of the Technique	Reference Number
Huleihel et al.	FTIR	Fibroblasts	+	161
Mossoba et al.	FTIR	Bacteria	+	162
Krishna et al.	Raman	Mixed cancer cells	+	163
Kuhnert and Thumser	Raman	Human living cells	+	164
Naumann	Raman and FTIR	Microbial cells	+	165
Dovbeshka et al.	FTIR	Tumour cells	n/a	166
Chan et al.	Raman	Individual cells	+	167
Bogomolny et al.	FTIR	Cells in culture	+	168

components responsible for the spectral dissimilarities, they found considerable differences between DNA of normal and cancerous cells [161].

In a proof-of-concept study, M. M. Mossoba et al. made an investigation on printing microarrays of bacteria for identification by infrared microscopy. They used the technique as a tool for rapid bacteria identification and could demonstrate its effectiveness [162].

C. M. Krishna et al. used micro-Raman spectroscopy to investigate randomly mixed cancer cell populations, including human promyelocytic leukaemia, human breast cancer, human uterine sarcoma, as well as their respective pure cell lines. In this study, the efficiency of micro-Raman spectroscopy to identify a cell type in a randomly distributed mixed cell population was assessed. According to the results, cells from different origins can display variances in their spectral signatures and the technique can be used to identify a cell type in a mixed cell population via its spectral signatures [163].

In a research article presented by N. Kuhnert and A. Thumser, Raman microspectroscopy with a diode laser at 785 nm or an argon ion laser at 512 nm was used for detection of vibrationally labelled compounds in living human cells and positive results were obtained. They suggested that future research should concentrate on sensitivity and experimental setup in order to achieve better detection limits [164].

According to a published paper by D. Naumann, FTIR and FT-NIR Raman spectra of intact microbial cells are highly specific, fingerprint-like signatures that can be used to (i) discriminate between diverse microbial species and strains; (ii) detect in situ intracellular components or structures such as inclusion bodies, storage materials, or endospores; (iii) detect and quantify metabolically released CO_2 in response to various different substances; and (iv) characterise growth-dependant phenomena and cell–drug interactions. Particularly interesting applications arise by means of a light microscope coupled to the spectrometer. FTIR spectra of microcolonies containing less than 103 cells can be obtained from a colony replica by a stamping technique that transfers microcolonies growing on culture plates to a special IR sample holder. FTIR and FT-NIR Raman spectroscopy can also be used in tandem to characterise medically important microorganisms [165].

G. I. Dovbeshko et al. carried out an FTIR reflectance study on surface-enhanced IR absorption of nucleic acids from tumour cells. The application of this method to nucleic acids isolated from tumour cells revealed some possible peculiarities of their structural organization, namely, the appearance of unusual sugar and base conformations, modification of the phosphate backbone, and redistribution of the H-bond net. The spectra of the RNA from the tumour cells showed more sensitivity to the grade of malignancy than the spectra of the DNA. After application of the anticancer drug doxorubicin to sensitive and resistant strains, the DNA isolated from these strains had different spectral features, especially in the region of the phosphate I and II bands [166].

Chan et al. worked on detection of individual neoplastic and normal hematopoietic cells using micro-Raman spectroscopy. The potential application of confocal micro-Raman spectroscopy as a clinical tool for single-cell cancer detection based on intrinsic biomolecular signatures was demonstrated. They showed that this method can discriminate among different kinds of unfixed lymphocytes, and single-cell Raman spectra provide a highly reproducible biomolecular fingerprint of each cell type. Characteristic peaks, mostly due to different DNA and protein concentrations, allowed for discerning normal lymphocytes from transformed ones with high confidence ($p \ll 0.05$). The method was shown to have a sensitivity of 98.3% for cancer detection, with 97.2% of the cells being correctly classified as belonging to the normal or transformed type [167].

Bogomolny et al. analysed the early spectral changes accompanying malignant transformation of cells in culture by FTIR microspectroscopy [168]. The cells were infected by the murine sarcoma virus (MuSV), which induces malignant transformation, and the spectral measurements were taken at various postinfection time intervals. The results indicate that the first spectral changes are detectable much earlier than the first morphological signs of cell transformation. It was found out that the first spectral signs of malignant transformation are observed on the first and third day of postinfection, while the first visible morphological alterations are observed only on the third and seventh day, respectively.

Analysing Saliva

Interestingly, Raman has been successfully used to study saliva and there have been a number of studies reported in the literature (Table 4.14). Farguharson et al. measured the chemotherapeutic drug 5-fluorouracil in saliva using SERS. A silver-doped sol-gel provided SERS and also some chemical selectivity. 5-fluorouracil and physiological thiocyanate produced SERS, whereas large biochemicals, such as enzymes and proteins, did not, supporting the expectation that the larger molecules do not diffuse through the sol-gel to any appreciable extent. In addition, 5-fluorouracil samples of $2\ \mu gml^{-1}$ were easily measured, and an estimated limit of detection of $5\ \mu gml^{-1}$ in 5 minutes should provide sufficient sensitivity to perform pharmacokinetic studies and to monitor and regulate patient dosage [169].

TABLE 4.14

Article on Saliva

Research Group	Research Method	Investigated Tissue or Sample	Effectiveness of the Technique	Reference Number
Farguharson et al.	Raman	Saliva	+	169

Other Spectroscopic Studies

A summary of some interesting samples investigated by vibrational spectroscopic methods is provided in Table 4.15. Characterisation of conformational
changes on guanine-cytosine and adenine-thymine oligonucleotides induced
by amiooxy analogues of spermidine using Raman spectroscopy was accomplished by A. J. Ruiz-Chica et al. These analogues resulted from the substitution of the two terminal aminomethylene groups of spermidine, $^+NH_3CH_2$,
by an amnooxy one, H_2NO. The spectra demonstrated the existence of strong
differences in the oligonucleotide–analogue interactions depending on base
sequences. Different spectral features were observed. This fact supported the
idea that the two amino terminal groups of spermidine could have different
roles in the interaction of this biogenic polyamine with DNA [170].

　　K. J. Jalkanen et al. used vibrational spectroscopy to study protein and
DNA structure, hydration, and binding of biomolecules, as a combined
theoretical and experimental approach. The systems studied systemati-
cally were the amino acids, peptides, and a variety of small molecules. The
goal was to interpret the experimentally measured vibrational spectra for
these molecules to the greatest extent possible, and to understand the structure, function, and electronic properties of these molecules in their various
environments. It was also believed that the application of different spectroscopic methods to biophysical and environmental assays is expanding, and
therefore a true understanding of the phenomenon from a rigorous theoretical basis is required [171].

TABLE 4.15

Articles on Different Samples

Research Group	Research Method	Investigated Tissue or Sample	Effectiveness of the Technique	Reference Number
Ruiz-Chika et al.	Raman	DNA oligonucleotides	n/a	170
Jalkanen et al.	Raman & FTIR	DNA	n/a	171
Vo-Dinh et al.	Raman	Cancer genes	+	172
Binoy et al.	Raman & FTIR	Anticancer drug	+	173
Faolin et al.	Raman & FTIR	Tissue processing	n/a	174
Sokolov et al.	Reflectance spectroscopy	Precancerous lesions	n/a	175
Chen et al.	Raman	Fluorescent agents	+	176
Sahu and Mordechai	FTIR	Cancer detection	+	177
Viehoever et al.	Raman	Organotypic raft cultures	+	178
Rabah et al.	Raman	Neuroblastoma and ganglioneuroma	+	179

Vo-Dinh et al. have described the development of the SERS method for cancer gene detection, using DNA gene probes based on SERS labels. Nanostructured metallic structures were used as SERS-active platforms. The surface-enhanced Raman gene (SERGen) probes were used to detect DNA targets via hybridisation to DNA sequences complementary to these probes. The details of the usefulness of the approach and its applications in cancer gene diagnosis (e.g., BRCA1 breast cancer gene and BAX gene) were discussed [172].

FTIR and NIR-FT Raman spectral features of the anticancer drug combretastatin-A4 were studied by J. Binoy et al. The vibrational analysis showed that the molecule exhibits similar geometric behaviour as *cis*-stilbene, and has undergone steric repulsion resulting in phenyl ring twisting with respect to the ethylenic plane [173].

E. O. Faolin et al. carried out a study examining the effects of tissue processing on human tissue sections using vibrational spectroscopy. This study investigated the effect of freezing, formalin fixation, wax embedding, and de-waxing. Spectra were recorded from tissue sections to examine biochemical changes before, during, and after processing with both Raman and FTIR spectroscopy. New peaks due to freezing and formalin fixation as well as shifts in the amide bands resulting from changes in protein conformation and possible cross-links were found. Residual wax peaks were observed clearly in the Raman spectra. In the FTIR spectra a single wax contribution was seen that may contaminate the characteristic CH_3 deformation band in biological tissue. This study confirmed that formalin-fixed paraffin-processed (FFPP) sections have diagnostic potential [174].

Sokolov et al. applied fluorescence and reflectance spectroscopy to assess tissue structure and metabolism in vivo in real time, providing improved diagnosis of precancerous lesions. Reflectance spectroscopy can probe changes in epithelial nuclei that are important in precancer detection, such as mean nuclear diameter, nuclear size distribution, and nuclear refractive index. Fluorescence spectroscopy can probe changes in epithelial cell metabolism, by assessing mitochondrial fluorophores, and epithelial–stromal interaction, by assessing the decrease in collagen cross-link fluorescence that occurs with precancer. Thus, it was believed that these techniques provide complementary information useful for precancer diagnosis [175].

To detect a subsurface tumour labelled by fluorescent contrast agents, Y. Chen et al. developed a phase cancellation imaging system for fast and accurate localisation of a fluorescent object embedded several centimetres deep inside the turbid media. The excitation wavelength was 780 nm and the fluorescence photons were collected through an 830 ± 10 nm band-pass filter. It was concluded that using this method, the localisation of fluorescent objects inside the scattering media could be accomplished. The localisation error for a 5-mm diameter sphere filled with one nanomole of fluorescent dye 3 cm deep inside the turbid media is about 2 mm. In addition, the accuracy of the localisation could suggest that this system can be helpful in guiding clinical fine-needle biopsy, and would benefit the early detection of breast tumour [176].

R. K. Sahu and S. Mordechai studied FTIR spectroscopy in cancer detection. The areas of focus were the distinction of premalignant and malignant cells and tissues from their normal state using specific parameters obtained from the spectra. It was concluded that while the method still requires pilot studies and designed clinical trials to ensure the applicability of such systems for cancer diagnosis, substantial progress has been made in incorporating advances in computation into the system to increase the sensitivity of the entire setup, making it an objective and sensitive technique suitable for automation to suit the demands of the medical community [177].

Viehoever et al. examined the use of organotypic raft as an in vitro model of in vivo tissue conditions in an attempt to overcome some of the limitations of previously used methods. In this study, organotypic raft cultures resembling normal and dysplastic epithelial cervical tissue were conducted and grown at an air–liquid interface for two weeks. Raman spectra of normal as well as dysplastic raft cultures were measured and compared with in vivo spectra from the corresponding tissue type. These investigations showed that the Raman spectra of the raft cultures are similar to the spectra acquired from the cervix in vivo for both normal and dysplastic tissues. It was concluded that this type of culture is an effective and useful tool for the cellular and biochemical analysis of tissue spectra [178].

Rabah et al. reported for the first time the evaluation of Raman spectroscopy in the diagnosis and classification of neuroblastoma in children [179]. A biopsy or resection of fresh tissue samples from normal adrenal glands, neuroblastomas, ganglioneuromas, nerve sheath tumours, and pheochromocytoma were equally divided between routine histology and spectroscopic studies and at least 12 spectra were collected from different regions of each sample. In total, 698 spectra were collected from 16 neuroblastomas, 5 ganglioneuromas, 3 normal adrenal glands, 6 nerve sheath tumours, and 1 pheochromocytoma. Raman spectroscopy differentiated between a normal adrenal gland, and neuroblastoma and ganglioneuroma with 100% sensitivity and 100% specificity. It was also able to differentiate a neuroblastoma from nerve sheath tumours and a pheochromocytoma with high sensitivity and specificity.

Summary

There are some key points in defining the corresponding functional groups of every peak. These points are described in Chapter 8 and can play an outstanding role in the process of characteristic peak analysis, and they are of outstanding importance in the process of having a clear understanding of the spectroscopic techniques used in the research work. There is a significant number of studies involving spectroscopy of cancer tissues and the use is growing rapidly. There is a growing need to identify the key spectral peaks and their correct assignment to the chemical structures. Therefore, it is of utmost importance to have a trustworthy spectral database that is widely available to researchers working with vibrational spectroscopic techniques.

The following are some spectral peaks that are considered extremely important in correctly assigning spectral peaks related to biological molecules:

1. Lipid bands mainly consist of three kinds of bands: C–H, C=C, and C–O [61].

2. Collagen, the major protein of connective tissue, contains approximately 33% glycine, and 20–25% praline, and hydroxyproline residues [63].

3. The vibrational modes of collagen are as follows:

   A. 1634 and 1654 cm^{-1}: amide I

   B. 1544 cm^{-1}: amide II

   C. 1450 and 1401 cm^{-1}: bending modes of methyl groups

   D. 1337, 1280, 1237, 1204 cm^{-1}: amide III and CH_2 wagging vibrations from glycine backbone and praline side chains

   E. 1082 and 1031 cm^{-1}: carbohydrate residues of collagen [63].

4. The vibrational modes of collagen type I are as follows:

   A. 1163 and 1174 cm^{-1}: tyrosine

   B. 937 cm^{-1}: proline

   C. 856 cm^{-1}: hydroxyproline

   D. 667 cm^{-1}: cysteine [93]

5. The protein amide I characteristics (peak frequency and width) are determined by conformational differences like α-helix, β-plates, and disordered structure [85].

6. The vibrational modes of lipids are as follows:

   A. 1750 cm^{-1}: C=O

   B. 1655 cm^{-1}: C=C

   C. 1437 cm^{-1}: CH_2 (1420–1450 cm^{-1}) [71]

   D. 1301 cm^{-1}: assigned for Parker (1300 cm^{-1}) [71]

   E. 1267 cm^{-1}: CH

   F. 1076 cm^{-1}: C–C [93]

7. Two possible assignments are available for the carbonyl band at about 1745 cm^{-1}:

   • Phospholipid band in the membrane

   • Ester group in fat (the carbonyl band of phospholipids) [94]

8. 1745 cm^{-1} can be found in normal tissues with fat cells. Thus, the relative intensity of the band can reliably identify fat content of tissue [94].

9. Protein bands:

   - 1640–80 cm^{-1}: amide I
   - 1220–300 cm^{-1}: amide III
   - 928–40 cm^{-1}: ν(C–C) stretching (probably in amino acids proline and valine) [71].

10. Protein band reflect the secondary protein structure [71].

11. The cellular cytoskeleton is composed of different kinds of protein fibres. Any depolymerization of these proteins would result in unravelling of the secondary structure and hence an increase in the NH$_3^+$ and COO$^-$ vibrations (1485–1550 cm^{-1} and 1560–600 cm^{-1}, respectively) [128].

12. Cell cytoplasm, fat, collagen, and cholesterol have many of the same functional groups (CH$_2$ bands, C–C stretches, etc.) [62].

13. 1121 cm^{-1}/1020 cm^{-1} ratio provides a measure of cellular RNA/DNA ratio and is higher in malignant tissues than in normal ones [71,116].

14. 1045 cm^{-1}/1545 cm^{-1} ratio gives an estimate of carbohydrate concentration in cells and is lower in malignant tissues than in normal ones [116].

15. Amide II region is not as sensitive to conformational changes as amide I [104].

16. The peak area of 1030 cm^{-1}/1080 cm^{-1} corresponds to the glycogen/phosphate ratio and is indicative of metabolic activity. This ratio has been used to differentiate neoplastic from non-neoplastic cells of different cancers [86].

17. The spectra of amide I vibrational modes of the proteins are highly sensitive to conformational changes in the secondary structure [60].

18. The frequencies and relative intensities of the characteristic protein vibrations (amide I, II, III) depend on the secondary structure that the protein assumes (e.g., helical, sheet, globular, or triple-helical) [32].

19. Both amide I and III have some overlapping with nucleic acid frequencies. Therefore, these spectral regions cannot be used to quantify the components in mixtures of protein and nucleic acids. The spectral features between 1000 cm^{-1} and 1150 cm^{-1} are reasonably specific for nucleic acids in the absence of glycogen [32].

20. It is thought that unsaturated lipids existing in nature predominantly take the cis configuration of the C=C double bond [180,181]. The cis configuration is known to increase the flexibility of lipid membranes, while the trans makes them stiff.

21. Raman and infrared spectroscopy are optical spectroscopic techniques that use light scattering (Raman) and light absorption (infrared) to probe the vibrational energy levels of molecules in tissue samples [143].

22. Spectroscopy cannot determine the molecules that cause spectral differences, if their tertiary or quaternary structures are involved [146].

There are two comprehensive review articles that can be of significant help in defining peak frequencies [182,183].

References

1. Quintas, G., Garrigues, A., de la Guardia, M. 2004. FT-Raman determination of Malathion in pesticide formulations. *Talanta* 63, 345–350.
2. Laska, J., and Widlarz, J. 2005. Spectroscopic and structural characterization of low molecular weight fractions of polyaniline. *Polymer* 46, 1485–1495.
3. Mazurek, S., and Szostak, R. 2006. Quantitative determination of captopril and prednisolone in tablets by FT-Raman spectroscopy. *Journal of Pharmaceutical and Biomedical Analysis* 40, 1225–1230.
4. Mazurek, S., and Szostak, R. 2006. Quantitative determination of diclofenac sodium and aminophylline in injection solutions by FT-Raman spectroscopy. *Journal of Pharmaceutical and Biomedical Analysis* 40, 1235–1242.
5. Singha, A., Ghosh, A., Roy, A., and Ray, N. R. 2006. Quantitative analysis of hydrogenated diamondlike carbon films by visible Raman Spectroscopy. *Journal of Applied Physics* 100(044910), 1–8.
6. McGoverin, C. M., Ho, L. C. H., Zeitler, J. A., Strachan, C. J., Gordon, K. C., and Rades, T. 2006. Quantification of binary polymorphic mixtures of ranitidine hydrochloride using NIR Spectroscopy. *Vibrational Spectroscopy* 41, 225–231.
7. Baklanova, N. I., Kolesov, B. A., and Zima, T. M. 2007. Raman study of yttria-stabilized zirconia interfacial coatings in Nicalon™ fibre. *Journal of the European Ceramic Society* 27, 165–171.
8. Romero-Torres, S., Perez-Ramon, J. D., Morris, K. R., and Grant, E. R. 2006. Raman spectroscopy for tablet coating thickness quantification and coating characterization in the presence of strong fluorescent interface. *Journal of Pharmaceutical and Biomedical Analysis* 41, 811–819.
9. Ravagnan, L., Bongiorno, G., Bandiera, D., Salis, E., Piseri, P., Milani, P., Lenardi, C., Coreno, M., de Simone, M., and Prince, K. C. 2006. Quantitative evaluation of sp/sp2 hybridization ratio in cluster-assembled carbon films by in situ near edge X-ray absorption fine structure spectroscopy. *Carbon* 44, 1518–1524.
10. Hwang, M.-S., Cho, S., Chung, H., and Woo, Y.-A. 2005. Nondestructive determination of the ambroxol content in tablets by Raman spectroscopy. *Journal of Pharmaceutical and Biomedical Analysis* 38, 210–215.
11. Kachrimanis, K., Braun, D. B., and Griesser, U. J. 2007. Quantitative analysis of paracetamol polymorphs in powder mixtures by FT-Raman spectroscopy and PLS regression. *Journal of Pharmaceutical and Biomedical Analysis* 43, 407–412.

12. Quintas, G., Garrigues, S., Pastor, A., and de la Guardia, M. 2004. FT-Raman determination of Mepiquet chloride in agrochemical products. *Vibrational Spectroscopy* 36, 41–46.
13. Yin, J.-H., Li, Z.-W., Tian, Y.-J., Sun, Z.-W., and Song, X.-L. 2005. A study on Raman scattering cross section of carbon tetrachloride at low concentrations. *Applied Physics*, B80, 573–576.
14. Agarwal, R., Tandon, P., and Gupta, V.D. 2006. Phonon dispersion in poly(dimethylsilane). *Journal of Organometallic Chemistry* 691, 2902–2908.
15. Sadezky, A., Muckenhuber, H., Grothe, H., Niessner, R., and Poschl, U. 2005. Raman microspectroscopy of soot and related carbonaceous materials: Apectral analysis and structural information. *Carbon* 43, 1731–1742.
16. Behrens, H,. Roux, J., Neuville, D. R., and Siemann, M. 2006. Quantification of dissolved H_2O in silicate glasses using confocal micro-Raman spectroscopy. *Chemical Geology* 229, 96–112.
17. Ortiz, C., Zhang, D., Xie, Y., Ribbe, A. E., and Ben-Amotz, D. 2006. Validation of the drop coating deposition Raman method for protein analysis. *Analytical Biochemistry* 353, 157–166.
18. Conroy, J., Ryder, A. G., Leger, M. N., Hennessey, K., and Madden, M. G. 2005. Qualitative and quantitative analysis of chlorinated solvents using Raman spectroscopy and machine learning. *Proc. SPIE – Int. Soc. Opt. Eng.* 5826, 131–142.
19. Schroeder, P. A., Melear, N. D., and Pruett, R. J. 2003. Quantitative analysis of anatase in Georgia kaolins using Raman spectroscopy. *Applied Clay Science* 23, 299–308.
20. Katainen, E., Elomaa, M., Laakkonen, U-M., Sippola, E., Niemela, P., Suhonen, J., and Jarninen, K. 2007. Qualification of the amphetamine content in seized street samples by Raman spectroscopy. *Journal of Forensic Science* 52(1), 88–92.
21. Niemela, P., Paallysaho, M., Harjunen, P., Koivisto, M., Lehto, V-P.,Suhonen, J., and Jarvinen, K. 2005. Quantitative analysis of amorphous content of lactose using CCD-Raman spectroscopy. *Journal of Pharmaceutical and Biomedical Analysis* 37, 907–911.
22. Coates, J. 2000. Interpretation of infrared spectra: A practical approach. In *Encyclopedia of Analytical Chemistry.* New York: John Wiley & Sons, 1–23.
23. Sacristan, J., Reinecke, H., Mijangos, C., Spells, S., and Yarwood, J. 2002. Surface modification of polystyrene films: Depth profiling and mapping by Raman microscopy. *Macromolecular Chemistry and Physics* 203, 678–685.
24. Sacristan, J., Mijangos, C., Reinecke, H., Spells, S., and Yarwood, J. 2000. Specific surface modification of PVC films as revealed by confocal Raman microspectroscopy. *Macromolecules* 33, 6134–6139.
25. Chittur, K. K. 1998. FTIR/ATR for protein adsorption to biomaterial surfaces. *Biomaterials* 19, 357–369.
26. Xue, G. 1997. Fourier transform Raman spectroscopy and its application for the analysis of polymeric materials. *Progress in Polymer Science* 22(2), 313–406.
27. Costa, J C. S., Sant Ana, A. C., Corio, P., and Temperini, M. L. A. 2006. Chemical analysis of polycyclic aromatic hydrocarbons by surface-enhanced Raman spectroscopy. *Talanta* 70, 1011–1016.
28. Liang, Y., Miranda, C. R., and Scandolo, S. 2006. Infrared and Raman spectra of silica polymorphs from an *ab initio* parameterized polarizable force field. *The Journal of Chemical Physics* 125, 194524, 1–9.

29. Wang, R. X., Xu, S. J., Fung, S., et al. 2005. Micro-Raman and photoluminescence studies of neutron-irradiated gallium nitride epilayers. *Applied Physics Letters* 87, 031906, 1–3.

30. Allen, A., Foulk, J., and Gamble, G. 2007. Preliminary Fourier-transform infrared spectroscopy analysis of cotton trash. *Journal of Cotton Science* 11, 68–74.

31. Edwards, H. G. M., Villar, S. E. J., Parnell, J., Cockell, C. S., and Lee, P. 2005. Raman spectroscopic analysis of cyanobacterial gypsum halotrophs and relevance for sulfate deposits on Mars. *Analysis* 130(6), 917–923.

32. Frost, R. L., Edwards, H. G. M., Duong, L., Kloprogge, J. T., and Martens, W. N. 2002. Raman spectroscopic and SEM study of cinnabar from Herod's palace and its likely origin. *Analyst* 127(2), 293–296.

33. Edwards, H. G. M., Nikhassan, N. F., Farwell, D. W., Garside, P., and Wyeth, P. 2006. Raman spectroscopic analysis of a unique linen artefact: The HMS Victory Trafalgar sail. *Journal of Raman Spectroscopy* 37, 1193–1200.

34. Kuptsovas, A. H. 1994. Applications of Fourier transform Raman spectroscopy in forensic sciences. *Journal of Forensic Sciences*, 39(2), 305–318.

35. Thomas, J., Buzzini, P., Massonnet, G., Reedy, B., and Roux, C. 2005. Raman spectroscopy and the forensic analysis of black/grey and blue cotton fibres: Investigation of the effects of varying laser wavelength. *Forensic Science International* 152(2–3), 189–197.

36. Mahadevan-Jansen, A., and Richards-Kortum, R. 1997. Raman spectroscopy for cancer detection, 19th International Conference IEEE EMBS, Chicago, IL, October 30–November 2.

37. Hanlon, E. B., Manoharan, R., Koo, T.-W., Shafer, K. E., Motz, J. T., Fitzmaurice, M., Kramer, J. R., Itzkan, I., Dasari, R. R., and Feld, M. S. 2000. Prospects for in vivo Raman spectroscopy. *Phys. Med. Biol.* 45, 1–59.

38. Dukor, R. K., 2002. Vibrational spectroscopy in the detection of cancer. *Biomedical Applications*, 3335–3359.

39. Choo-Smith, L.-P., Edwards, H. G. M., Endtz, H. P., Kros, J. M., Heule, F., Barr, H., Robinson, J. S., Jr., Bruining, H. A., and Pupells, G. J. 2002. Medical applications of Raman spectroscopy: From proof of principle to clinical implementation. *Biopolymers (Biospectroscopy)* 67, 1–9.

40. Swinson, B., Jerjes, W., El-Maaytah, M., Norris, P., and Hopper, C. 2006. Optical techniques in diagnosis of head and neck malignancy. *Oral Oncology* 42, 221–228.

41. Shaw, R. A., and Mantsch, H. H. 1999. Vibrational biospectroscopy: From plants to animals to humans. A historical perspective. *Journal of Molecular Structure.* 480–481, 1–13.

42. Barr, H., Dix, T., and Stone, N. 1998. Optical spectroscopy for the early diagnosis of gastrointestinal malignancy. *Lasers Medical Science* 13, 3–13.

43. Petrich, W. 2001. Mid-infrared and Raman spectroscopy for medical diagnostics. *Applied Spectroscopy Reviews* 36(2), 181–237.

44. Zeng, H., McWilliams, A., and Lam, S. 2004. Optical spectroscopy and imaging for early lung cancer detection: A review. *Photodiagnosis and Photodinamic Therapy* 1, 111–122.

45. Schultz, H., and Baranska, M. 2007. Identification and qualification of valuable plant substances by IR and Raman spectroscopy. *Vibrational Spectroscopy* 43, 13–25.

46. Moreira, L. M., Silveira, L., Santos, F. V., et al. 2008. Raman spectroscopy: A powerful technique for biochemical analysis and diagnosis. *Spectroscopy: An International Journal* 22(1), 1–19.

47. Krishna, C. M., Kurien, J., Mathew, S., et al. 2008. Raman spectroscopy of breast tissues. *Expert Review of Molecular Diagnostics* 8(2), 149–166.
48. Murali Krishna, C., Sockalingum, G. D., Vidyasagar, M. S., et al. 2008. An overview on applications of optical spectroscopy in cervical cancers. *Journal of Cancer Research and Therapeutics* 4(1), 26–36.
49. Wood, B. R., Quinn, M. A., Burden, F. R., and McNaughton, D. 1996. An investigation into FT-IR spectroscopy as a bio-diagnostic tool for cervical cancer. *Biospectroscopy* 2, 143–153.
50. Chiriboga, L., Xie, P., Yee, H., Vigorita, V., Zarou, D., Zakim, D., and Diem, M. 1998. Infrared spectroscopy of human tissue. I. Differentiation and maturation of epithelial cells in the human cervix. *Biospectroscopy* 4, 47–53.
51. Wood, B. R., Quinn, M. A., Tait, B., Ashdown, M., Hislop, T., Romeo, M., and McNaughton, D. 1998. FTIR microspectroscopic study of cell types and potential confounding variables in screening for cervical malignancies. *Biospectroscopy* 4, 75–91.
52. Utzinger, U. R. S., Heintzelman, D. L., Mahadevan-Jansen, A., et al. 2001. Near-infrared Raman spectroscopy for in vivo detection of cervical precancers. *Applied Spectroscopy* 55(8), 955–959.
53. Sindhuphak, R., Issaravanich, S., Udomprasertgul, V., et al. 2003. A new approach for the detection of cervical cancer in Thai women. *Gynecologic Oncology* 90, 10–14.
54. Mordechai, S., Sahu, R. K., Hammody, Z., et al. 2004. Possible common biomarkers from FTIR microspectroscopy of cervical cancer and melanoma. *Journal of Microscopy* 215(1), 86–91.
55. Chiriboga, L., Xie, P., Vigorita, V., Zarou, D., Zakin, D., and Diem, M. 1998. Infrared spectroscopy of human tissue. II. A comparative study of spectra of biopsies of cervical squamous epithelium and of exfoliated cervical cells. *Biospectroscopy* 4, 55–59.
56. Wong, P. T. T., Lacelle, S., Fung, M. F. K., Senterman, M., and Mikhael, N. Z. 1995. Characterization of exfoliated cells and tissues from human endocervix and ectocervix by FTIR and ATR/FTIR spectroscopy. *Biospectroscopy* 1(5), 357–364.
57. Fung, M. F. K., Senterman, M. K., Mikhael, N. Z., Lacelle, S., and Wong, P. T. T. 1996. Pressure-tuning Fourier transform infrared spectroscopic study of carcinogenesis in human endometrium. *Biospectroscopy* 2, 155–165.
58. Krishna, C. M., Prathima, N. B., Malini, R., et al. 2006. Raman spectroscopy studies for diagnosis of cancers in human uterine cervix. *Vibrational Spectroscopy* 41(1), 136–141.
59. Maheedhar, K., Bhat, R. A., Malini, R., et al. 2008. Diagnosis of ovarian cancer by Raman spectroscopy: A pilot study. *Photomedicine and Laser Surgery* 26(2), 83–90.
60. Mo, J. H., Zheng, W., Low, J., et al. 2008. In vivo diagnosis of cervical precancer using high wavenumber Raman spectroscopy. *Optics in Health Care and Biomedical Optics III* 6826, 195–599.
61. Jess, P. R. T., Smith, D. D. W., Mazilu, M., et al. 2007. Early detection of cervical neoplasia by Raman spectroscopy by Raman spectroscopy. *International Journal of Cancer* 121, 2723–2728.
62. El-Tawil, S. G., Rohana, A., Zaki, N. M., and Nor Hayati, O. 2008. Comparative study between Pap smear cytology and FTIR spectroscopy: A new tool for screening for cervical cancer. *Pathology* 40(6), 600–603.

63. Stone, N., Kendell, C., Shepherd, N., Crow, P., and Barr, H. 2002. Near-infrared Raman spectroscopy for the classification of epithelial pre-cancers and cancers. *J. Raman Spectroscopy.* 33, 564–573.

64. Stone, N., Kendall, C., Smith, J., Crow, P., and Barr, H. 2004. Raman spectroscopy for identification of epithelial cancers. *Faraday Discussion* 126, 141–157.

65. Wang, H. P., Wang, H.-C., and Huang, Y.-J. 1997. Microscopic FTIR studies of lung cancer cells in pleural fluid. *The Sci. Total Envir.* 204, 283–287.

66. Yano, K., Ohoshima, S., Grotou, Y., Kumaido, K., Moriguchi, T., and Katayama, H. 2000. Direct measurement of human lung cancerous and noncancerous tissues by Fourier transform infrared microscopy: Can an infrared microscope be used as a clinical tool? *Analytical Biochemistry* 287, 218–225.

67. Yang, Y., Sule-Suso, J., Sockalingum, G. D., Kegelaer, G., Manfait, M., and El Haj, A. J. 2005. Study of tumour cell invasion by Fourier transform infrared microspectroscopy. *Biopolymers* 78, 311–317.

68. Min, Y.-K., Yamamoto, T., Kohoda, E., Ito, T., and Hamaguchi, H. 2005. 1064-nm near-infrared multichannel Raman spectroscopy of fresh human lung tissues. *J. Raman Spectroscopy.* 36, 73–76.

69. Kaminaka, S., Yamazaki, H., Ito, T., Kohoda, E., and Hamaguchi, H. 2001. Near-infrared Raman spectroscopy of human lung tissues: Possibility of molecular-level cancer diagnosis. *J. Raman Spectroscopy.* 32, 139–141.

70. Huang, Z., McWilliams, A., Lui, M., McLean, D. I., Lam, S., and Zeng, H. 2003. Near-infrared Raman spectroscopy for optical diagnosis of lung cancer. *International Journal of Cancer* 107, 1047–1052.

71. Alfano, R. R., Tang, G. C., Pradhan, A., Lam, W., Choy, D. S. J., and Opher, E. 1987. Optical spectroscopic diagnosis of cancer and normal breast tissues. *IEEE J. Quantum Electronics* 6(5), 1015–1023.

72. Huang, Z., McWilliams A., Lam, S., English, J., McLean, D., Lui, H., Zeng, H. 2003. Effect of formalin fixation on the near-infrared Raman spectroscopy of normal and cancerous human bronchial tissues. *International Journal of Oncology* 23, 649–655.

73. Short, M. A., Lam, S., McWilliams, A., et al. 2008. Development and preliminary results of an in-vivo Raman probe for early lung cancer detection. *Biomedical Optical Spectroscopy* 6853, 530.

74. Eckel, R., Huo, H., Guan, H.-W., Hu, X., Che, X., and Huang, W.-D. 2001. Characteristic infrared spectroscopic patterns in the protein bands of human breast cancer tissue. *Vibrational Spectroscopy* 27, 165–173.

75. Kline, N. J., and Treado, P. J. 1997. Raman chemical imaging of breast tissue. *J. Raman Spectroscopy.* 28, 119–124.

76. Shafer-Peltier, K. E., Haka, A. S., Fitzmaurice, M., Crowe, J., Dasar, R. R., and Feld, M. S. 2002. Raman microspectroscopic model of human breast tissue: implications for breast cancer diagnosis *in vivo*. *J. Raman Spectroscopy.* 33, 552–563.

77. Ronen, S. M., Stier, A., and Degani, H. 1990. NMR studies of the lipid metabolism of T47D human breast cancer spheroids. *FEBS Letter* 266, 147–149.

78. Frank, C. J., McCreery, R. L., and Redd, D. C. B. 1995. Raman spectroscopy of normal and diseased human breast tissues. *Anal. Chem.* 67, 777–783.

79. Fabian, H., Jackson, M., and Murphy, L., et al. 1995. A comparative infrared spectroscopic study of human breast tumours and breast tumour cell xenografts. *Biospectroscopy* 1(1), 37–45.

80. Tam, K., Armstrong, R. S., Carter, E. A., Lay, P. A., Mountford, C., Dowd, S., Himmelreich, U., and Russell, P. 2004. Raman spectroscopy for breast cancer detection: A sample processing study. In Proceedings of the XIX International Conference on Raman Spectroscopy, Victoria, Australia: CSIRO Publishing, 457–457.

81. Abramczyk H., Placek I., Brozek-Pluska B., et al. 2008. Human breast tissue cancer diagnosis by Raman spectroscopy. *Spectroscopy: An International Journal* 22(2–3). 113–121.

82. Kast, R. E., Serhatkulu, G. K., Cao, A., et al. 2008. Raman spectroscopy can differentiate malignant tumours from normal breast tissue and detect early neoplastic changes in a mouse model. *Biopolymers* 89, 134–241.

83. Marzullo, A. C. D., Neto, O. P., Bitar, R., et al. 2007. FT-Raman spectra of the border of infiltrating ductal carcinoma lesions. *Photomedicine and Laser Surgery* 25, 455–460.

84. Sukuta, S., and Bruch, R. 1999. Factor analysis of cancer Fourier transform infrared evanescent wave fibreoptical (FTIR-FEW) spectra. *Lasers in Sur. and Med.* 24, 382–388.

85. Cheng, W.-T., Liu, M.-T., Liu, H.-N., and Lin, S.-Y. 2005. Micro-Raman spectroscopy used to identify and grade human skin pilomatrixoma. *Microscopy Research and Technique* 68, 75–79.

86. Kaminaka, S., Ito, T., Yamazaki, H., Kohoda, E., and Hamaguchi, H. 2002. Near-infrared multichannel Raman spectroscopy toward real-time in vivo cancer diagnosis. *J. Raman Spectrosc.* 33, 498–502.

87. Sigurdsson, S., Philipsen, P. A., Hansen, L. K., Laesen, L., Gniadecka, M., and Wulf, H. C. 2004. Detection of skin cancer by classification of Raman spectra. *IEEE Transcriptions on Biomedical Engineering* 51(10), 1784–1793.

88. Gniadecka, M., Wulf, H. C., Mortensen, N. N., Nielsen, O. F., and Christensen, D. H. 1997. Diagnosis of basal cell carcinoma by Raman spectroscopy. *J. Raman Spectroscopy.* 28, 125–129.

89. Panjehpour, M., Julius, C. E., Phan, M. N., and Vo-Dinh, T. 2002. Laser-induced fluorescence spectroscopy for in vivo diagnosis of non-melanoma skin cancers. *Lasers in Sur. and Med.* 31, 367–373.

90. Wong, P. T., Goldstein, S. M., Grekin, R. C., et al. 1993. Distinct infrared spectroscopic patterns of human basal cell carcinoma. *Cancer Res.* 53, 762–765.

91. Lucassen, G. W., Van Veen, G. N., and Jansen, J. A. 1998. Band analysis of hydrated human skin stratum corneum attenuated total reflectance Fourier transform infrared spectra in vivo. *J. Biomed. Opt.* 3, 267–280.

92. McIntosh, I. M., Jackson, M., Mantsch, H. H., et al. 1999. Infrared spectra of basal cell carcinomas are distinct from non-tumour-bearing skin components. *J. Invest. Dermatol.* 112, 951–956.

93. Barry, B. W., Edwards, H. G. M., and Williams, A. C. 1992. Fourier transform Raman and infrared vibrational study of human skin: Assignment of spectral bands. *J. Raman Spectroscopy.* 23, 641–645.

94. Huang, Z., Lui, H., McLean, D. I., Korbelik, M., and Zeng, H. 2005. Raman spectroscopy in combination with background near-infrared autofluorescence enhances the in vivo assessment of malignant tissues. *Photochemistry and Photobiology* 81, 1219–1226.

95. Caspers, P. J., Lucassen, G. W., Carter, E. A., Bruining, H. A., and Puppels, G. J. 2001. In vivo confocal Raman microspectroscopy of the skin: Non-invasive

determination of molecular concentration profiles. *Journal of Investigative Dermatology* 116(3), 434–442.

96. Ly, E., Piot, O., Wolthuis, R., et al. 2008. Combination of FTIR spectral imaging and chemometrics for tumour detection from paraffin-embedded biopsies. *Analyst* 133(2), 197–205.

97. Lieber, C. A., Majumder, S. K., Billheimer D., et al. 2008. Raman microspectroscopy for skin cancer detection in vitro. *Journal of Biomedical Optics* 13(2), 024013.

98. Hammody, Z., Argov, S., Sahu, R. K., Cagnano E., Moreh R., and Mordechai S. 2008. Distinction of malignant melanoma and epidermis using IR microspectroscopy and statistical methods. *Analyst* 133(3), 372–378.

99. Fujioka, N., Morimoto, Y., Arai, T., and Kikuchi, M. 2004. Discrimination between normal and malignant human gastric tissues by Fourier transform infrared spectroscopy. *Cancer Detection & Prevention* 28, 32–36.

100. Tan, Y.-Y., Shen, A.-G., Zhang, J.-W., Wu, N., Feng, L., Wu, Q.-F., Ye, Y., and Hu, J.-M. 2003. Design of auto-classifying system and its application in Raman spectroscopy diagnosis of gastric carcinoma. 2nd International Conference Machine Learning & Cybernetics. November 2–5.

101. Weng, S.-F., Ling, X.-F., Song, Y.-Y., et al. 2000. FT-IR fibre optics and FT-Raman spectroscopic studies for the diagnosis of cancer. *American Clinical Laboratory* 19(7), 20.

102. Mordechai, S., Salman, A., Argov, S., et al. 2000. Fourier-transform infrared spectroscopy of human cancerous and normal intestine. *Proc. SPIE* 3918, 66–77.

103. Shetty, G., Kedall, C., Shepherd, N., Stone, N., and Barr, H. 2006. Raman spectroscopy: Evaluation of biochemical changes in carcinogenesis of oesophagus. *British Journal of Cancer* 94, 1460–1464.

104. Li, Q.-B., Sun, X.-J., Xu, Y.-Z., et al. 2005. Diagnosis of gastric inflammation and malignancy in endoscopic biopsies based on Fourier transform infrared spectroscopy. *Clinical Chemistry* 51(2), 346–350.

105. Xu, Y. Z., Yang, L. M., Xu, Z., et al. 2005. Distinguishing malignant from normal stomach tissues and its in vivo, in situ measurement in operating process using FTIR Fibre-Optic techniques. *Science in China Series B-Chemistry* 48(2), 167–175.

106. Suzuki, T., Hattori, Y., Katagiri, T., et al. 2008. In situ Raman study of the instant spectral changes observed in a pancreatic tumour tissue in living and dead model mice. *Biomedical Optical Spectroscopy* 6853, 85314.

107. Kawabata, T., Mizuno, T., Okazaki, S., et al. 2008. Optical diagnosis of gastric cancer using near-infrared multichannel Raman spectroscopy with a 1064-nm excitation wavelength. *Journal of Gastroenterology* 43(4), 283–290.

108. The, S. K., Zheng, W., Ho, K. Y., et al. 2008. Diagnostic potential of near-infrared Raman spectroscopy in the stomach: Differentiating dysplasia from normal tissue. *British Journal of Cancer* 98(2), 457–465.

109. Pandya, A. K., Serhatkulu, G. K., Cao, A., et al. 2008. Evaluation of pancreatic cancer with Raman spectroscopy in a mouse model. *Pancreas* 36(2), E1–E8.

110. Hu, Y. G., Shen, A. G., Jiang, T., et al. 2008. Classification of normal and malignant human gastric mucosa tissue with confocal Raman microspectroscopy and wavelet analysis. *Spectrochemica Acta Part A—Molecular and Biomolecular Spectroscopy* 69(2), 378–382.

111. Maziak, D. E., Do, M. T., Shamji, F. M., Sundaresan, S. R., Perkins, D. G., and Wong, P. T. T. 2007. Fourier-transform infrared spectroscopic study of characteristic molecular structure in cancer cells of esophagus: An exploratory study. *Cancer and Prevention* 31(3), 244–253.

112. Xu, Y. Z., Zhao, Y., Ling, X. F., et al. 2007. Fourier transform mid-infrared spectroscopy (FTIR) used for the rapid intraoperative diagnosis of gallbladder diseases. *Chemical Journal of Chinese Universities* 28(4), 645–648.

113. Kondepati, V. R., Keese, M., Heise, H. M., and Backhaus J. 2006. Detection of structural disorders in pancreatic tumour DNA with Fourier-transform infrared spectroscopy. *Vibrational Spectroscopy* 40(1), 33–39.

114. Sun, X. J., Li, Q. B., Shi, J. S., Ren, Y., Zhang, Y. F., Xu, D. F., and Wu, J. G. 2004. FTIR spectroscopic study on gastric tissue samples from gastroscopy. *Spectroscopy and Spectral Analysis* 24(8), 933–935.

115. Choo, L.-P., Mansfield, J. R., Pizzi, N., et al. 1995. Infrared spectra of human central nervous system tissue: Diagnosis of Alzheimer's disease by multivariate analyses. *Biospectroscopy* 1(2), 141–148.

116. Dovbeshko, G. I., Gridina, N. Y., Kruglova, E. B., and Pashchuk, O. P. 2000. FTIR spectroscopy studies of nucleic acid damage. *Talana* 53, 233–246.

117. Lehnhardt, F.-G., Rohn, G., Ernestus, R.-I., Grune, M., and Hoehn, M. 2001. [1]H- and ^{31}P-MR spectroscopy of primary and recurrent human brain tumours *in vitro*: Malignancy-characteristic profiles of water soluble and lipophilic spectral components. *NMR Biomed.* 14, 307–317.

118. Howe, F. A., and Opstad, K. S. 2003. [1]H MR spectroscopy of brain tumours and masses. *NMR Biomed.* 16, 123–131.

119. Yoshida, S., Miyazaki, M., Sakai, K., et al. 1997. Fourier transform infrared spectroscopic analysis of rat brain microsomal membranes modified by dietary fatty acids: Possible correlation with altered learning behavior. *Biospectroscopy* 3(4), 281–290.

120. Lakshmi, R. J., Kartha, V. B., Krishna, C. M., Solomon, J. G. R., Ullas, G., and Uma Devi, P. 2002. Tissue Raman spectroscopy for the study of radiation damage: Brain irradiation of mice. *Radiation Research* 157, 175–182.

121. Krafft, C., Neudert, L., Simat, T., and Salzer, R. 2005. Near-infrared Raman spectra of human brain lipids. *Spectrochimia Acta, Part A* 61, 1529–1535.

122. Fukuyama, Y., Yoshida, S., Yanagisawa, S., and Shimizu, M. 1999. A study on the differences between oral squamous cell carcinomas and normal oral mucosas measured by Fourier transform infrared spectroscopy. *Biospectroscopy* 5, 117–126.

123. Malini, R., Venkatakrishma, K., Kurien, J., et al. 2006. Discrimination of normal, inflammatory, premalignant, and malignant oral tissue: A Raman spectroscopy study. *Biopolymers* 81(3), 179–193.

124. Wu, J.-G., Xu, Y.-Z., Sun, C.-W., Soloway, R., Xu, D.-F., Wu, Q.-G., Sun, K.-H., Weng, S.-F., and Xu, G.-X. 2001. Distinguishing malignant from normal oral tissues using FTIR fibre-optic techniques. *Biopolymer (Biospectroscopy)* 62, 185–192.

125. Majumder, S. K., and Ghosh, N. 2005. Relevance vector machine for optical diagnosis of cancer. *Lasers in Sur. and Med.* 36, 323–333.

126. Lau, D. P., Huang, Z., Lui, H., Man, C. S., Berean, K., Morrison, M. D., and Zeng, H. 2003. Raman spectroscopy for optical diagnosis in normal and cancerous tissue of the nasopharynx—preliminary findings. *Lasers in Sur. and Med.* 32, 210–214.

127. Lau, D. P., Huang, Z., Lui, H., Anderson, D. W., Beren, K., Morrison, M. D., Shen, L., and Zeng, H. 2005. Raman spectroscopy for optical diagnosis in the larynx: Preliminary findings. *Lasers in Sur. and Med.* 37, 192–200.

128. Wu, J. G., Xu, Y. Z., Sun, C. W., et al. 2001. Distinguishing malignant from normal oral tissues using FTIR fibre-optic techniques. *Biopolymers* 62(4), 185–192.

129. The, S. K., Zheng W., Lau D. P., et al. 2008. Raman spectroscopy for optical diagnosis of laryngeal cancer. *Photonic Therapeutics and Diagnostics IV* 6842, 8421.

130. Wang, B. B., Pan, Q. H., Zhang, Y. F., et al. 2007. Application of Fourier transform infrared spectroscopy to non-invasive detection of pleomorphic adenoma of salivary gland in vivo. *Spectroscopy and Spectral Analysis* 27(12), 2427–2431.

131. Li, X., Lin, J., Ding, J., Wang, S., Liu, Q., and Qing, S. 2004. Raman spectroscopy and fluorescence for the detection of liver cancer and abnormal liver tissue. Annual International Conference IEEE EMBS, San Francisco, CA, September 1–5.

132. Chiang, H.-P., Song, R., Mou, B., Li, K. P., Chiang, P., Wang, D., Tse, W. S., and Ho, L. T. 1999. Fourier transform Raman spectroscopy of carcinogenic polycyclic aromatic hydrocarbons in biological systems: Binding to heme proteins. *J. Raman Spectroscopy.* 30, 551–555.

133. Van de Poll, S. W. E., Romer, T. J., Volger, O. L., et al. 2001. Raman spectroscopic evaluation of the effects of diet and lipid-lowering therapy on atherosclerotic plaque development in mice. *Arterioscler. Thromb. Vasc. Biol.* 21, 1630–1635.

134. Duarte, J., Pacheco, M. T. T., Machado, R. Z., Silveira, L., Jr., Zangaro, R. A., and Villaverde, A. B. 2002. Use of near-infrared Raman spectroscopy to detect IgG and IgM antibodies against *Toxoplasma gondii* antibodies in serum samples of domestic cats. *Cell Mol. Bio.* 48(5), 585–589.

135. Rohleder, D., Kiefer, W., and Petrich, W. 2004. Quantitative analysis of serum and serum ultrafiltrate by means of Raman spectroscopy. *Analyst* 129, 905–911.

136. Silveira, L., Sathaiah, S., Zangaro, R. A., et al. 2002. Correlation between near-infrared Raman spectroscopy and the histopathological analysis of atherosclerosis in human coronary arteries. *Lasers in Surgery and Medicine* 30, 290–297.

137. Andrus, P. G. L., and Strickland, R. D. 1998. Cancer grading by Fourier transform infrared spectroscopy. *Biospectroscopy* 4, 37–46.

138. Puppels, G. J., Garritsen, H. S. P., Kummer, J. A., and Greve, J. 1993. Carotenoids located in human lymphocyte subpopulations and natural killer cells by Raman microspectroscopy. *Cytometry* 14, 251–256.

139. Deng, J. L., Wei, Q., Zhang, M. H., Wang, Y. Z., and Li, Y. Q. 2005. Study of the effect of alcohol on single human red blood cells using near-infrared laser tweezers Raman spectroscopy. *Journal of Raman Spectroscopy* 36, 257–261.

140. Li, X., and Jin, H. 2001. Raman spectroscopy of serum for cancer detection. 23rd Annual EMBS International Conference, October 25–28, Istanbul, Turkey, 3221–3223.

141. Mordechai, S., Mordechai, J., Ramesh, J., Levi, C., Huleihel, M., Erukhimovitch, V., Moser, A., and Kapelushnik, J. 2001. Application of FTIR microspectroscopy for the follow-up of childhood leukaemia chemotherapy. *Proc. SPIE Subsurface and Surface Sensing Technologies and Applications III* 4491, 243–250.

142. Andrus, P. G. 2006. Cancer monitoring by FTIR spectroscopy. *Technology in Cancer Research and Treatment* 5(2), 157–167.

143. Isabelle M., Stone N., Barr H., et al. 2008. Lymph node pathology using optical spectroscopy in cancer diagnostics. *Spectroscopy: An International Journal* 22(2–3), 97–104.

144. Chan, J., Taylor, D. S., Lane, S. M., et al. 2008. Nondestructive identification of individual leukemia cells by laser trapping Raman spectroscopy. *Analytical Chemistry* 80(6), 2180–2187.

145. Khanmohammadi, M., Nasiri R., Ghasemi K., et al. 2007. Diagnosis of basal cell carcinoma by infrared spectroscopy of whole blood samples applying soft independent modeling class analogy. *Journal of Cancer Research and Clinical Oncology* 133, 1001–1010.

146. Pichardo-Molina, J. L., Frausto-Reyes, C., Barbosa-Garcia, O., et al. 2007. Raman spectroscopy and multivariate analysis of serum samples from breast cancer patients. *Lasers in Medical Science* 22, 229–236.

147. Huleihel, M., Karpasas, M., Talyshansky, M., Souprun, Y., Doubijanski, Y., and Erukhimovitch, V. 2005. MALDI-TOF and FTIR microscopy analysis of blood serum from diarrhea patients. *Spectroscopy: An International Journal* 19(2), 101–108.

148. Rehman, S., Movasaghi, Z., Darr, J. A., and Rehman, I. U. 2010. Fourier transform infrared spectroscopic analysis of breast cancer tissues: Identifying differences between normal breast invasive ductal carcinoma, and ductal carcinoma in situ of the breast; *Applied Spectroscopy Reviews* 45, 5, 355–368.

149. Rehman, S., Movasaghi, Z., Tucker, A. T., Joel, S., Darr, J. A., and Rehman, I. U. 2007. Raman spectroscopic analysis of breast cancer tissues: Identifying differences between normal breast invasive ductal carcinoma, and ductal carcinoma in situ of the breast, *Journal of Raman Spectroscopy*, 38, 10, 1345–1351.

150. Gazi, E., Dwyer, J., Gardner, P., Ghanbari-Siakhani, A., Wde, A. P., Lockyer, N. P., Vickerman, J. C., Clarke, N. W., Shanks, J. H., Scott, L. J., Hart, C. A., Brown, M., 2003. Applications of Fourier transform infrared microspectroscopy in studies of benign prostate and prostate cancer: A pilot study. *J. Pathol.* 201, 99–108.

151. Paluszkiewicz, C., and Kwiatek, W. M., 2001. Analysis of human cancer prostate tissues using FTIR microscopy and SXIXE techniques. *J. Mol. Structures* 565–566, 329–334.

152. Argov, S., Sahu, R. K., Bernshtain, E., Salam, A., Shohat, G., Zelig, U., and Mordechai, S. 2004. Inflammatory bowel diseases as an intermediate stage between normal and cancer: A FTIR-microspectroscopy approach. *Biopolymers* 75, 384–392.

153. Richter, T., Steiner, G., Abu-Id, M. H., Salzer, R., Gergmann, R., Rodig, H., and Johannsen, B. 2002. Identification of tumour tissue by FTIR spectroscopy in combination with positron emission tomography. *Vibrational Spectroscopy* 28, 103–110.

154. Rigas, B., Morgello, S., Goldman, I. S., and Wong, P. T. T. 1999. Human colorectal cancers display abnormal Fourier-transform infrared spectra. *Proc. Natl. Acad. Sci. USA* 87, 8140–8144.

155. Rigas, B., and Wong, P. T. T. 1992. Human colon adenocarcinoma cell lines display infrared spectroscopic features of malignant colon tissues. *Cancer Res.* 52, 84–88.

156. Pinzaru, S. C., Andronie, L. M., Domsa, I., et al. 2008. Bridging biomolecules with nanoparticles: Surface-enhanced Raman scattering from colon carcinoma and normal tissue. *Journal of Raman Spectroscopy* 39(3), 331–334.

157. Widjaja, E., Zheng, W., and Huang, Z. W. 2008. Classification of colonic tissues using near-infrared Raman spectroscopy and support vector machines. *International Journal of Oncology* 32(3), 653–662.

158. Kondepati, V. R., Heise, H. M., Oszinda, T., Mueller, R., Keese, M., and Backhaus, J. 2008. Detection of structural disorders in colorectal cancer DNA with Fourier-transform infrared spectroscopy. *Vibrational Spectroscopy* 46(2), 150–157.

159. Li, Q. B., Xu, Z., Zhang, N. W., et al. 2005. In vivo and in situ detection of colorectal cancer using Fourier transform infrared spectroscopy. *World Journal of Gastroenterology* 11(3), 327–220.

160. Sahu, R. K., Argov, S., Bernshtain, E., Salman, A., Walfisch, S., Goldstein, J., and Mordechai, S. 2004. Detection of abnormal proliferation in histologically "normal" colonic biopsies using FTIR-microspectroscopy. *Scandinavian Journal of Gastroenterology* 39(6), 557–566.

161. Huleihel, M., Salman, A., Erukhimovich, V., Ramesh, J., Hammody, Z., and Mordechai, S. 2002. Novel optical method for study of viral carcinogenesis in vitro. *J. Biochem. Biophys. Methods* 50, 111–121.

162. Mossoba, M. M., Al-Khaldi, S. F., Kirkwood, J., Fry, F. S., Sedman, J., and Ismail, A. A. 2005. Printing microarrays of bacteria for identification by infrared microspectroscopy. *Vibrational Spectroscopy* 38, 229–235.

163. Krishna, C. M., Sockalingum, G. D., Kegelaer, G., Rubin, S., Kartha, V. B., and Manfait, M. 2005. Micro-Raman spectroscopy of mixed cancer cell populations. *Vibrational Spectroscopy* 38, 95–100.

164. Kuhnert, N., and Thumser, A. 2004. An investigation into the use of Raman microscopy for the detection of labelled compounds in living human cells. *J. Label Compd. Radiopharm.* 47, 493–500.

165. Naumann, D. 1998. Infrared and NIR Raman spectroscopy in medical microbiology. *Proc. SPIE* 3257, 245–257.

166. Dovbeshka, G. I., Chegel, V. I., Gridina, N. Y., et al. 2002. Surface enhanced IR absorption of nucleic acids from tumour cells: FTIR reflectance study. *Biopolymer(Biospectroscopy)* 67, 470–486.

167. Chan, J. W., Taylor, D. S., Zwerdling T., Lane, S. T., Ihara, K., and Huser, T. 2006. Micro-Raman spectroscopy detects individual neoplastic and normal hematopoietic cells. *Biophysical Journal* 90, 648–656.

168. Bogomolny, E., Huleihel, M., Suproun, Y., et al. 2007. Early spectral changes of cellular malignant transformation using Fourier transform infrared microspectroscopy. *Journal of Biomedical Optics* 12(2), 024003.

169. Farguharson, S., Shende, C., Inscore, F. E., Maksymiuk, P., and Gift, A. 2005. Analysis of 5-fluorouracil in saliva using surface-enhanced Raman spectroscopy. *Journal of Raman spectroscopy* 36, 208–212.

170. Ruiz-Chica, A. J., Medina, M. A., Sanchez-Jimenez, F., and Ramirez, F. J. 2004. Characterization by Raman spectroscopy of conformational changes on guanine-cytosine and adenine-thymine oligonucleotides induced by aminooxy analogues of spermidine. *J. Raman Spectroscopy.* 35, 93–100.

171. Jalkanen, K. J., Jurgensen, V. W., Claussen, A., et al 2006. Use of vibrational spectroscopy to study protein and DNA structure, hydration, and binding of biomolecules: A combined theoretical and experimental approach. *J. Quantum Chem.* 106, 1160–1198.

172. Vo-Dinh, T., Allian, L. R., and Stokes, D. L. 2002. Cancer gene detection using surface-enhanced Raman scattering (SERS). *Journal of Raman Spectroscopy* 33, 511–516.

173. Binoy, J., Abraham, J. P., Joe, I. H., Jayakumar, V. S., Petit, G. R., and Nielsen, O. F. 2004. NIR-FT Raman and FT-IR spectral studies and *ab initio* calculations of the anti-cancer drug combretastatin-A4. *Journal of Raman Spectroscopy* 35, 939–946.

174. Faolain, E. O., Hunter, M. B., Byrne, J. M., Kelehan, P., McNamer, M., Byrne, H. J., and Lyng, F. M. 2005. A study examining the effects of tissue processing on human tissue sections using vibrational spectroscopy. *Vibrational Spectroscopy* 38, 121–127.

175. Sokolov, K., Follen, M., and Richards-Kortum, R. 2002. Optical spectroscopy for detection of neoplasia. *Current Opinion in Chemical Biology* 6, 651–658.

176. Chen, Y., Mu, C., Intes, X., Blessington, D., and Chance, B. 2003. Near-infrared phase cancellation instrument for fast and accurate localization of fluorescent heterogeneity. *Reviews of Scientific Instruments* 74(7), 3466–3473.

177. Sahu, P. K., and Mordechai, S. 2005. Fourier transform infrared spectroscopy in cancer detection. *Future Oncology* 1, 635–647.

178. Viehoever, A. R., Anderson, D., Jansen, D., and Mahadevan-Jansen, A. 2003. Organotypic raft cultures as an effective in vitro tool for understanding Raman spectral analysis of tissues. *Photochemistry and Photobiology* 78(5), 517–524.

179. Rabah, R., Weber, R., Serhatkulu, G. K., et al. 2008. Diagnosis of neuroblastoma and ganglioneuroma using Raman spectroscopy. *Journal of Pediatric Surgery* 43, 171–176.

180. Onogi C., Motoyama M., and Hamaguchi H. 2008. High concentration trans form unsaturated lipids detected in a HeLa cell by Raman microspectroscopy. *Journal of Raman Spectroscopy* 39(5), 555–556.

181. Gunstone, F. D., Harwood, J. L., Dijkstra, A. J., 2007. *The Lipid Handbook*, 3rd ed. Boca Raton, FL: CRC Press.

182. Movasaghi, Z., Rehman, S., and Rehman, I,U. 2008. FTIR spectroscopy of biological samples. *Applied Spectroscopy Reviews* 43(1), 1–46.

183. Movasaghi, Z., Rehman, S., and Rehman, I. U. 2007. Raman spectroscopy of biological samples. *Applied Spectroscopy Reviews* 42(2), 493–541.

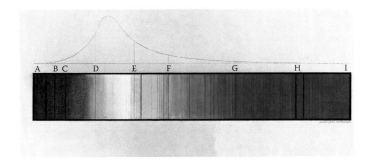

FIGURE 1.1
Solar spectrum drawn and coloured by Fraunhofer with the dark lines named after him, 1817. (Courtesy Deutsches Museum, Munich, Germany. With permission.)(1–3).

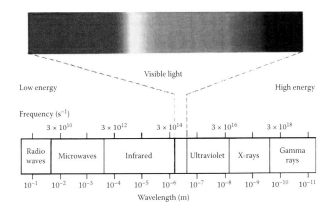

FIGURE 1.3
Electromagnetic spectrum.

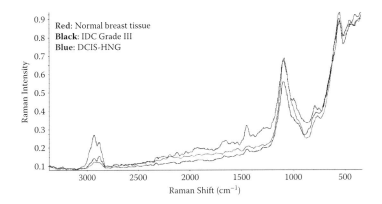

FIGURE 5.1
Raman spectra of normal, DCIS (HNG), and IDC (GIII) breast tissues.

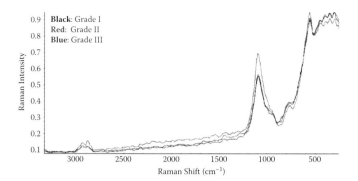

FIGURE 5.2
Raman spectra of different IDC grades.

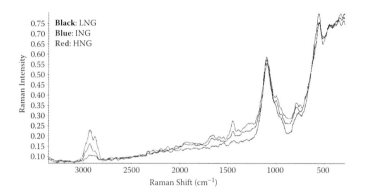

FIGURE 5.3
Raman spectra of different DCIS grades.

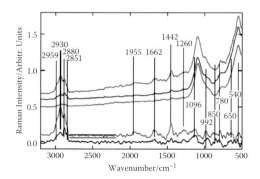

FIGURE 5.4
Raman spectra of normal (red), DCIS (HNG) (blue), and IDC (GIII) (black) breast tissues. Bottom traces are differences between IDC (black) and control, and DCIS (blue) and control tissues. For calculating the differences, spectra have been normalised to the peak at 1093 cm⁻¹. Horizontal double lines indicate the level of noise in order to facilitate justification of peak significance.

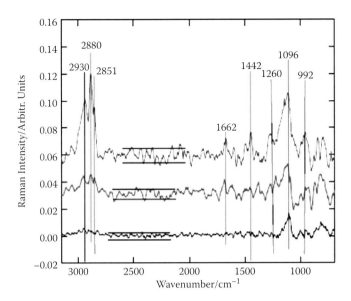

FIGURE 5.5
Raman difference spectra among various grades of IDC and control tissues. Grade I minus control, Grade II minus control, and Grade III minus control are in black, red, and blue, respectively. Horizontal double lines indicate the level of noise in order to facilitate justification of peak significance.

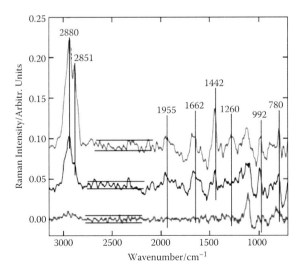

FIGURE 5.7
Raman difference spectra among various grades of DCIS and control tissues. LNG grade minus control, ING grade minus control, and HNG grade minus control are in red, black and blue, respectively. Horizontal double lines indicate the level of noise in order to facilitate justification of peak significance.

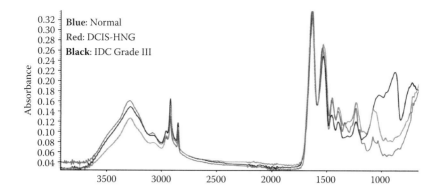

FIGURE 5.9
FTIR spectra of normal, DCIS (HNG), and IDC (GIII) breast tissues.

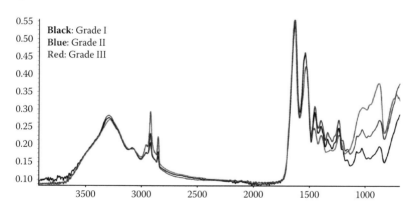

FIGURE 5.10
FTIR spectra of different IDC grades.

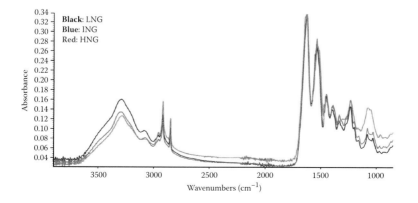

FIGURE 5.11
FTIR spectra of different DCIS grades.

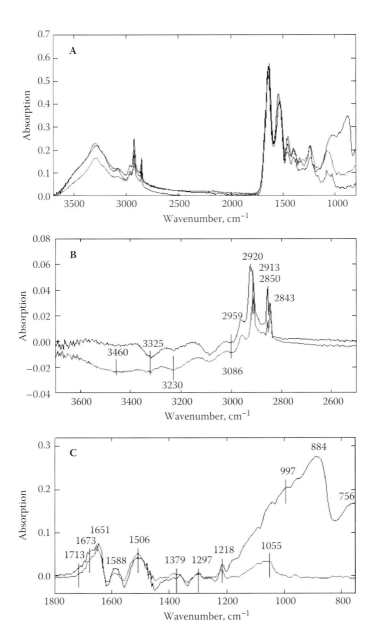

FIGURE 5.12
A: FTIR spectra of normal breast tissue (blue curve) NG, DCIS (red curve) of IDC (GIII) (black curve). B and C: Difference FTIR spectra. Red curve is a difference between DCIS and normal, and the black curve is a difference between IDC (GIII) and normal breast tissue.

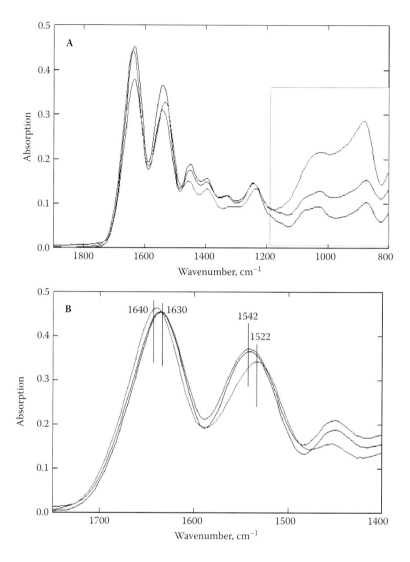

FIGURE 5.13
A: FTIR differences among IDC grades (I, in black; II in blue; and III, in red). B: Peak shifting in protein amide I and II regions.

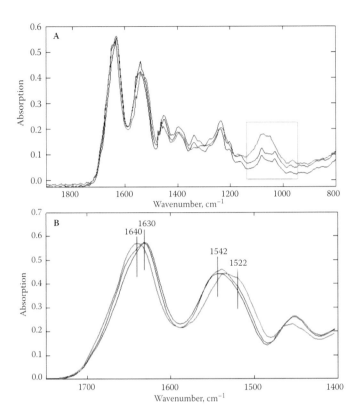

FIGURE 5.15
A: FTIR spectra of sections from DCIS, (LNG in black, ING in blue, and HNG in red). B: Variation in intensity of the peaks in the spectral region 1700–1400 cm⁻¹.

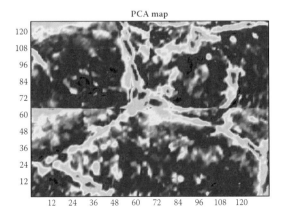

FIGURE 5.17
Red region of the image illustrates the area that gives rise to greatest shift and the blue region shows the minimum shift in the IR spectrum.

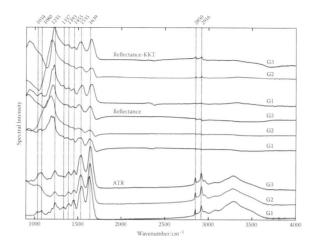

FIGURE 5.19
Spectra of three grades of IDC samples.

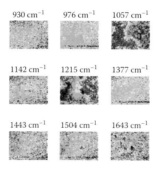

FIGURE 5.21
False-colour images based on the second-derivative intensity at the indicated wave numbers. Blue indicates high intensity (large negative second derivative) while red indicates low intensity.

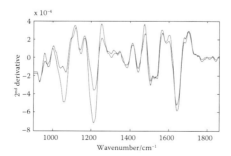

FIGURE 5.23
The blue spectrum is from the left side and the green spectrum is from the right. The difference in the relative intensities at 1057 and 1215 cm⁻¹ is obvious.

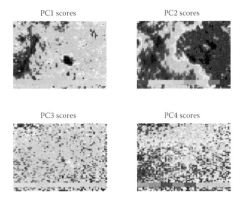

FIGURE 5.25
The second principal component (PC2) provides the same contrast as the difference between the bands at 1057 and 1215 cm^{-1}. The first PC identifies a small region near the centre of the image.

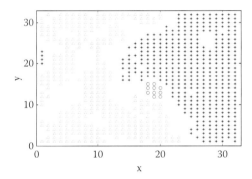

FIGURE 5.31
Rectangular regions around the samples; spatial (x, y) coordinates calculated and plotted.

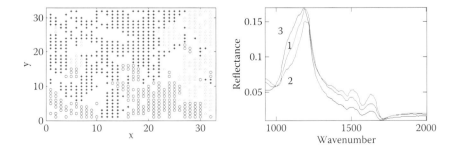

FIGURE 5.41
GII: spatial coordinates of points in three clusters.

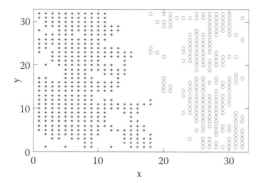

FIGURE 5.43
GI: regions containing the clusters, and the clusters are plotted with their (x, y) coordinates.

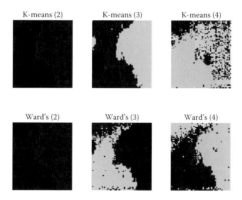

FIGURE 5.49
K-means and hierarchical clustering (Ward's method) into 2, 3, and 4 clusters.

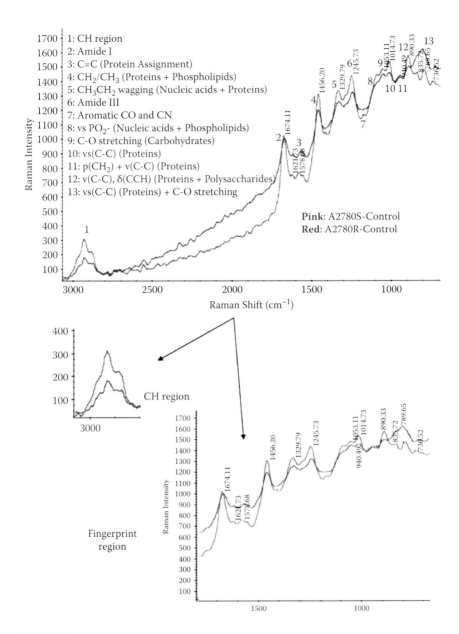

FIGURE 6.5
Spectra of A2780 R and S (control samples).

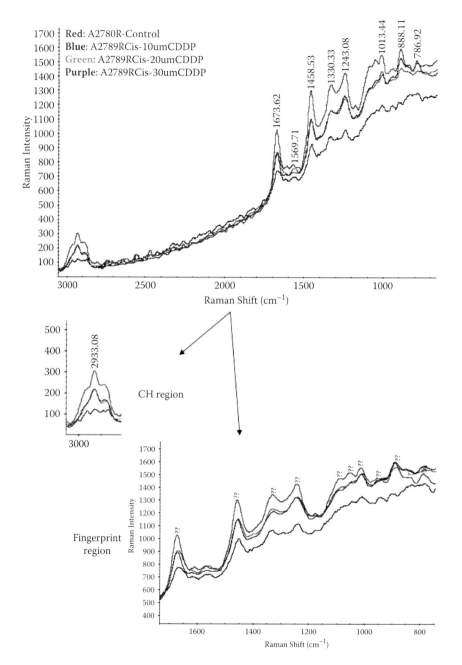

FIGURE 6.6
Spectra of R samples treated with different concentrations of cisplatin.

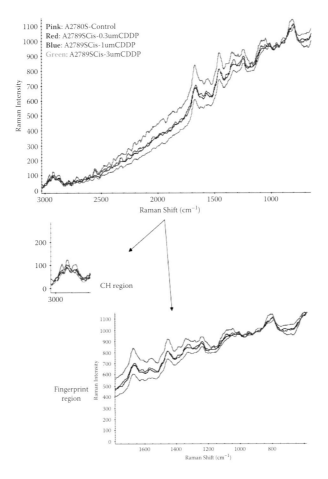

FIGURE 6.7
Spectra of S samples treated with different concentrations of cisplatin.

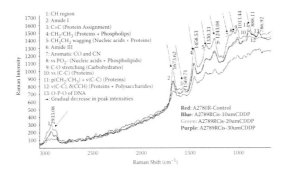

FIGURE 6.8
Changes in spectra of treated R samples.

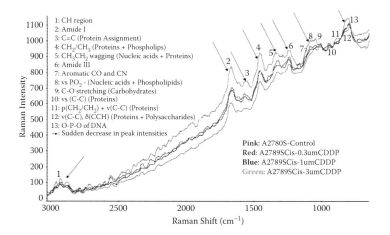

1: CH region
2: Amide I
3: C=C (Protein Assignment)
4: CH₂/CH₃ (Proteins + Phospholips)
5: CH₃CH₂ wagging (Nucleic acids + Proteins)
6: Amide III
7: Aromatic CO and CN
8: vs PO₂ - (Nucleic acids + Phospholipids)
9: C-O stretching (Carbohydrates)
10: vs (C-C) (Proteins)
11: p(CH₂/CH₂) + v(C-C) (Proteins)
12: v(C-C), δ(CCH) (Proteins + Polysaccharides)
13: O-P-O of DNA
→: Sudden decrease in peak intensities

Pink: A2780S-Control
Red: A2789SCis-0.3umCDDP
Blue: A2789SCis-1umCDDP
Green: A2789SCis-3umCDDP

FIGURE 6.9
Changes in spectra of treated S samples.

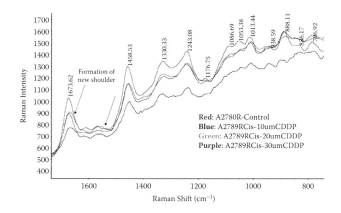

Red: A2780R-Control
Blue: A2789RCis-10umCDDP
Green: A2789RCis-20umCDDP
Purple: A2789RCis-30umCDDP

FIGURE 6.10
Treated R spectra, fingerprint region.

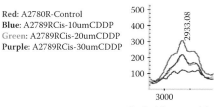

Red: A2780R-Control
Blue: A2789RCis-10umCDDP
Green: A2789RCis-20umCDDP
Purple: A2789RCis-30umCDDP

Gradual decrease of the
peak intensities of the R
samples in the CH region

FIGURE 6.11
Treated R spectra, CH region.

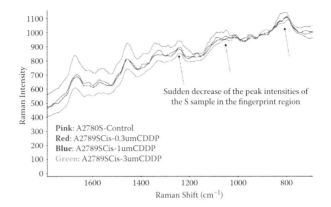

Pink: A2780S-Control
Red: A2789SCis-0.3umCDDP
Blue: A2789SCis-1umCDDP
Green: A2789SCis-3umCDDP

Sudden decrease of the peak intensities of
the S sample in the fingerprint region

FIGURE 6.12
Treated S spectra, fingerprint region.

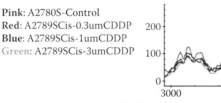

Pink: A2780S-Control
Red: A2789SCis-0.3umCDDP
Blue: A2789SCis-1umCDDP
Green: A2789SCis-3umCDDP

Sudden change in the
peak intensities of the S
samples in the CH region

FIGURE 6.13
Treated S spectra, CH region.

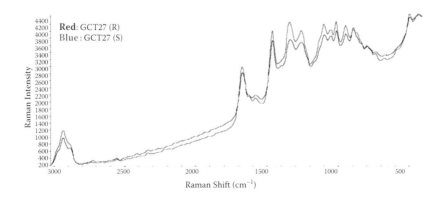

Red: GCT27 (R)
Blue : GCT27 (S)

FIGURE 6.14
Comparative image of GCT27 spectra.

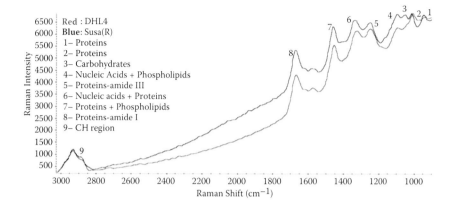

FIGURE 6.16

Peak definitions of spectra of DHL4 and Susa (R).

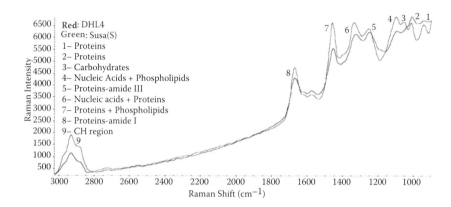

FIGURE 6.17

Peak definitions of spectra of DHL4 and Susa (S).

5

Infrared and Raman Spectroscopy and Imaging of Breast Cancer

Introduction

The incidence of breast cancer is increasing worldwide, with the highest rates reported in affluent Western societies (the risk ratio is approximately 1 in 8 to 1 in 12 affluent to nonaffluent Western societies). According to the latest data of Cancer Research, United Kingdom, it accounts for 12,471 deaths per year in the UK, the lifetime incidence is 9%, and the average incidence is 65 per 100,000 persons per year. Breast cancer remains the leading cause of cancer death in adult women under 54 years of age and the second most common cause after age 54. Among women of all ages, breast cancer is second only to lung cancer as the leading cause of cancer deaths in women [1–7].

Several techniques are currently used for breast cancer diagnosis. These techniques include mammography, ultrasound, fine-needle aspiration cytology (FNAC) and magnetic resonance imaging (MRI). However, most of these methods have considerable shortcomings and inaccuracies [8–14]. In mammography, for instance, 10–14% of clinically diagnosable cancers will not be detected [14] and ultrasound is not as sensitive or specific as mammography. As a result, in a considerable majority of cases, different biopsy methods (i.e., core and localization biopsy) must be applied. MRI is sensitive for the visualisation of both ductal carcinoma in situ (DCIS) and invasive ductal carcinoma (IDC) and can detect these only when they are clinically and mammographically occult [15].

In this chapter, Raman and Fourier transform infrared (FTIR) spectra of the normal breast tissue were compared with malignant tissue. Different grades of DCIS and IDC were analysed to evaluate precisely how spectroscopy could differentiate between these grades and any biochemical changes taking place.

Materials and Methods

Sample Preparation

The breast cancer specimens were accessed from the Histopathology Department, Royal Free and University College Medical School, London, archives between 1987 and 1997. A total of 67 individual cases were studied, details of which are tabulated in Table 5.1. All patients were diagnosed and clinically managed at the Royal Free Hospital, London. Age at presentation ranged from 42–75 years (median 56 years). All patients were treated by mastectomy or local excision with or without radiotherapy. Formalin-fixed, paraffin-wax-embedded sections from each case were 20-μm thick and were dewaxed in xylene and rinsed in graded alcohols by immersion in 50%, 70%, and 100% absolute alcohol for 5 minutes each. Five sections in each case were mounted on slides. A total of seven normal breast tissues and ten cases each of low nuclear grade (LNG), intermediate nuclear grade (ING), and high nuclear grade (HNG) DCIS, as well as ten cases each of grade I, II, and III IDC were studied (Table 5.1). DCIS was classified into LNG, ING, and HNG according to the recommendations of the National Co-ordinating Group for Breast Screening Pathology, UK [16]. IDC was classified into well-differentiated grade I, intermediate-differentiated grade II, and poorly differentiated grade III according to the Elston and Ellis grading system [17]. The classification was done by an accredited pathologist experienced in breast pathology.

Raman Spectroscopy of Breast Tissue Samples

Collecting the Spectra

Raman spectra were recorded using a Thermo Nicolet Almega Raman spectrophotometer, equipped with a microscope. A 786-nm laser was used.

TABLE 5.1

Number of Breast Tissue Samples Studied with Raman and FTIR Spectroscopy

Number	Histology	Total Number of Cases
1	Normal	7
2	LNG	10
3	ING	10
4	HNG	10
5	G I	10
6	G II	10
7	G III	10
Total Number		**67**

The low laser power in the Raman spectrophotometer makes it possible to analyse biological tissues without the loss of sensitivity due to fluorescence of organics.

The tissue samples were analysed by mounting them on slides. The spectra were obtained by using 4 exposures, 16 scans, and at 4 cm^{-1} resolution. The tissue samples were mapped across the section.

Infrared spectra were obtained using a Nicolet 800™ spectrometer in conjunction with a Nicplan™ microscope equipped with a liquid nitrogen–cooled MCTA detector. An attenuated total reflectance (ATR) slide on objective equipped with diamond crystal (Spectra-Tech™, USA) was mounted on the microscope. This ATR objective provides two modes of operation; one is for viewing the specimen and the other for infrared (IR) analysis. The view mode allows viewing of the area of interest and the IR mode allows spectral data acquisition. A zoom on aperture was used to mask areas of interest.

Spectra were obtained at 4 cm^{-1} resolution, averaging 256 scans. Three randomly selected spots were analysed for each specimen taking fifteen spectra for each tissue. Spectra of the normal tissues were also obtained for comparison.

In this study, the spectra were obtained by using 4 exposures, 16 scans, and 4 cm^{-1} resolution. The tissue sample was mapped across the section. Figure 5.1 shows the Raman spectra of normal, DCIS (HNG), and IDC (GIII) breast tissues. Spectra of IDC grades (GI, GII, and GIII) are shown in Figure 5.2, and the spectra of DCIS grades (HNG, ING, and LNG) appear in Figure 5.3.

Raman spectroscopy has the potential to identify markers associated with malignant changes. Normal and neoplastic tissues have distinct biochemical

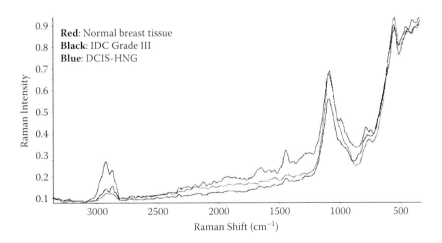

FIGURE 5.1 (See colour insert.)
Raman spectra of normal, DCIS (HNG), and IDC (GIII) breast tissues.

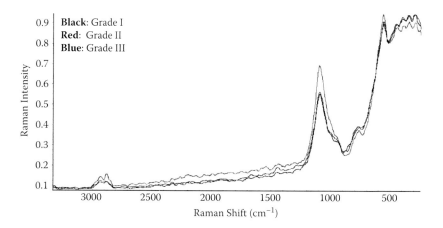

FIGURE 5.2 (See colour insert.)
Raman spectra of different IDC grades.

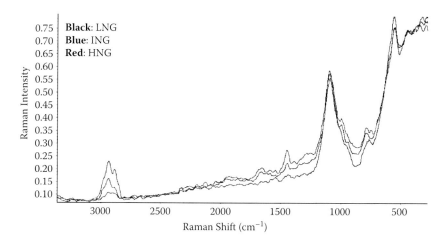

FIGURE 5.3 (See colour insert.)
Raman spectra of different DCIS grades.

features that are manifestations of unique biological processes. The direct detection of these biochemical constituents may provide a new methodology for noninvasive detection of disease processes, together with information for classifying, grading, and evaluating the progression of malignant neoplasms to provide a potential diagnostic tool. The spectrum of the normal tissue has well-defined peaks in the region 3500–650 cm^{-1}. Assignment of the peaks is presented in Table 5.2.

TABLE 5.2

Peak Assignments of the Raman Spectra of Breast Tissue [18]

Peak Assignment	Normal	DCIS HNG	DCIS ING	DCIS LNG	IDC GI	IDC GII	IDC GIII
3300–2800 cm⁻¹ (C–H stretching vibrations of methyl (CH$_3$) and methylene (CH$_2$) groups of phospholipids, cholesterol and creatine[46,60])							
CH$_3$ (ν_{as}CH$_3$ of lipids, DNA, and proteins[32])	2959	2958	2958	2957	2959	2957	2957
CH$_2$ (ν_{as} CH$_2$ of lipids[24,32,38])	2930	2930	2930	2930	2930	2930	2930
	2884	2882	2881	2880	2882	2880	2880
CH$_2$ (ν_sCH$_2$ of lipids[24,32,38])	2851	2851	2850	2850	2849	2849	2849
Vibrational bands for lipids, proteins, collagen, etc. 1670–750 cm⁻¹							
C=C (amide I (C=O stretching mode of proteins, α-helix conformation)/C=C lipid stretch[44])	1662	1660	1658	1658	1660	1659	1658
CH$_2$ (overlapping asymmetric CH$_3$ bending & CH$_2$ scissoring (is associated with elastin, collagen & phospholipids)[29,30,61])	1442	1452	1452	1451	1452	1452	1452
C–N–H (ν(CN), δ(NH) amide III, α-helix conformation collagen, tryptophan; and PO$_2^-$ asymmetric (Phosphate I)[38,39,41])	1260	1261	1261	1260	1261	1260	1260
O–P–O (stretching PO$_2^-$ symmetric (Phosphate II) of phosphodiesters[31,38,62,63])	1096	1096	1095	1095	1095	1095	1095
C–C (ν_s(C–C)), Symmetric ring breathing of phenylalanine (protein assignment) band[39])	–	992	992	992	992	992	992
C–C (C–C stretch or proline, hydroxyproline and tyrosine and ν_2PO$_2^-$ stretch of nucleic acids bands[39,41,58])	780	795	795	795	792	792	792
C–C (Symmetric breathing of tryptophan – protein assignment[39,41,44])	650 540	751	750	–	–	–	–

Source: S. Rehman, Z. Movasaghi, A. T. Tucker, S. P. Joel, J. A. Darr, A. V. Ruban, and I. U. Rehman. Raman spectroscopic analysis of breast cancer tissues: Identifying differences between normal, invasive ductal carcinoma and ductal carcinoma in situ of the breast tissue. *Journal of Raman Spectroscopy* 38(2007): 1345–1351. With permission.

Raman Differences among Normal, DCIS, and IDC Tissues

Raman spectra of normal breast tissue, a case of DCIS HNG, and a case of IDC GIII are shown in Figure 5.4. The most notable feature of each spectrum is the well-defined peaks in the 1800–550 cm^{-1} region of the spectra. Obvious spectroscopic differences exist in both the absolute and relative intensities of the peaks in the spectra (bottom traces).

The peak present at 1662 cm^{-1} in the spectrum of normal breast tissue is due to amide I band of tissue proteins [19–22]. These bands are highly sensitive to the conformational changes in the secondary structure. The amide I band is due to the in-plane stretching of the C=O bond of the peptide backbone [22], weakly coupled to stretching of the C–N and in-plane bending of the N–H bond [21]. A noticeable point about this peak is the fact that it has been used the most for structural studies because of its high sensitivity to small changes in molecular geometry and hydrogen bonding of the peptide group [22]. The intensity of the bands varies with the degree of fatty-acid unsaturation and depends on the lipid to protein ratio. The peak located at 1442 cm^{-1} is a protein assignment attributed to $\nu(CH_2)$ and $\nu(CH_3)$ of collagen [23], or a lipid assignment attributed to $\nu(CH_2)$, scissoring, phospholipids [23], and is of diagnostic significance [24]. The spectral differences in the regions 960–800 cm^{-1} and 1400–1080 cm^{-1} further confirm the increase of protein content and relative decrease in the lipids/acylglyceride content in the cancerous tissues. The bands at 1662, 1442, and 1260 cm^{-1} are due to β-sheet structure of amide I[21], arising from C=O stretching vibrations of collagen and elastin [23,25] and C=C of lipids [26]; CH$_2$ scissoring and CH$_3$ bending in lipids

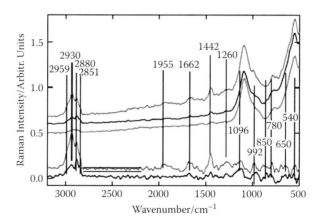

FIGURE 5.4 (See colour insert.)
Raman spectra of normal (red), DCIS (HNG) (blue), and IDC (GIII) (black) breast tissues. Bottom traces are differences between IDC (black) and control, and DCIS (blue) and control tissues. For calculating the differences, spectra have been normalised to the peak at 1093 cm^{-1}. Horizontal double lines indicate the level of noise in order to facilitate justification of peak significance.

and proteins [23,27]; and amide III and CH_2 wagging vibrations from glycine backbone and proline sidechains, as well as asymmetric PO_2^- (phosphate I) [27–31] vibrations, respectively. The presence of amide I bands further confirms the increase in protein content in the cancer tissue as compared to the normal tissue.

Raman Differences among IDC Grades (I, II, and III)

Figures 5.5 and 5.6 illustrate the differences among grades I, II, and III of IDC. The Raman spectrum of the normal tissue is weaker in intensity than that of the IDC tissue and the intensity of the peaks increases with increasing grades; grade III has the highest intensity, grade I possesses the minimum, and grade II lies in between (Figure 5.7). The increase in intensity of the OH–NH–CH peaks suggests a change in the lipids, proteins, and DNA contents. The absolute intensity of each of the OH–NH–CH peaks in the 3500–2700 cm^{-1} region [31] varies with increasing grade (I, II, and III), indicating varying concentrations of fatty acyl chains [32], phospholipids, cholesterol, and creatine [25], proteins [32,33] and nucleic acids [21,25].

Similarly, an increase in intensity and shifting of peaks in the area of 1660–600 cm^{-1}, peaks of nucleic acids due to the base carbonyl stretching and ring breathing mode [34], and amide absorption having some overlap with the

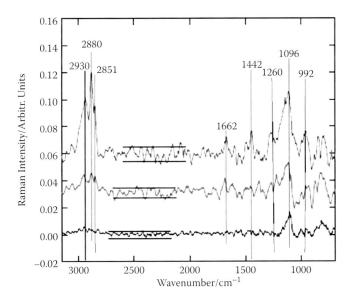

FIGURE 5.5 (See colour insert.)
Raman difference spectra among various grades of IDC and control tissues. Grade I minus control, Grade II minus control, and Grade III minus control are in black, red, and blue, respectively. Horizontal double lines indicate the level of noise in order to facilitate justification of peak significance.

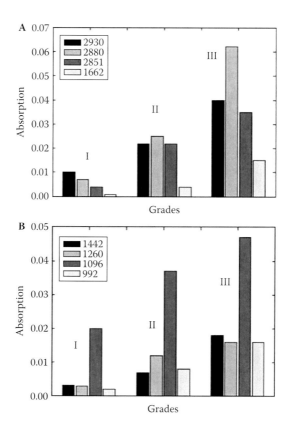

FIGURE 5.6
Plots of the amplitudes of selected peaks, from the Raman difference spectra presented as a function of the IDC grade of the breast tissues.

carbonyl stretching modes of nucleic acids [34] and C=C stretch of lipids [35], confirm the changes in the chemical structure of the breast tissue and this change in intensity provides useful information in differentiating among the IDC grades.

Raman Differences among DCIS Grades (LNG, ING, and HNG)

Raman spectra from LNG, ING, and HNG DCIS tissues normalised with respect to maximum absorption intensity are shown in Figures 5.7 and 5.8.

There are significant spectroscopic differences observed between the three DCIS grades in both the relative and absolute intensities of absorption bands in the spectra. The absolute intensity of each of the OH–NH–CH peaks, in the 3500–2700 cm^{-1} region [31], increases with increasing grade (low/intermediate/high), indicating an increase in concentrations of fatty acyl chains (32), phospholipids, cholesterol, and creatine [25], proteins [32,33] and nucleic

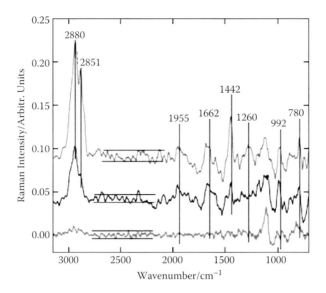

FIGURE 5.7 (See colour insert.)
Raman difference spectra among various grades of DCIS and control tissues. LNG grade minus control, ING grade minus control, and HNG grade minus control are in red, black and blue, respectively. Horizontal double lines indicate the level of noise in order to facilitate justification of peak significance.

acids [21,25]. In addition, the intensity of the absorption bands at 2958 cm^{-1} (asymmetric CH$_3$ of lipids, DNA, and proteins) [36], 2920 cm^{-1} (asymmetric CH$_2$ of lipids) [31,36,37] and 2851 cm^{-1} (symmetric CH$_2$ of lipids) [31,36,37] is highest in HNG, progressing to the lowest relative intensity in LNG. This indicates that the tissue section giving rise to the spectrum from HNG is lipid- and acylglyceride-rich and the one giving rise to the spectrum from LNG is protein-rich, while the one giving rise to the ING spectrum contains significant amounts of both lipids/acylglycerides and proteins. This complements the FTIR data reported in the previous section.

Similarly, the variation in intensity of the peaks in the spectral region 1660–700 cm^{-1} (phosphodiester stretching bands region, for absorbances due to collagen and glycogen [38] and porphyrin ring of heme proteins [39]) also confirms that the HNG spectrum is rich in acylglyceride and the LNG spectrum is rich in protein contents. The amide I band centred at 1660 cm^{-1} (arising from C=O and C=C stretching vibrations) [25,26,40] not only varies in intensity, but also in shape, since HNG has a higher intensity as compared to ING and LNG, confirming the varying protein contents with respect to the grades.

FTIR Spectroscopy of Breast Tissue Samples

In this study, FTIR microspectroscopy was conducted using an ATR slide on an objective equipped with diamond crystal (Spectra-Tech™, USA) on

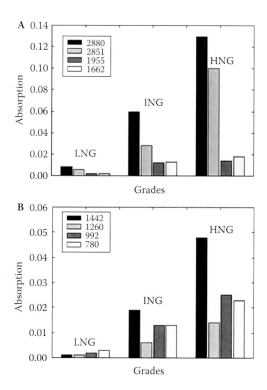

FIGURE 5.8
Plots of the amplitudes of selected peaks, from the Raman difference spectra presented as a function of the DCIS grade of breast tissues.

the microscope. This ATR objective provides two modes of operation; one is for viewing the specimen and the other for IR analysis. The view mode allows viewing of the area of interest and the IR mode for spectral data acquisition. A zoom on aperture was used to mask areas of interest. Spectra were obtained from three different areas at 4 cm^{-1} resolution, averaging 256 scans for each region.

Figure 5.9 shows the FTIR spectra of normal, DCIS (HNG), and IDC (GIII) breast tissues. Figure 5.10 represents the FTIR spectra of different IDC grades, and Figure 5.11 shows the FTIR spectra of different DCIS grades of breast tissue.

A summary of the peak assignments and definitions is presented in Table 5.3.

FTIR Differences among Normal, DCIS, and IDC

FTIR spectra of normal breast tissue, cases of HNG, DCIS, and a case of GIII IDC are given in Figure 5.12. The most noticeable feature of each spectrum is the complexity, with prominent absorption in almost all regions of the spectra.

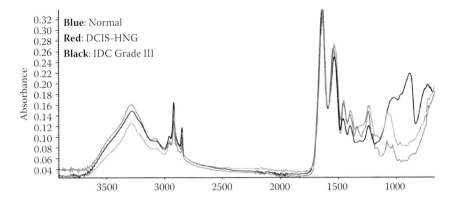

FIGURE 5.9 (See colour insert.)
FTIR spectra of normal, DCIS (HNG), and IDC (GIII) breast tissues.

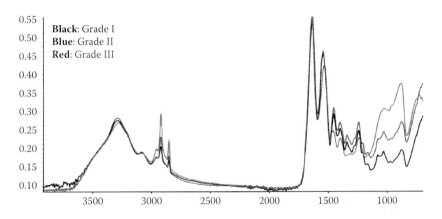

FIGURE 5.10 (See colour insert.)
FTIR spectra of different IDC grades.

In summary, the spectral data reveal the presence of DNA, proteins, lipids, and fatty acids. Spectral bands in the region of 1700 to 900 cm^{-1} arise from biochemical groups (such as C=O, CH$_2$, CH$_3$, C–O–C, and O–P–O) or linkages, belonging to phospholipids, proteins, carbohydrates, collagen, and amino acids, and an overall shift of peaks is observed in the fingerprint region (1700 to 800 cm^{-1}). Differences in the intensity and positioning of peaks in the spectra may be attributed to the compositional changes between normal and cancerous tissue (Figure 5.12 A, B, and C). The increase in intensity of the C–H peaks (3000–2800 cm^{-1}) [31,36,37] may be attributed to the increase in lipids, proteins, and DNA. Similarly, increase in intensity and shifting of peaks in the area of 1700 to 1500 cm^{-1} (peaks of amide I and II absorption) [20,28,29] confirm the changes in the chemical composition and protein structure of the normal tissue.

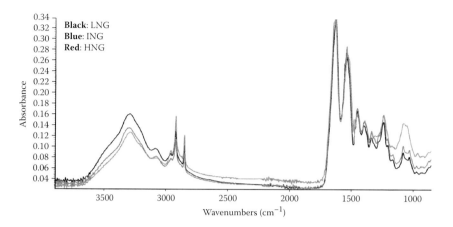

FIGURE 5.11 (See colour insert.)
FTIR spectra of different DCIS grades.

TABLE 5.3

Peak Assignments of the FTIR Spectra of Normal Breast Tissue and Breast Carcinoma

Approximate Peak Intensity	Assignment	Ref. No
3290 cm⁻¹	N–H asymmetric stretching modes in proteins and nucleic acids	21,31
3085 cm⁻¹	N–H symmetric stretching modes in proteins and nucleic acids	21,31
2960 cm⁻¹	$v_{as}CH_3$ of lipids, DNA, and proteins	20
2920 cm⁻¹	$v_{as}CH_2$	31,36,37
2850 cm⁻¹	v_sCH_2	31,36,37
1630 cm⁻¹	Amide I (in-plane stretching of the C=O bond of the peptide backbone, weakly coupled to stretching of the C–N and in-plane bending of the N–H bond)	20,41,42
1540 cm⁻¹	Amide II (primarily N–H bending coupled to C–N stretching, or C–N stretching and C–N–H bending vibrations)	20,29,37,41,42
1452 cm⁻¹	Overlapping asymmetric CH_3 bending & CH_2 scissoring (is associated with elastin, collagen, and phospholipids	28,43
1372 cm⁻¹	CH_3 (Symmetric CH_3 bending)	37,42
1310 cm⁻¹	Amide III, Symmetric stretch	44,45,46
1260 cm⁻¹	Amide III, $v(CN)$, $\delta(NH)$	44,45,46
1216 cm⁻¹	C–O (C–OH groups of serine, threosine and tyrosine of proteins)	46,47
1080 cm⁻¹	O–P–O (Symmetric stretching of phosphate groups of phosphodiester linkages)	29,42
1030 cm⁻¹	O–P–O (Phosphate groups of phosphorylated proteins & cellular nucleic acids	41

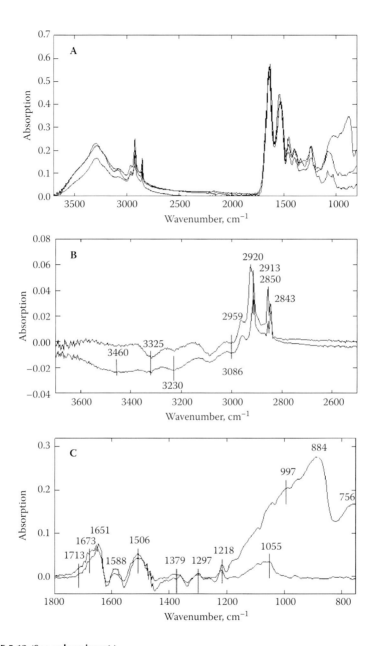

FIGURE 5.12 (See colour insert.)
A: FTIR spectra of normal breast tissue (blue curve) NG, DCIS (red curve) of IDC (GIII) (black curve). B and C: Difference FTIR spectra. Red curve is a difference between DCIS and normal, and the black curve is a difference between IDC (GIII) and normal breast tissue.

FTIR Differences among IDC Grades (I, II, and III)

The most dramatic changes in the FTIR spectra of cancerous tissues occur in the spectral range of 1100 to 1000 cm⁻¹ (phosphodiester stretching bands region) [48,49] and show significant differences between the spectra of IDC grades, as peak positioning is shifted in addition to changes in absorbance of the peaks (Figure 5.13, outlined, boxed area). This indicates that the chemical

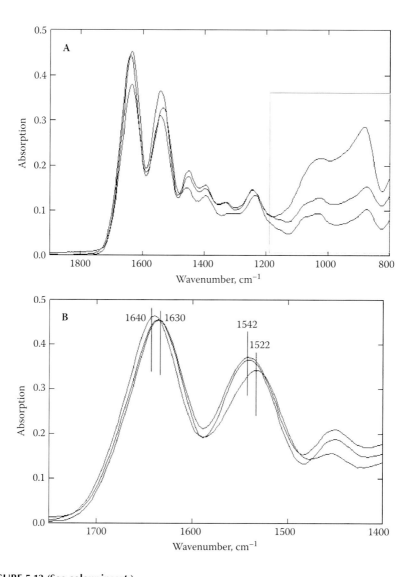

FIGURE 5.13 (See colour insert.)
A: FTIR differences among IDC grades (I, in black; II in blue; and III, in red). B: Peak shifting in protein amide I and II regions.

structure is changing with the formation of new peaks, suggesting the formation of new groups, as two small peaks in the region of 1200–1000 cm^{-1} merge together to form a broad band. This appears to be due to an increase in the substances in the cancerous cells.

A pronounced difference is also observed in the bands appearing at 1080 and 1030 cm^{-1}, which can be assigned to the $v(PO_2^-)$ symmetric stretching of phosphodiesters [45,46,50], and phosphate groups of phosphorylated proteins and cellular nucleic acids [50], respectively. Grade I has a well-defined doublet, similar to the one observed in the FTIR spectrum of the normal tissue. The peak centred at 1029 cm^{-1}, increases in intensity with increasing grades (I to III) and the new peek appears centred at 1055 cm^{-1} with increasing grades (I to III) (Figure 5.13). Change in the peak shape and intensity indicates that significant biochemical changes are taking place. The group of bands at 1055, 997, 890, and 765 cm^{-1} also display a steady rise as the cancer progresses from grade I to III (Figure 5.14A).

In the protein amide I and II regions at 1630 and 1540 cm^{-1}, significant shifts of the maxima can be observed, in particular for grade III (Figure 5.13). While amide I upshifts toward 1640 cm^{-1}, amide II downshifts to 1522 cm^{-1}. This is a clear indication of secondary structure alterations, most likely due to the elimination of the β-sheets and formation or replacement with the alpha or random coil structure. The quantitative analysis of the changes in the a Amide I and II regions is presented in Figure 5.14B.

The gradual alteration in CH of lipids, DNA, and proteins is quantified and shown in Figure 5.14C by measuring the amplitudes of 2920, 2896, 2962, and 2850 cm^{-1} components for different cancer grades.

FTIR Differences among DCIS Grades (LNG, ING, and HNG)

FTIR spectra of DCIS tissues provide prominent absorption bands to distinguish among the three grades. FTIR spectra of sections from LNG, ING, and HNG DCIS normalised with respect to maximum absorption intensity are given in Figures 5.15 and 5.16. Significant difference is observed in both the relative and absolute intensities of absorption bands in the spectra.

The absolute intensity of each of the N–H [49,51] peaks in the 3400 to 3100 cm^{-1} region decreases with increasing grade (low–intermediate–high), whilst the amplitude of 3000 to 2800 cm^{-1} (C–H region) increases, indicating increasing concentrations of proteins and nucleic acids [52,53] and lipids (the lipid band has a tendency toward higher levels of energy, while the protein band's tendency is toward lower energy levels) [33]. The relative intensity of the absorption at around 2960 to 2922 cm^{-1} is highest in HNG, progressing to the lowest relative intensity in LNG. These show the relative concentrations of $v(as)$ CH of lipids, DNA, and proteins [25,36], respectively(Figure 5.16A).

The amide I band centred at 1630 cm^{-1} [28,29] not only varies in intensity, but varies in shape, indicating that different tumour grades have varying protein contents with different types of structure (Figures 5.15 and 5.16B).

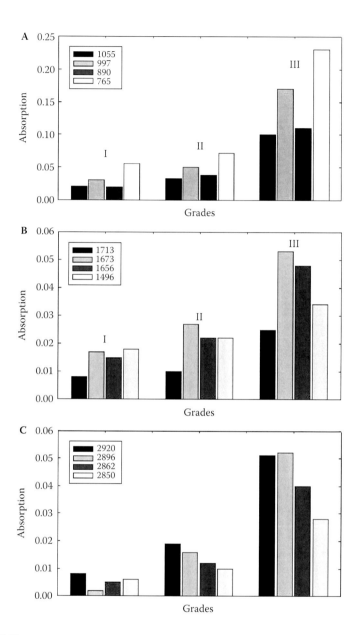

FIGURE 5.14

A: Plots of amplitudes of selected peeks from the new group of bands at 1055, 997, 890, and 765 cm^{-1} appear and display a steady rise as the cancer progresses from Grade I to III. B: Plots of amplitudes representing the quantitative analysis of amide I and amide II regions in different cancer grades. C: Plots of the amplitudes of the gradual alteration in CH content in different cancer grades.

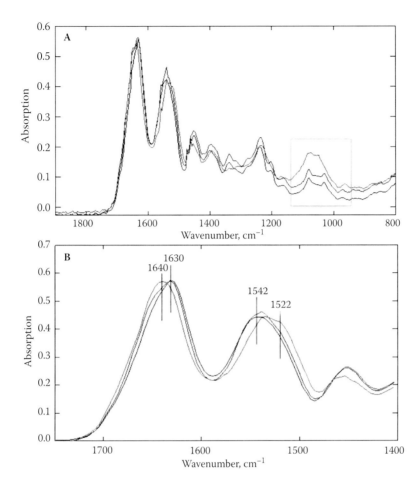

FIGURE 5.15 (See colour insert.)
A: FTIR spectra of sections from DCIS, (LNG in black, ING in blue, and HNG in red). B: Variation in intensity of the peaks in the spectral region 1700–1400 cm⁻¹.

This is due to the fact that the infrared absorption band for amide I groups is sensitive to protein conformation; therefore, this band can be employed to grade the DCIS tissue, in addition to relative intensities.

In the amide I and II area, a pattern similar to IDC samples was observed in DCIS samples (Figures 5.15B and 5.13B). In both cases, whilst amide I upshifts toward 1640 cm⁻¹, amide II downshifts to 1522 cm⁻¹, indicating secondary structure alterations.

The presence of peaks centred at 1310–1300 cm⁻¹ (amide III band of proteins) [46], in the 1700–900 cm⁻¹ region can be attributed to type I collagen (Figure 5.15 and 5.16C). The difference in intensity and shape of these peaks indicates the varying type and amount of type I collagen present in the three nuclear-grade tissues (Figure 5.16C). Therefore, combining the three

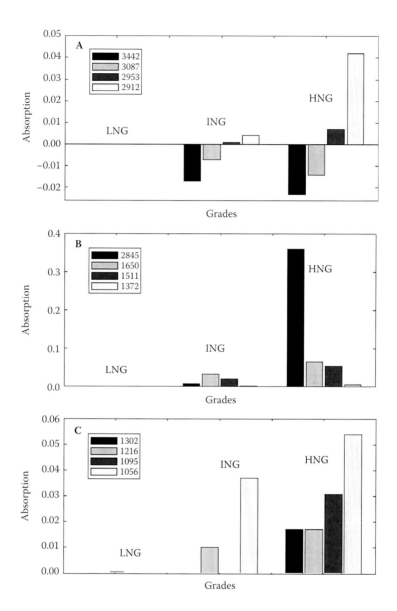

FIGURE 5.16
Plots of amplitudes of selected peaks. A: The intensity of each of the N–H peaks in the 3400 to 3100 cm^{-1} region decreases with increasing grade (low–intermediate–high), while the amplitude of 3000 to 2800 cm^{-1} (C–H region) increases, indicating increasing concentrations of proteins and nucleic acids and lipids. B: Plots of amplitudes of the bands at CH and amide I and II regions (indicating that different tumour grades have varying protein contents) C: Plots of amplitudes of some other peaks.

parameters of lipids/acylglyceride, proteins, and collagen together helps in classifying three different nuclear grades (high, intermediate, and low) more precisely and consistently. All of these results have been backed by our published review works in *Applied Spectroscopy Reviews* [54,55].

Analysis of Cancerous Tissues Using FTIR Rapid Scan Imaging

The basic question is whether FTIR-MS (micro-spectroscopy) imaging with a Focal Plane Array (FPA) camera can be useful in cancer diagnosis and grading by analysis of tissue samples, e.g., biopsies or excised tumours. Its potential advantages are that it requires little sample preparation and is therefore rapid; that a single image can contain contrasts that would normally take several different types of stains to obtain; and that it can be implemented more objectively than a standard histological analysis.

There appear to have been a few tens of studies dealing with FTIR-MS imaging of cancer-related samples published in the last few years. Several reviews have been published recently [56–58].

FTIR-MS imaging produces three-dimensional data: an intensity quantity (e.g., absorbance or reflectance) as a function of two spatial coordinates (x and y) and wavelength (λ). This type of data is sometimes called a *hyperspectral image*. Since an entire infrared spectrum is recorded for every pixel, hyperspectral images potentially contain a lot more information than can be obtained with a visible-light microscope.

Most of the studies published to date analyse only one or a few samples, and seek to distinguish between types of tissue in the imaged area, with the aim of revealing information about the physical structure of the sample. The infrared image is processed to give a false-colour image in which chemically similar cells are represented by the same colour. To put it another way, the hyperspectral image is reduced in dimensionality in an optimal way so that it can be interpreted. The dimensionality reduction can be achieved in a purely unsupervised (or nearly so) manner [59] by hierarchical or K-means clustering algorithms. However, attaching meaning to the false-colour images thus produced generally requires comparison of the image with a visible-light image of the same sample that has been examined by a pathologist. Potentially, this information could be obtained by comparison of the IR spectra with reference spectra, but this comparison can be difficult because the different kinds of tissue have remarkably similar spectra.

These studies with small sample sizes have shown that FTIR-MS imaging can produce information equivalent to several different types of optical examination [60]. However, a much larger set of samples is necessary to show that the method can be useful for diagnosis. A few studies have used large sets of data, e.g., 57 or 20 samples [61–64]. These studies were all by the same group, and sought to replace or complement traditional histological methods for brain cancer grading with an automated classification algorithm (more details are given in the following text).

The most ambitious and comprehensive work to date appears to be that of Bhargava et al. [58], who sought to establish a very large calibration set using tissue microarrays (see below).

Size-Scale/Resolution Issues and Instrumentation

Most human cells are on the order of 10 μm in diameter. Mid-infrared wavelengths (λ) are 2.5–10 μm. The best attainable (diffraction-limited) spatial resolution with an infrared microscope is approximately $2\lambda/3$ [65], but this depends on the design of the instrument and may be limited by the aperture (for mapping instruments) or the pixel size (for imaging instruments; typically ~5–10 μm). The field of view at the sample depends on the magnifying optics; some examples for 64 × 64 arrays are 11 × 11 mm [66] (176 μm square pixels), 4 mm × 4 mm (63 μm square pixels) and 500 μm × 500 μm (7.8 μm pixels). The field of view must be appropriate to the problem at hand. Sometimes individual cells, or small groups of cells, are of interest, and high spatial resolution is required [58]. In other applications, it is necessary to image a larger tissue sample [66]. The field of view can be effectively increased by using a motorised sample stage and tiling several images, allowing a large sample to be imaged at high resolution. A synchrotron source can be used with a point detector, providing better signal-to-noise ratio with small apertures [65].

Tissue Sampling

All the studies discussed here have used thin sections (4–10 μm) cut from a larger sample by microtome. The section is then mounted on either an IR-transparent substrate (e.g., CaF_2, BaF_2) for transmission measurements or an IR-reflecting substrate (glass slide coated with silver and tin oxide, or *low-E slide*) for reflection–absorption measurements (less common, but apparently gaining in popularity) [66–69]. One paper discussed tissue microarrays [58] in which core biopsies (500-μm diameter) from many samples are implanted in a wax block, which is then sliced. This can be thought of as a high-throughput approach to microscopic tissue imaging.

It is common for parallel sections (another thin slice adjacent to the IR sample) to be taken and mounted on glass slides for standard histological analysis, in which contrast is provided by various kinds of staining. When low-E slides are used, the same sample may be used for IR and visible-light examination, since these slides are transparent to visible light.

Many studies use samples that have been snap-frozen prior to sectioning; others use samples that have been archived in wax (paraffin) blocks (these are more readily available, since hospitals often keep them). Paraffin-embedded samples are usually de-waxed with xylene prior to analysis.

Ó Faoláin et al. compared the effects of various physical and chemical treatments [70] and concluded that tissue degradation as a result of freezing or wax-embedding can lead to spectral differences. It may be challenging to create protocols that can deal with multiple different sample treatments.

While sectioning is necessary for transmission or reflection–absorption measurements, reflectance and ATR methods can potentially be used to measure the spectra of fresh samples (possibly even in vivo with fibre optics), but reflectance may be more prone to artefacts and instrumentation for ATR studies may require customisation.

FTIR imaging has allowed analysis of the same specimen used for histopathological evaluation, as described in Table 5.1. Breast cancer tissues were analysed to demonstrate that chemical changes taking place in biological tissue can be identified precisely and the results are reproducible.

The analysis of natural tissues and biomaterials by spectroscopic means has had great impact on the development of specific structure–property relationship in materials. For biomedical applications, the composition and structure of these materials is of great importance since interactions with biological environments has a significant impact in predicting long-term behavior.

As described earlier, FTIR spectroscopy is a widely used tool for structural and compositional analysis of natural materials due to its relatively simple, nondestructive capability and allows analysis of characteristic group frequencies in a spectrum that provides qualitative estimates of chemical composition in natural materials. Structural factors such as linear chains, branching, or cross-linking can also be measured. In addition, the nature and quantity of any imperfections can also be determined.

With new technological advances in the field of spectroscopy, applications of FTIR spectroscopy have increased a great deal and numbers of new clinical studies reported in the literature are increasing at a rapid pace due to the advances in FTIR and perhaps more importantly due to the better understanding and interpretation of the spectral data. FTIR rapid scan imaging is a prime example, where a cancer sample, as described earlier, can be effectively imaged and the chemical distribution of various components extracted from the image in a very short time span. It is well-established fact now that cancer is a genetic disease resulting from various mutations in specific genes. These mutations, which may either abrogate gene function or increase gene function, drive the cell toward the unregulated and irreversible cellular proliferation that is the hallmark of cancer. Over the last 15 to 20 years, considerable progress has been made in defining and understanding these changes at the molecular level. There is a need to understand the chemical structure of cancer tissue and compare it with healthy tissue to monitor the chemical changes precisely and accurately. This will help to understand the chemical pathways to cancer formation, as FTIR rapid scan imaging has made it possible to obtain the visual image and relate it to the chemical structure profile of the samples.

Spectroscopic Analysis

Infrared spectra were obtained using a Digilab rapid-scan FTIR spectrometer Excalibur FTS 3000 with the UMA-600 infrared microscope equipped with a liquid nitrogen–cooled MCTA detector and a mercury cadmium telluride (MCT) focal plane array detector. The samples were mounted on glass slides and collected using infrared reflectance techniques. A gold mirror was used as a background. Spectra were collected at 8 cm^{-1} resolution, averaging 256 scans. A variety of techniques, such as imaging and mosaic images, were employed for samples with areas varying from 350 × 350 µm to a maximum area of 1.4 × 1.4 mm. The collection time for a single image was 5 minutes with 256 scans. For the larger image at 64 scans, the mosaic was collected in 17 minutes.

The infrared spectra exhibited signs of specular reflectance, which were indicated by restrahlen bands in the spectral region 2000–900 cm^{-1}. The spectra were converted to percentage reflectance and the Kramers-Krönig correction applied to correct for specular reflectance.

A spectral region is defined and the shift of the centre of gravity of this peak relative to a baseline can be imaged. In the image presented in Figure 5.17, the red region of the image illustrates the greatest shift and the blue region shows the minimum shift. The image in Figure 5.17 was collected using a 64 × 64 MCT focal plane array and was collected at 8 cm^{-1} resolution. This technique was used for imaging all the biological tissue provided.

Strong absorption bands in the region of 1700 to 1600 cm^{-1} arising from C=O vibrations of the amide groups of polypeptide chain, tend to be among the strongest and the intensity of the carbonyl band makes the detection of this

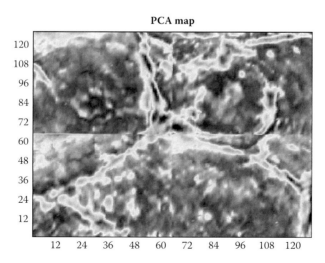

PCA map

FIGURE 5.17 (See colour insert.)
Red region of the image illustrates the area that gives rise to greatest shift and the blue region shows the minimum shift in the IR spectrum.

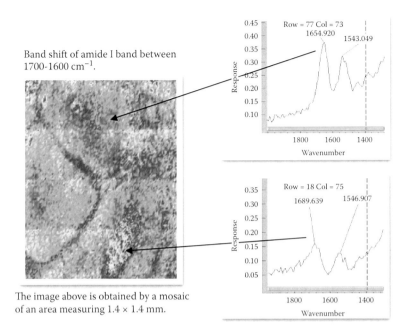

Band shift of amide I band between 1700-1600 cm^{-1}.

The image above is obtained by a mosaic of an area measuring 1.4 × 1.4 mm.

FIGURE 5.18
FTIR spectra of cancerous tissue, amide I band of cancer, and normal tissue.

group simpler. The peak centred at 1632 cm^{-1} is due to bonded C=O and the band at 1538 cm^{-1} confirms the presence of C–N–H moieties, see Figure 5.18.

Monosubstituted amides usually exist with the –NH and C=O bonds. The hydrogen bonded stretch is seen near 3290 cm^{-1} with a weaker band near 3070 cm^{-1} assigned to a Fermi resonance–enhanced overtone of the 1536 cm^{-1} band. This band at 1536 cm^{-1} involves both C–N stretch and C–N–H in-plane bend in the amide II band. This band is a characteristic of monosubstituted amides. Amide III bands absorb more weakly and are confirmed by the presence of peaks at 1310 and 1235 cm^{-1}. A band centred at 1160 cm^{-1} is due to C–O stretching modes of the C–OH of cell proteins and the C–O group of carbohydrates.

Spectral bands in the region of 1700 to 900 cm^{-1} arise from C=O, CH$_2$, CH$_3$, C–O–C, and O–P–O groups, confirming the presence of phospholipids, proteins, carbohydrates, collagen, and amino acids. Differences between normal and cancerous tissues spectra are observed. Spectra of normal and breast cancer tissue are complex and possess well-defined and prominent spectral bands at the 1700 to 700 cm^{-1} and 3500 to 2700 cm^{-1} regions. Difference in intensity and positioning of peaks in the spectra are attributed to the compositional changes between the normal and cancer tissues.

The intensity increase of the C–H peaks may be attributed to the increase in the lipids, proteins, and DNA contents indicating the increment of fatty acids. Similarly, increase in intensity and shifting of peaks in the area of 1700 to 1500 cm^{-1} (carbonyl stretching and amide-bending vibrations in the DNA) confirm the changes in the chemical structure of normal tissue. An overall shift of peaks in the fingerprint region (1700 to 800 cm^{-1}) is observed in the spectrum of breast cancer tissue. Figure 5.18 shows the shift in these bands as characterised by an infrared image of the amide I band between 1700–1600 cm^{-1}. The deeper the red colour, the greater the infrared shift in this spectral region.

Useful information pertaining to carbonyl stretching and amide-bending vibrations in the DNA can be extracted. This data can be extremely useful in defining infrared images that show the shift of amide bands between 1700–1500 cm^{-1}. The images extracted show that this shift is significant, sometimes varying by as much as 20 cm^{-1}. In addition, principal components analysis (PCA) can be applied to the images to extract further data producing images based on the principal components.

Data Handling Requirements and Spectral Pretreatments

A single image usually consists of $64 \times 64 = 4096$ spectra, each consisting of measurements at several hundred to a few thousand wavelengths. Assuming 1000 wavelengths, there are roughly 4 million data points in one image. This equates to about 30 MB. Many images may be tiled together to image a larger sample, and when images from many specimens from many samples are combined, datasets can be several GB. Computations (particularly hierarchical clustering) on the dataset as a whole can then be challenging, and data-reduction methods (such as PCA or clustering) may be applied to individual images prior to analysing the data set as a whole.

Sometimes, the spectra are analysed without any pretreatments [67]. More commonly, however, varieties of numerical treatments are used to reduce the influence of artefacts. First [71,72] and second [68] derivatives are used to reduce baseline effects and to enhance small features. An alternative approach to removing baselines is to subtract a linear function directly [59,63]. Often, some kind of scaling is used as well, such as an absorbance ratio [59,61,63] (often applied to the amide I band or to the difference between the minimum and maximum absorbance values) or the standard normal variate (SNV) transform [60,68], in which each spectrum is scaled to have zero mean and unit variance. The purpose of scaling is to emphasise differences in the shapes, rather than intensities, of the spectra, and it is particularly justified when there are variations in sample thickness due to imperfections in sample preparation.

Spectra are usually screened for certain quality criteria such as having acceptable absorbance levels or signal-to-noise ratios. Liquid nitrogen–cooled FPA cameras are apparently prone to developing pixel defects due to thermal cycling, and these pixels should be identified and excluded from the analysis.

It is also common to select a wavenumber range that is known or suspected to pertain to the problem at hand, rather than using the entire spectrum. The most commonly used range is the "fingerprint" region (~1000–1800 cm^{-1}); the C–H stretching region (~2800–3000 cm^{-1}) has also been used.

Classification Approaches and Algorithms

Most studies involve assigning a particular tissue type to every pixel in the image via some kind of classification algorithm. K-means [60,74], fuzzy C-means [59], and hierarchical clustering using Ward's method [59,68,71,72] are popular unsupervised methods. Lasch [74] compared all three and reported that hierarchical clustering was the most effective. In all of these methods, the clusters are generated automatically, and then the resulting colour-coded images are compared to the visible-light image of the same sample. By searching for correspondences, clusters are assigned to tissue types. These studies have focused on establishing the ability of infrared microspectroscopy to discriminate between different kinds of tissue, but have not looked at its use for actual diagnosis.

Studies that have attempted to use FTIR-MS imaging for diagnosis have done so by using supervised classification methods, with a training set that has been manually classified by a histological analysis. Several algorithms have been used, including linear discriminant analysis (LDA) [61,62], artificial neural networks (ANNs) [71,72], soft independent modelling by class analogy (SIMCA) [63], and a fast Bayesian classification scheme [58,64]. These studies seek to develop predictive models that can be used to classify future samples. In general, it seems to be relatively easy to distinguish between tumour tissue and other types of tissue, but harder to distinguish between different types of tumours [61]. However, there have been some examples of successful tumour classification [63,72].

In some cases, PCA was used for data reduction prior to clustering [60,73]. In terms of data handling, this does not alleviate the main obstacle to clustering, which is the large number of objects (spectra). Clustering based on PCA scores gives very similar results to using the raw spectra, unless very few principle components (PCs; ~4 or fewer) are retained.

Some studies have used band-fitting approaches [57]. These are probably quite labour-intensive and may be better suited for analysis of sample structural features within single images rather than for automated classification.

Fabian et al. [72] used a two-stage ANN method, in which the first neural network classifies a spectrum as being cancer or noncancer, and the second network further classifies the cancer spectra. This approach is sensible because the differences between cancer and noncancer are greater than the differences among the different grades of cancer.

There are a number of factors that need to be considered for the set of data reported here.

First, the spectra were measured in reflectance mode, whereas a number of other researchers have used either transmission or reflection–absorbance (reflectance of a thin film mounted on a mirror), which is essentially equivalent to transmission. In theory, the reflectance spectrum should contain the same information as the transmission spectrum. However, reflectance may be more prone to artefacts due to scattering. More importantly, it is hard to compare reflectance spectra to transmission spectra because the peaks are shifted, and not by a constant amount. Using reflectance instead of transmission is justified, however, if there is a significant advantage in doing so; for example, does it reduce sample preparation time or enable analysis of important samples that cannot be sectioned thinly enough for transmission? If so, however, ATR would probably give the same benefits and would also be more comparable to transmission data.

Second, the data reported here was run blind, and did not have detailed histological information. In most of the studies reported in the literature, samples have been analysed by a pathologist using light-microscopy techniques. This provides *Y-block* information such as which regions of the sample correspond to which kinds of tissue, and greatly facilitates the development of classification models. The only Y-block information available in this study was the overall tumour grade of the sample. In principle, it was only possible to compare the obtained spectra (e.g., the cluster centroids) to spectra from the literature to try to identify the tissue, but this may be extremely difficult with reflectance spectra.

Obviously, a method that does not require further histological analysis has advantages in terms of data acquisition time. But the challenge is to make effective use of the imaging aspect of the data.

Analysis of the Reflectance Spectral Data

Structure of the Data

Table 5.4 lists the number of samples of different types of breast cancer. In total, there were 38 samples, including 7 DCIS samples and 30 IDC samples. There is only one sample of healthy or normal tissue.

TABLE 5.4

Sample Numbers Inferred from the File Names

Tissue		Number of Samples
Healthy		1
Ductal carcinoma in situ	Low nuclear grade	2
	Intermediate nuclear grade	1
	High nuclear grade	4
Invasive ductal carcinoma	Grade I	6
	Grade II	14
	Grade III	10

Each sample has an infrared image consisting of $64 \times 64 = 4096$ spectra over the wavelength range 900–4000 cm^{-1}. The spectra were originally recorded in absorbance mode, that is,

$$A = -\log10(R)$$

where R is the reflectance relative to a gold background. This transformation may not be correct for specular reflectance spectra; therefore, transform these back to reflectance mode. One transformation that can be useful is the Kramers-Krönig transform (KKT). This calculates the imaginary part of the refractive index, which can be compared more easily to absorbance spectra.

Spectra of the three grades of IDC samples are presented in Figure 5.19, measured in three ways. The reflectance and reflectance-KKT spectra were obtained by choosing a sample of the appropriate type and taking the average of all its spectra. A number of points of comparison can be made:

- There are differences between the three grades of cancer in the ATR spectra and these differences are reproducible. The full dataset from that paper would probably be amenable to quite a straightforward statistical classification analysis, and would be large enough to provide reasonable validation.

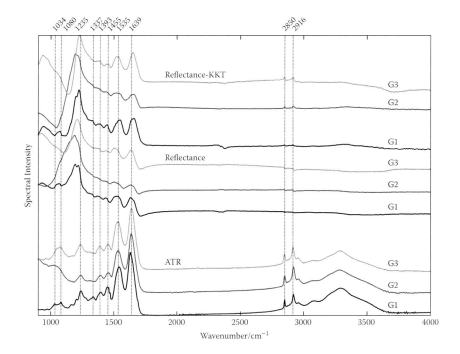

FIGURE 5.19 (See colour insert.)
Spectra of three grades of IDC samples.

- Note the distorted (sigmoidal) bands in the reflectance spectra, particularly visible as the dip in reflectance to a higher wavenumber of the amide I band at ~1639 cm^{-1}. In theory, this effect should shift the band maxima to a lower wavenumber (by a few tens of wavenumbers), but this does not seem to be the case here, as the positions are very similar to the ATR positions. The amide II peak (~1535 cm^{-1}) is shifted to a lower wavenumber, as expected. In the KKT spectra, the amide II peak is in a position similar to the ATR spectra.

- Overall, it can be stated that the KKT transform does not make a consistent improvement in the spectra, so the best option is to use the reflectance spectra.

- There are substantial differences between the ATR and reflectance spectra at wavenumbers lower than about 1350 cm^{-1}.

Analysis of Individual Images

There are a number of methods that can be employed to analyse individual and complex images. In this section, these methods are described using the samples described in Table 5.4.

Single-Wavelength Images

To reduce the effects of baselines, the second derivative of the spectra can be been taken. One way to visualise the images is to do false-colour plots based on the intensity at particular wavenumbers corresponding to peak maxima (which become minima in second-derivative spectra).

Using one of the spectrums as an example, nine peaks can be chosen from the mean spectrum (Figure 5.20). The second derivative intensity at these

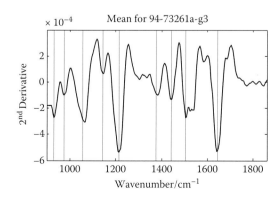

FIGURE 5.20
Peaks selected for false-colour images.

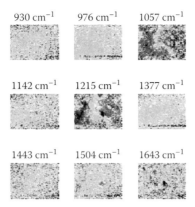

FIGURE 5.21 (See colour insert.)
False-colour images based on the second-derivative intensity at the indicated wave numbers. Blue indicates high intensity (large negative second derivative) while red indicates low intensity.

wavenumbers can then be plotted as images (Figure 5.21). These plots reveal a few interesting features.

- The peaks at 1057 and 1215 cm^{-1} clearly contrast two regions of the sample. They appear to be complementary to each other: high intensity at 1057 cm^{-1} is correlated to low intensity at 1215 cm^{-1}. This implies a significant physical difference between these two regions of the sample.

- There is a horizontal line of pixels near the bottom of the image that stands out, particularly at 976 cm^{-1}. This is presumably due to a defect of the detector (it appears in the images of other samples as well). Specifically, it affects rows 5–8. These rows appear to be noisier than the rest of the pixels. A plot of the Root Mean Square (RMS) noise levels presented in Figure 5.22 confirms this, along with a vertical line toward the right side and a few other isolated noisy pixels.

- There also appears to be a small region near the centre of the image that has quite low intensity for most bands (e.g., 1643, 1504, 1142, 930 cm^{-1}).

The mean spectra for the left 20 columns and right 20 columns of the image are plotted in Figure 5.23.

It is also possible to look at the frequency distributions for the intensity at these wave numbers.

The distributions for 1057 and 1215 cm^{-1} are clearly bimodal, corresponding to the two regions identified in Figure 5.24. Most of the other bands appear fairly normally distributed.

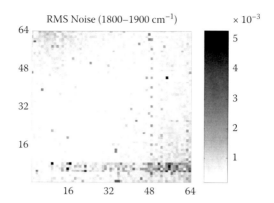

FIGURE 5.22
Noise levels along with a vertical line toward the right side and other isolated noisy pixels.

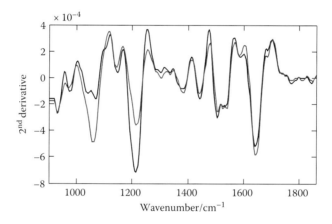

FIGURE 5.23 (See colour insert.)
The blue spectrum is from the left side and the green spectrum is from the right. The difference in the relative intensities at 1057 and 1215 cm^{-1} is obvious.

Principal Components Analysis (PCA)

False-colour images can also be generated by plotting the principal component scores. From Figure 5.25, it can be seen that pixels in this region have quite low intensity for most bands.

An alternative method is to plot the scores on one PC against the scores on another and to look for groups that form. Some plots relating to this method and some analyses of several samples are presented in Figures 5.26 and 5.27. The two plots in these figures are of PCA scores (T1 vs. T2, and T2 vs. T3). From these, some samples were identified as forming clusters, and were investigated further.

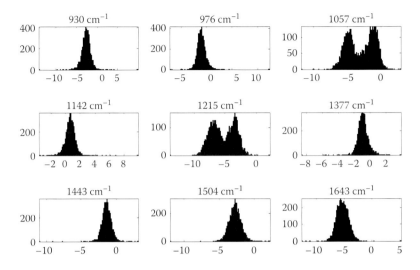

FIGURE 5.24
Histograms for the second-derivative intensity at the indicated wavelengths. The x axis units are arbitrary.

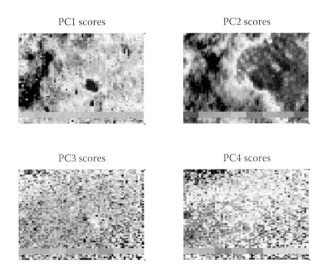

FIGURE 5.25 (See colour insert.)
The second principal component (PC2) provides the same contrast as the difference between the bands at 1057 and 1215 cm^{-1}. The first PC identifies a small region near the centre of the image.

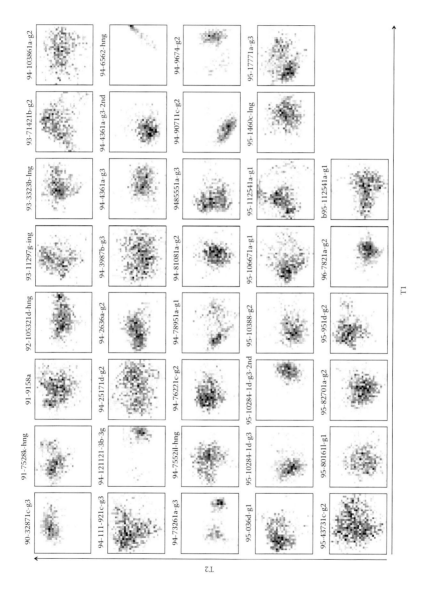

FIGURE 5.26
Scores on PC1 versus scores on PC2 for all the images.

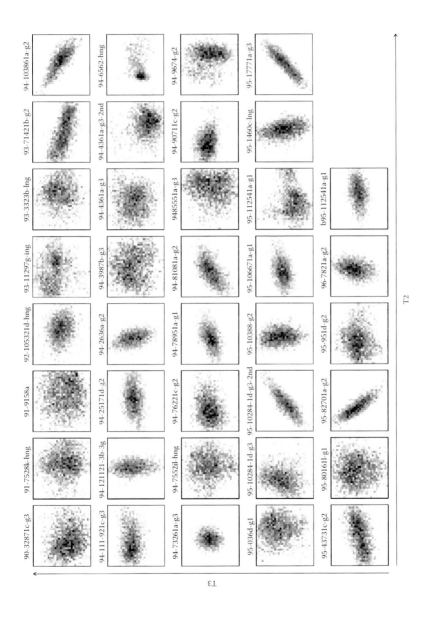

FIGURE 5.27
Scores on PC2 versus scores on PC3 for all the images.

The samples of GI, GII, GIII, and HNG types were chosen for analysis. In each case, the steps were as follows:

1. Average in 2 × 2 squares to reduce size and reduce noise
2. Second derivative via Savitzky-Golay filter with mild smoothing
3. Selection of wavelength range ~1000–2000 cm^{-1}
4. Principal component decomposition using singular value decomposition (SVD)
5. Selection of clusters from score plots
6. Spatial plot of clusters and mean cluster spectra

IDC Grade III

This section will help in understanding how calculations are carried out and the way these help in performing statistical analysis. The plot in Figure 5.28 of the singular values suggests that two PCs explain a large percentage of the variance in the spectra. The PC1/PC2 score plot suggests two clusters and is presented in Figure 5.29. The clusters of these two PCs are more obvious if the data are plotted as a histogram instead, as described in Figure 5.30.

It is interesting to see whether these clusters correspond to different spatial regions of the sample. To test this, points in rectangular regions around the samples were selected and their spatial (x, y) coordinates calculated and plotted (Figure 5.31).

The blue asterisks correspond to the upper cluster in step 5 above, the green triangles to the lower one, and the red circles to the one on the far right. It can be seen that the blue points are generally on the right side of the image (large x) while the red are on the left.

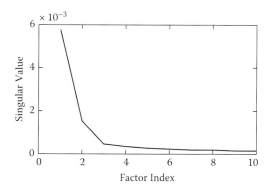

FIGURE 5.28
Two PCs explaining a large percentage of the variance in the spectra.

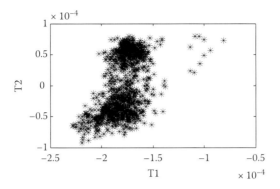

FIGURE 5.29
Two PCs (PC1/PC2) score plot.

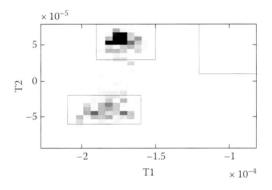

FIGURE 5.30
The clusters of these two PCs.

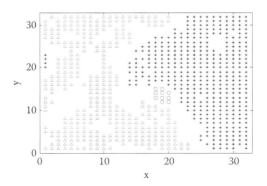

FIGURE 5.31 (See colour insert.)
Rectangular regions around the samples; spatial (x, y) coordinates calculated and plotted.

Since the main difference between the clusters is in the values of the PC2 score and the x coordinate, plotting these should further illustrate the trend.

In another example of GIII, the same preprocessing outlined in steps 1–6 was used. The SVD indicates that there are three factors that explain most of the data. The histogram for T1 and T2 is presented in Figure 5.32.

The Matlab function inpolygon was used to find points in polygonal regions containing the clusters, and the clusters are plotted with their (x, y) coordinates given in Figure 5.33. Again, the clusters are evident in the spatial coordinates as well as the PC coordinates. Plotting spectra averaged at $x = 5$ and $x = 25$ (Figure 5.34).

The same kinds of differences at ~1000 cm^{-1} and ~1200 cm^{-1} can be observed in Figure 5.35. The plot of all the spectra shows the differences as well in a bimodal distribution for T2 presented in Figure 5.36.

A plot some of the original spectra to see what changes in the spectra are responsible for this behaviour is given in Figure 5.37. The easiest way to do

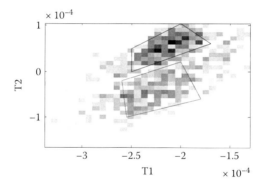

FIGURE 5.32
Histogram for T1 and T2.

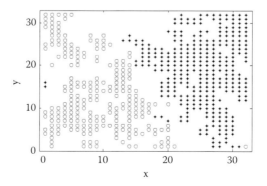

FIGURE 5.33
Polygonal regions containing the clusters, and the clusters are plotted with their (x, y) coordinates.

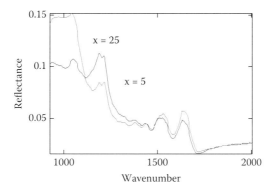

FIGURE 5.34
Spatial and PC coordinates plotting average spectra.

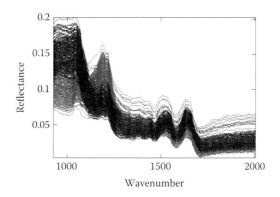

FIGURE 5.35
Differences at specific spectral bands, ~1000 cm^{-1} and ~1200 cm^{-1}.

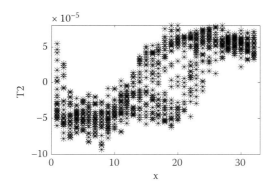

FIGURE 5.36
A bimodal distribution for T2.

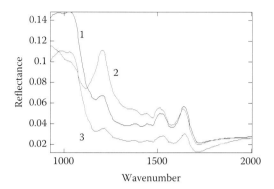

FIGURE 5.37
A plot of some of the original spectra (wavenumber cm^{-1}).

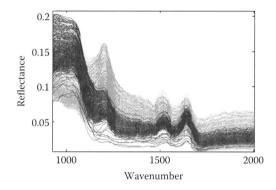

FIGURE 5.38
Differences at spectral bands between ~1000 cm^{-1} and ~1200 cm^{-1} in all the different grades of spectra.

this is to average over all the spectra in each cluster. Plotting all the spectra, colour-coded by cluster, is presented in Figure 5.38, indicating that the spectra are well represented by the means.

There are occasions where second-derivative spectra can provide a wealth of information. The large bands at ~1000 cm^{-1} and ~1200 cm^{-1} are the main difference among the three spectra. The spectra for clusters 1 and 3 are quite similar, but differ in intensity. This is clearer in the second-derivative spectra as shown in Figure 5.39.

IDC Grade II

Grade II does not display any obvious clustering in T1 versus T2. However, there is slightly more evidence for it in T2 versus T3; both of the clusters are given in Figure 5.40. The spatial coordinates of points in these three clusters are plotted in Figure 5.41. Where blue asterisks are the largest (bottom) cluster,

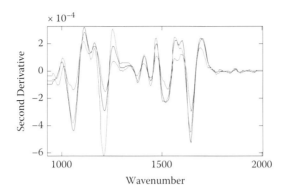

FIGURE 5.39
GIII: second-derivative spectra for clusters 1 and 3, bands at ~1000 and ~1200 cm^{-1}.

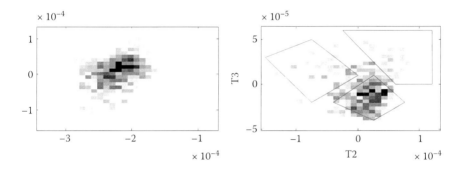

FIGURE 5.40
GII: clustering for T1 and T2 and T2 and T3.

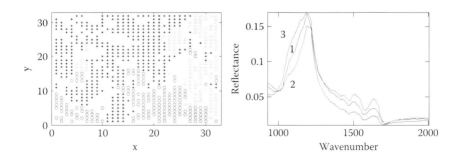

FIGURE 5.41 (See colour insert.)
GII: spatial coordinates of points in three clusters.

red circles are the cluster on the left, and green triangles are the cluster at the top right. Again, points with similar score values are likely to have similar spatial coordinates.

Since the differences are not found in the first two factors (but require both the second and third), the differences in the spectra are expected to be smaller than in the previous examples, and this is indeed the case.

Presumably, this means that the chemical differences in the tissue types are not as extreme as in the case of GIII, indicating that the spectral differences among the spectra of GII are not that apparent.

IDC Grade I

In the case of the Grade I type of the cancer, the histogram as presented in Figure 5.42, shows two clusters that are not strongly separated. Once again, the two clusters correspond to the left and right halves of the sample (blue asterisks = cluster 1, at top; red circles = cluster 2, at bottom), polygonal regions containing the clusters, and the clusters are plotted with their (x, y) coordinates in Figure 5.43.

Plotting x against T2 shows a smooth transition, in keeping with the poor differentiation of the clusters and suggesting a continuous change in tissue type. Plotting the original spectra averaged over each cluster shows the same trend as previously seen. Plotting all of the spectra, instead of just the averages, shows that the averages are a good representation of all the spectra. If the cluster edges are chosen to be closer together, the separation in the spectra is not that great. This can be demonstrated by the histograms presented in Figure 5.44.

DCIS High Nuclear Grade (HNG)

This sample has an unusual histogram for T1 and T2 and the three clusters correspond to horizontal bands across the sample as shown in Figure 5.45.

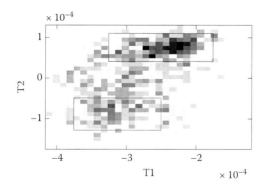

FIGURE 5.42
GI: clustering histogram for T1 and T2.

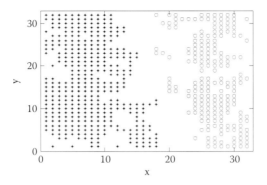

FIGURE 5.43 (See colour insert.)
GI: regions containing the clusters, and the clusters are plotted with their (x, y) coordinates.

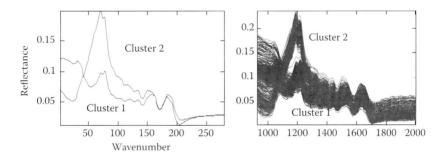

FIGURE 5.44
GI: a plot of some of the original spectra (wavenumber cm^{-1}).

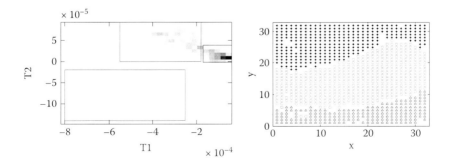

FIGURE 5.45
HNG, clustering histogram for T1 and T2, and the clusters plotted with their (x, y) coordinates.

The HNG spectra appear to differ more in intensity than in shape, which is clear from Figure 5.46.

Healthy Normal Tissue

The histogram for normal tissue indicates two poorly separated clusters and there is a spatial basis to the clustering as presented in Figure 5.47. There appears to be a significant difference in the spectra at about 1000 cm^{-1}, but plotting the second-derivative spectra reveals that this is mostly due to the sloping baseline rather than any differences among the samples (Figure 5.48).

These differences could be due to variations in the tissue, but they are certainly not as dramatic as those observed for some of the earlier samples of GI, GII, GIII, and HNG. These spectra, being representative of healthy tissue, may be helpful in assigning clusters found earlier as either healthy or cancerous.

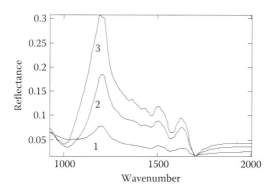

FIGURE 5.46
HNG: a plot of some of the original spectra (wavenumber cm^{-1}).

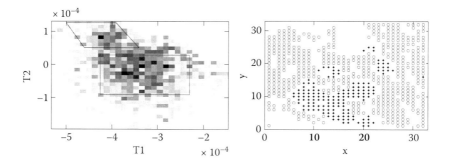

FIGURE 5.47
Normal: clustering histogram for T1 and T2 and the clusters plotted with their (x, y) coordinates.

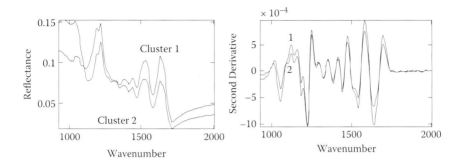

FIGURE 5.48

Normal: a plot of some of the original spectra (wavenumber cm^{-1}) and second-derivative spectra for clusters 1 and 3, bands at ~1000 and ~1200 cm^{-1}.

Cluster Analysis

Another way to generate false-colour images is through clustering. K-means or hierarchical clustering can also be used to separate the spectra automatically into a specified number of classes. The example given in Figure 5.49 compares K-means and hierarchical clustering (Ward's method) into 2, 3, and 4 clusters.

With two clusters, the methods give fairly similar results and correlate well with the single-wavelength and PCA images plotted previously. For three clusters, the results are somewhat different. The mean spectra for each cluster (Figure 5.50) can also be calculated, which can potentially help with assignment.

A number of researchers have used different protocols to analyse cancer tissues. Steller et al. [59] used IR microspectroscopic imaging to identify different types of tissue in squamous cell cervical carcinoma samples. Many images (a total of 122) were collected in absorbance mode from a 10-µm sample on a CaF$_2$ slide. They used fuzzy C-means clustering on the full dataset, then used the cluster centroids (average of all the spectra in a cluster) as input to an hierarchical clustering algorithm (which would have been too time-consuming with the full dataset), which they used to generate clusters equal in number to the number of different types of tissue they expected (based on light-microscopic analysis of a parallel sample). Their basic goal was to generate a false-colour image showing the physical structure of the sample in terms of tissue types. They mention that the long-term goal is to "calibrate" the method so that it can be used on new samples without the parallel histological analysis.

Zhang et al. [73] worked with breast-cell suspensions rather than tissue samples. The suspensions were spread onto BaF$_2$ plates and measured in transmission mode. They used PCA (8 factors) for data reduction prior to K-means clustering (i.e., unsupervised) to divide the image into *cell* and *noncell* regions. They used an ANN to further divide the cells into cancerous and normal, with partial success. Their spectra looked slightly unusual.

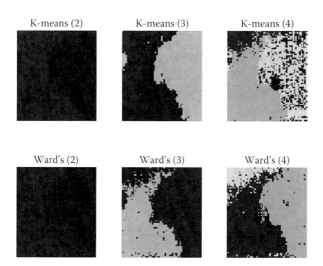

FIGURE 5.49 (See colour insert.)
K-means and hierarchical clustering (Ward's method) into 2, 3, and 4 clusters.

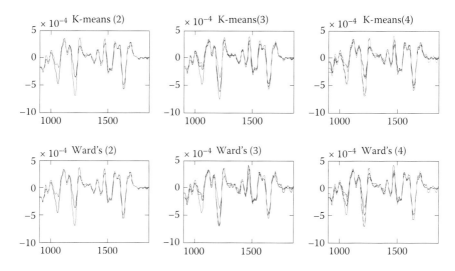

FIGURE 5.50
Mean spectra for each cluster.

Fabian et al. [71,72] present some studies comparing *mapping* (point-by-point) with *imaging* (FPA camera) for both benign and malignant breast tumours. The samples were 7-μm slices on BaF$_2$ windows, and the spectra were measured in transmission. They describe FTIR microspectroscopy as an imaging method that uses "infrared absorbance as an intrinsic contrast mechanism." [71] They used hierarchical clustering (with Ward's method for linkage) to generate false-colour images, which were compared with the conventional microscopic images.

A special issue of *Vibrational Spectroscopy* devoted to the conference Shedding Light on Disease: Optical Diagnostics for the New Millennium included several papers dealing with infrared microspectroscopic imaging (including those described in the next three paragraphs).

Beleites et al. [61] used FTIR microspectroscopic imaging to distinguish among different types of gliomas (brain tumours). Ten-micron slices (area ~0.5 cm²) were cut from frozen samples and mounted and dried on CaF₂ windows. Parallel slices were mounted and stained on slides for histological evaluation. They used linear discriminant analysis (on the whole dataset of 59 samples) in conjunction with a genetic algorithm for wavelength selection. The model was assessed by cross-validation. The classification for each sample was obtained by aggregating the predictions for its individual spectra. Their method was able to distinguish between cancer and noncancer with ~95% accuracy, but the prediction of the grade of cancer was much worse.

Romeo and Diem [68] used a different sampling method. To reduce costs, they used reflection-absorption spectroscopy with thin samples (6 µm) mounted on silver-coated glass microscope slides (low-e slides—cheaper than IR windows). This approach can result in reflectance artefacts (superimposed dispersion-shaped bands in the spectra), particularly around the edges of the sample where it may not adhere as well to the slide. They propose an interesting (and very simple) correction method that appears to work quite well [69].

Ó Faoláin et al. [70] discuss the effects of tissue processing. They compared spectra from fresh frozen samples to those from wax-block archived samples using both Raman and FTIR spectroscopy (not imaging). They list sample types as: fresh, frozen, air-dried, formalin-fixed, and de-waxed formalin-fixed paraffin preserved. They found that there are changes in the spectra depending on the treatment: bands due to formalin/wax were evident even after standard de-waxing procedures. In addition, there were more subtle changes in the spectra, for which they provided some partial explanations.

Amharref et al. [60] studied rat brain tumours with FTIR microspectroscopic imaging. Cryosectioned samples were mounted on ZnSe windows. First derivative and SNV preprocessing was used, and K-means clustering. They were able to distinguish between several different kinds of brain tissue based on the infrared spectra. In fact, the infrared map contained information equivalent to several separate samples stained with different methods, but they do not present any cross-sample results as described by Beleites et al. [61].

Fabian et al. [72] used IR microspectroscopy to distinguish between benign (fibro-adenoma, FA) and malignant (DCIS) breast tumours, as well as adipose and connective tissues. They had a large set of spectra from 22 patients, and spectra were classified histologically to provide the Y-block information. An ANN was used for classification, and worked fairly well. Initially, there was some misclassification among the tumour types. The ANN was replaced with two ANNs: one to classify spectra as either cancer or not cancer, and another to classify the cancers as either FA or DCIS. This improved the results substantially, but there was one sample from which most of the spectra were misclassified, highlighting the important problem of interpatient variability.

Krafft et al. [63] used SIMCA classification of FTIR microspectroscopic images to identify the primary tumours of brain metastases. They had 20 samples, split 5/15 into training and validation sets. There was one training sample for each type of primary tumour. They used min-max normalisation to account for thickness differences. The reclassification of the training data was >90% for each sample. The classification of the independent samples was quite good, with all samples but one having >50% of their spectra correctly classified; the average correct classification percentage was 79%.

Bhargava et al. and Fernandez et al. [58,64] present results of a study on *tissue microarrays*, arrays formed by embedding small core biopsies of a sample in a wax block, which is then microtomed and mounted on a slide (transparent window, but they mention reflective slides as well). Their samples were of prostate tissue. They suggest that standard chemometric data processing and classification–clustering algorithms are too slow for the large datasets produced in their study. They use a different approach based on Bayesian classification of spectral metrics. The *spectral metrics* are things like peak positions, widths, and heights, and are isolated from representative spectra manually by a spectroscopist. This appears to be simply a way to reduce the number of variables while retaining specific, local information. The set of spectral metrics is further reduced, algorithmically, to only the useful ones. They defined the classification task as determining the probability that an unknown spectrum originated from a particular class of tissue. Their classification approach is similar to LDA, but utilises Bayes's theorem to derive discriminant functions automatically. They state that the approach relies heavily on having a very large dataset available, since the weights in the discriminant functions are overly determined by the underlying distributions of components.

Bambery et al. [66] studied a malignant glioma tumour in a rat brain via FTIR imaging. They used reflection-absorbance. The sampling area covered nearly the entire rat brain section (11 mm × 11 mm; 4-μm thick), and the tumour was a few milimeters in diameter. Hierarchical clustering was easily able to identify the tumour region. Comparison was made to a healthy rat brain and to histological examination of the brain with the tumour.

Summary

To summarise, both the Raman and FTIR spectra are well known for their sensitivity to composition and three-dimensional structure of biomolecules. The biochemical changes in the subcellular levels developing in abnormal cells, including a majority of cancers, manifest themselves in different optical signatures, which can be detected by spectroscopy. In this research, this technique was employed to investigate whether morphological changes in DCIS and IDC grades translate into compositional differences.

The results obtained in this study have demonstrated that it is possible to identify differences between the normal and the cancerous tissue accurately, and also differentiate between the different nuclear grades of DCIS and IDC with confidence. The differences in the intensity and positioning of peaks in the spectra are attributed to the compositional changes between the different tumours. The variation in intensity of these peaks is due to the variation in the amount of lipids, proteins, and DNA contents present in the breast carcinoma.

Currently a number of clinical diagnostic techniques, such as imaging by computed tomography (CT) and MRI are widely used. However, it is extremely difficult to obtain chemical structural information in vivo. There is a need to develop a clinical diagnostic technique that can offer chemical analysis of the tissue, thus helping clinicians to not only detect the cancer, but its type and classification of nuclear grades.

A standardised method of analysis using spectroscopy could only be achieved by consensus of pathologists and spectroscopists. A combination of both would further enhance the diagnosis by removing some of the interobserver variation of individual pathologists. Although the technique is quite specialised, a comprehensive computer database, minimal training, and expertise will be required.

Spectroscopy appears to be a valuable technique in the armoury of pathologists for clinical diagnosis of cancerous tissue. Microspectroscopic methods can be applied to breast tissue samples mounted on slides. This is a nondestructive technique and even after the spectroscopic analysis, samples can be used for further investigations. The technique could also be valuable in studying archived samples, ultimately helping in the continuous monitoring of the patients for years. Spectroscopy could be a valuable clinical evaluation system for regular monitoring of the effects of therapies (chemo- and radio-therapy) and to assist in predicting the type and progression of the disease.

Future Work

Some interesting ideas could be introduced to carry out the next steps of this research:

- The technique could be tried on fresh tissue. It would be extremely useful to compare the spectral data both from fresh and embedded tissues because comparable results in both these type of cases would mean that prospective studies could be carried out.

- Spectroscopy can be applied to living tissue (during surgeries or biopsies, for instance).

- A comprehensive database could be made using spectral information received from similar research on larger numbers of samples. This database can be combined with statistical methods, and an automated instrument in the form of software can be established. This can be used as an alternative in laboratories, and as no professional knowledge is required, the method can easily be used by clinicians and technicians. In other words, the spectra of different tissue samples are given to the software, and the software determines whether the sample is normal tissue or a malignancy. The same method can be used for diagnosing different grades of breast malignancies (DCIS and IDC). We believe this idea can revolutionize spectroscopy and its application in studies related to biological samples.

References

1. Various Contributors, *The Health of the Nation.* Report, 1993.
2. Eles, R. A., and Ponder, B. A. J. 1992. Familial breast cancer, RSM current medical literature. *Breast and Prostate Cancer* 5, 67–70.
3. Kelsey, J. L., and Berkowitz, G. S. 1998. Breast cancer epidemiology. *Cancer Research* 48, 5615–5623.
4. Kalache, A., Maguire, A., and Thompson, S. G. 1993. Age at last full term pregnancy and risk of breast cancer. *Lancet* 341(8836), 33–36.
5. McPherson, K., Vessey, N., Neil, A., et al. 1987. Early oral contraceptive use and breast cancer: Results of another case control study. *British Journal of Cancer* 56, 653–660.
6. Van Leeuwen, F. E., and Rookus, M. A. 1989. The role of exogenous hormones in the epidemiology of breast, ovarian and endometinal cancer. *European Journal of Cancer and Clinical Oncology* 25, 1961–1972.
7. Steinberg, K. K., Thacker, S. B., et al. 1991. Meta analysis of the effect of oestrogen replacement therapy on the risk of breast cancer. *Journal American Medical Association* 265, 1985–1990.
8. Tobar L., Duffy, S. W., and Krusemo, U. B. 1987. Detection method, tumour size and node metastasis in breast cancers diagnosed during a trial of breast cancer screening. *European Journal of Clinical Oncology* 23, 959–962.
9. Bassett, L. W., Liu, T. H., Giuliano, A. E., and Gold, R. H. 1991. The prevalence of carcinoma in palpable versus impalpable mammographically detected lesions. *American Journal of Roentgenology* 157, 21–24.

10. Rosen, P. P., Senie, R. T., Farr, G. H., et al. 1979. Epidemiology of breast carcinoma: Age, menstrual status and exogenous hormone usage in patients with lobular carcinoma in situ. *Surgery* 85: 219–224.

11. Carter, C. L., Allen, C, and Henson, D. E. 1989. Relation of tumour size, lymph node status and survival in 24,740 breast cancer cases. *Cancer* 63, 181–187.

12. Day, N. E. 1991. Screening for breast cancer. *British Medical Bulletin* 47, 400–415.

13. McKinna, J. A., Davey, J. B., Walsh, G. A., et al. 1992. The early diagnosis of breast cancer: A 20-year experience at the Royal Marsden Hospital. *European Journal of Cancer* 28A, 911–916.

14. Nystrom, L., Rutqvist, L. E., Wall, S., et al. 1993. Breast cancer screening with mammography: Overview of Swedish randomized trials. *Lancet* 341, 973–978.

15. Orel, S. G. 2000. MR imaging of the breast. *Radiologic Clinics of North America* 38, 899–913.

16. National Co-ordinating Group for Breast Screening Pathology. 1995. *Pathology Reporting in Breast Cancer Screening*, 2nd ed. NHBSP Publications, no. 3.

17. Elston, C. W., and Ellis, I. O. 1991. Pathological prognostic factors in breast cancer: The value of histological grade in breast cancer: Experience from a large study with long term follow-up. *Histopathology* 19, 403–410.

18. Rehman, S., Movasaghi, Z., Tucker, A. T., Joel, S. P., Darr, J. A., Ruban, A. V., and Rehman, I.U. 2007. Raman spectroscopic analysis of breast cancer tissues: Identifying differences between normal, invasive ductal carcinoma and ductal carcinoma in situ of the breast tissue. *Journal of Raman Spectroscopy* 38, 1345–1351.

19. Rehman, S., Movasaghi, Z., Darr, J. A., and Rehman, I. U. 2010. Fourier transform infrared spectroscopic analysis of breast cancer tissues: Identifying differences between normal breast, invasive ductal carcinoma, and ductal carcinoma in situ of the breast. *Applied Spectroscopy Reviews* 45(5), 355–368.

20. Sigurdsson, S., Philipsen, P. A., Hansen, L. K., Laesen, L., Gniadecka, M., and Wulf , H. C. 2004. Detection of skin cancer by classification of Raman spectra. IEEE T. *Bio-Med. Eng.* 51, 10.

21. Eckel, R., Huo, H., Guan, H. W., Hu, X., Che, X., and Huang, W. D. 2001. Characteristic infrared spectroscopic patterns in the protein bands of human breast cancer tissue. *Vibrational Spectroscopy* 27, 165–173.

22. Dukor, R. K. 2002. Vibrational spectroscopy in the detection of cancer. *Biomed. Appl.* 5, 3335–3359.

23. Huang, Z., McWilliams, A., Lui, M., McLean, D. I., Lam, S., and Zeng, H. 2003. Near-infrared Raman spectroscopy for optical diagnosis of lung cancer. *International Journal of Cancer* 107, 1047–1052.

24. Lau, D. P., Huang, Z., Lui, H., Man, C. S., Berean, K., Morrison, M. D., and Zeng, H. 2003. Raman spectroscopy for optical diagnosis in normal and cancerous tissue of the nasopharynx: Preliminary findings. *Lasers Surg. Med.* 32, 210–214.

25. Huleihel, M., Salman, A., Erukhimovich, V., Ramesh, J., Hammody, Z., and Mordechai, S. 2002. Novel optical method for study of viral carcinogenesis in vitro. *J. Biochem. Biophys. Methods* 50, 111–121.

26. Malini, R., Venkatakrishna, K., Kurien, J., et al. 2006. Discrimination of normal, inflammatory, premalignant, and malignant oral tissue: A Raman spectroscopy study. *Biopolymers* 81(3), 179–193.

27. Lau, D. P., Huang, Z., Lui, H., Anderson, D. W., Beren, K., Morrison, M. D., Shen, L., and Zeng, H. 2005. Raman spectroscopy for optical diagnosis in the larynx: Preliminary findings. *Lasers in Sur. and Med.* 37, 192–200.

28. Wood, B. R., Quinn, M. A., Burden, F. R., and McNaughton, D. 1996. An investigation into FTIR spectroscopy as a biodiagnostic tool for cervical cancer. *Biospectroscopy* 2, 143–153.

29. Wood, B., Quinn, M., Tait, B., Ashdown, M., Hislop, T., Romeo, M., and McNaughton, D. 1998. FTIR microspectroscopic study of cell types and potential confounding variables in screening for cervical malignancies. *Biospectroscopy* 4, 75–91.

30. Pichardo-Molina, J. L., Frausto-Reyes, C., Barbosa-Garcia, O., et al. 2007. Raman spectroscopy and multivariate analysis of serum samples from breast cancer patients. *Lasers in Medical Science* 22, 229–236.

31. Dovbeshko, G. I., Gridina, N. Y., Kruglova, E. B., and Pashchuk, O. P. 2000. FTIR spectroscopy studies of nucleic acid damage. *Talana* 53, 233–246.

32. Wu, J.-G., Xu, Y.-Z., Sun, C.-W., Soloway, R. D., Xu, D.-F., Wu, Q.-G., Sun, K.-H., Weng, S.-F., and Xu, G.-X. 2001. Distinguishing malignant from normal oral tissues using FTIR fiber-optic techniques. *Biopolymer (Biospectroscopy)* 62, 185–192.

33. Kline, N. J., and Treado, P. J. 1997. Raman chemical imaging of breast tissue. *J. Raman Spectroscopy* 28, 119–124.

34. Chiriboga, L., Xie, P., Yee, H., Vigorita, V., Zarou, D., Zakim, D., and Diem, M. 1998. Infrared spectroscopy of human tissue. I. Differentiation and maturation of epithelial cells in the human cervix. *Biospectroscopy* 4, 47–53.

35. Tan, Y.-Y., Shen, A.-G., Zhang, J.-W., Wu, N., Feng, L., Wu, Q.-F., Ye, Y., and Hu, J.-M. 2003. Design of auto-classifying system and its application in Raman spectroscopy diagnosis of gastric carcinoma. 2nd International Conference Machine Learning & Cybernetics, November 2–5.

36. Fung, M. F. K., Senterman, M. K., Mikhael, N. Z., Lacelle, S., and Wong, P. T. T. 1996. Pressure-tuning FTIR spectroscopic study of carcinogenesis in human endometrium. *Biospectroscopy* 2, 155–165.

37. Paluszkiewicz, C., and Kwiatek, W. M. 2001. Analysis of human cancer prostate tissues using FTIR microscopy and SXIXE techniques. *J. Mol. Structures* 565–566, 329–334.

38. Andrus, P. G. L., and Strickland, R. D. 1998. Cancer grading by Fourier transform infrared spectroscopy. *Biospectroscopy* 4, 37–46.

39. Alfano, R. R., Tang, G. C., Pradhan, A., Lam, W., Choy, D. S. J., and Opher, E. 1987. Optical spectroscopic diagnosis of cancer and normal breast tissues. *IEEE J. Quantum Electronics*, QE-23(10), 1806.

40. Richter, T., Steiner, G., Abu-Id, M. H., Salzer, R., Gergmann, R., Rodig, H., and Johannsen, B. 2002. Identification of tumor tissue by FTIR spectroscopy in combination with positron emission tomography. *Vibrational Spectroscopy* 28, 103–110.

41. Hwang, Z., McWilliams, A., Lui, H., McLean, D. I., Lam, S., and Zeng, H. 2003. Near-infrared Raman spectroscopy for optical diagnosis of lung cancer. *International Journal of Cancer* 107, 1047–1052.

42. Fabian, H., Jackson, M., Murphy, L., et al. 1995. A comparative infrared spectroscopic study of human breast tumor and breast tumor cell xenografts. *Biospectroscopy* 1(1), 37–45.

43. Wong, P. T. T., Papavassiliou, E. D., and Rigas, B. 1991. Phosphodiester stretching bands in the infrared spectra of human tissues and cultured cells. *Applied Spectroscopy* 45, 1563–1567.
44. McIntosh, I. M., Jackson, M., Mantsch, H. H., et al. 1999. Infrared spectra of basal cell carcinomas are distinct from non-tumor-bearing skin components. *Journal Invest. Dermatology* 112, 951–956.
45. Wong, P. T., Goldstein, S. M., Grekin, R. C., et al. 1993. Distinct infrared spectroscopic patterns of human basal cell carcinoma of the skin. *Cancer Research* 53, 762–765.
46. Fujioka, N., Morimoto, Y., Arai, T., and Kikuchi, M. 2004. Discrimination between normal and malignant human gastric tissues by Fourier transform infrared spectroscopy. *Cancer Detection & Prevention* 28, 32–36.
47. Lucassen, G. W., van Veen. G. N. A., and Jansen, J. A. J. 1998. Band analysis of hydrated human skin stratum corneum attenuated total reflectance Fourier transform infrared spectra in vivo. *Journal of Biomedical Optics* 3, 267–280.
48. Yang, Y., Sule-Suso, J., Sockalingum, G. D., Kegelaer, G., Manfait, M., and El Haj, A. J. 2005. Study of tumor cell invasion by Fourier transform infrared microspectroscopy. *Biopolymers* 78, 311–317.
49. Dovbeshka, G. I., Chegel, V. I., Gridina, N. Y., et al. 2002. Surface enhanced IR absorption of nucleic acids from tumor cells: FTIR reflectance study. *Biopolymer (Biospectroscopy)* 67, 470–486.
50. Teller, J., Wolf, M., Keel, M., et al. 2000. Can near-infrared spectroscopy of the liver monitor tissue oxygenation? *European Journal of Pediatrics* 159, 549.
51. Binoy, J., Abraham, J. P., Joe, I. H., Jayakumar, V. S., Pettit, G. R., and Nielsen, O. F. 2004. NIR-FT and FT-IR spectral studies and ab initio calculations of the anti-cancer drug combretastatin-A4. *Journal of Raman Spectroscopy* 35, 939–946.
52. Chiang, H. P., Song, R., Mou, B., et al. 1999. Fourier transform Raman spectroscopy of carcinogenic polycyclic aromatic hydrocarbons in biological systems: Binding to heme proteins. *Journal of Raman Spectroscopy* 30, 551–555.
53. Jackson, M., Mansfield, J. R., et al. 1999. Classification of breast tumours by grade and steroid receptor status using pattern recognition analysis of infrared spectra. *Cancer Detection and Prevention* 23, 245–253.
54. Movasaghi, Z., Rehman, S., and Rehman, I. U. 2008. FTIR spectroscopy of biological samples. *Applied Spectroscopy Reviews* 43(1), 1–46.
55. Movasaghi, Z., Rehman, S., and Rehman, I. U. 2007. Raman spectroscopy of biological samples. *Applied Spectroscopy Reviews* 42(2), 493–541.
56. Levin, I. W., and Bhargava, R., 2005. Fourier transform infrared vibrational spectroscopic imaging: Integrating microscopy and molecular recognition. *Annual Review of Physical Chemistry* 56(1), 429–474.
57. Petibois, C., and Deleris, G. 2006. Chemical mapping of tumor progression by FT-IR imaging: Towards molecular histopathology. *Trends in Biotechnology* 24(10), 455–462.
58. Bhargava, R., et al., 2006. High throughput assessment of cells and tissues: Bayesian classification of spectral metrics from infrared vibrational spectroscopic imaging data. *Biochimica et Biophysica Acta—Biomembranes* 1758(7), 830–845.
59. Steller, W., et al. 2006. Delimitation of squamous cell cervical carcinoma using infrared microspectroscopic imaging. *Analytical and Bioanalytical Chemistry* 384(1), 145–154.

60. Amharref, N., et al. 2006. Brain tissue characterisation by infrared imaging in a rat glioma model. *Biochimica et Biophysica Acta—Biomembranes* 1758(7), 892–899.
61. Beleites, C., et al. 2005. Classification of human gliomas by infrared imaging spectroscopy and chemometric image processing. *Vibrational Spectroscopy* 38(1–2), 143–149.
62. Krafft, C., et al. 2006. Classification of malignant gliomas by infrared spectroscopy and linear discriminant analysis. *Biopolymers* 82(4), 301–305.
63. Krafft, C., et al. 2006. Identification of primary tumors of brain metastases by SIMCA classification of IR spectroscopic images. *Biochimica et Biophysica Acta (BBA)—Biomembranes* 1758(7), 883–891.
64. Fernandez, D. C., et al. 2005. Infrared spectroscopic imaging for histopathologic recognition. *Nature Biotechnology* 23(4), 469–474.
65. Miller, L. M., and Smith, R. J. 2005. Synchrotrons versus globars, point-detectors versus focal plane arrays: Selecting the best source and detector for specific infrared microspectroscopy and imaging applications. *Vibrational Spectroscopy* 38(1–2), 237–240.
66. Bambery, K. R., et al. 2006. A Fourier transform infrared micro spectroscopic imaging investigation into an animal model exhibiting glioblastoma multiforme. *Biochimica et Biophysica Acta—Biomembranes* 1758(7), 900–907.
67. Gough, K., et al. 2005. Choices for tissue visualization with IR microspectroscopy. *Vibrational Spectroscopy* 38(1–2), 133–141.
68. Romeo, M. J., and Diem, M. 2005. Infrared spectral imaging of lymph nodes: Strategies for analysis and artifact reduction. *Vibrational Spectroscopy* 38(1–2), 115–119.
69. Romeo, M., and Diem, M. 2005. Correction of dispersive line shape artifact observed in diffuse reflection infrared spectroscopy and absorption/reflection (transflection) infrared micro-spectroscopy. *Vibrational Spectroscopy* 38(1–2), 129–132.
70. Ó Faoláin, E., et al. 2005. A study examining the effects of tissue processing on human tissue sections using vibrational spectroscopy. *Vibrational Spectroscopy* 38(1–2), 121–127.
71. Fabian, H., et al. 2003. Infrared microspectroscopic imaging of benign breast tumor tissue sections. *Journal of Molecular Structure* 661, 411–417.
72. Fabian, H., et al. 2006. Diagnosing benign and malignant lesions in breast tissue sections by using IR-microspectroscopy. *Biochimica et Biophysica Acta—Biomembranes* 1758(7), 874–882.
73. Zhang, L., et al. 2003. Classification of Fourier transform infrared microscopic imaging data of human breast cells by cluster analysis and artificial neural networks. *Applied Spectroscopy* 57(1), 14–22.
74. Lasch, P., et al. 2004. Imaging of colorectal adenocarcinoma using FT-IR microspectroscopy and cluster analysis. *Biochimica et Biophysica Acta—Molecular Basis of Disease* 1688, 176–186.

6

Detecting and Monitoring of Cancer at the Cellular Level

Introduction

Spectroscopy can be employed to study cancer cell lines. Raman spectroscopy has been applied to analyse the testicular, lymphoma, and ovarian cancer cell lines. In this chapter, a study describing the spectral differences between resistant and sensitive subtypes of the 833k testicular cancer cell line samples is reported. Raman spectroscopy allowed reproducible and quantitative analysis of the specimen and illustrated the chemical specifications of the samples precisely. Six pairs of 833k testicular cancer cell lines were studied and the findings were backed by statistical methods, i.e., partial least squares discriminant analysis (PLS-DA). In addition, two other testicular cancer cell lines (GCT27 and Susa), a lymphoma cell line (DHL4), and an ovarian cancer cell line (A2789) treated with chemotherapeutic medications are analysed.

Raman spectroscopy can objectively differentiate between resistant and sensitive cell lines. It can also analyse the effect of chemotherapeutic medications on cancer cell lines, as well as identify spectral differences between cell lines of different organs. These results suggested that in the future it may be possible to use cell lines and diagnostic Raman spectroscopy for preoperative classification of biological molecules.

Analysis of Resistant and Sensitive Subtypes of Testicular Cancer Cell Lines

Raman spectroscopy is an established technique for analysis of human tissue samples, and several research groups have focused on this newly emerging field [1–4]. However, most of the available literature is on applications of the technique on human tissue samples. A few articles have used cancer cell lines for spectroscopic investigations [5–7], and this field is relatively new in spectroscopy.

In general, testicular cancer is not very common: only 1 to 2 of every 100 cancers diagnosed in men are testicular cancers [8]. It is most common among males aged 15–40 years [9] and is the most common cancer affecting young men between 20 and 39 years [8]. Only about 1 in 7 (14%) are diagnosed in men over 50 years [8]. Over the lifetime of a male, the risk of developing testicular cancer is roughly 1 in 250 (four tenths of one percent, or 0.4%) [9,10]. Around 2,100 men are diagnosed in the United Kingdom and 7,500 to 8,000 diagnoses of testicular cancer are made in the United States each year [8,9]. Testicular cancer has one of the highest cure rates of all cancers [8,9] and chemotherapy offers a cure rate of at least 85% today [10,11]. Similar to other malignancies, this type of cancer can be sensitive or resistant to chemotherapeutic medications. It would be greatly beneficial if clinicians could deduce if cancer strains are sensitive or resistant to chemotherapy prior to a main course of chemotherapy treatment as this could help determine the best approach to use.

Spectral Acquisition

In this study, the Raman spectroscopic technique was employed to analyse testicular cancer cell line samples. A total of 12 samples, which included six resistant and six sensitive (to chemotherapeutic medications) subtypes, were analysed.

Raman spectra of the samples were recorded using an Almega (Thermo Fisher Scientific, Madison Wisconsin, USA), equipped with two types of lasers. The first laser was a 532-nm (green) diode-pumped solid-state laser, with a laser power of 25 mW TEM00 continuous wave beam, and power stability of <5%. The second laser, which was also the main one employed in this work, was a 785-nm near-infrared (NIR) high-power diode laser, with a laser power of 300 mW, multimode continuous wave beam, and power stability of <5%. It must be mentioned that the laser power can be changed from 15 to 100%, according to the requirements of experiments. The machine also has an Andor DV420 open electrode CCD detector, with a working temperature of −50°C. All of the spectra were collected in the range 96–3430 cm^{-1} using 10X objectives and more than an average of 64 scans.

The samples used in this research are ATTC cancer cell lines. After defrosting, the cell lines are cultured in culturing media and incubated. Culture medium is then removed, and an adequate quantity of 10% trypsin is added to detach the cells. The cells (in trypsin) are transferred into a 50-mL centrifuge tube containing an equal volume of fresh culture medium to neutralise the trypsin. After centrifuging, the supernatant is tipped off and the pellet resuspended in fresh culture medium. In this stage, the cells are counted by trypan blue method and 3×10^6 cells are calculated and transferred into a 15-mL centrifuge tube. After centrifuging another time, the pellet is resuspended in 5 mL of 70% ethanol and frozen until ready to analyse. Samples were prepared for spectroscopic investigation by removing the ethanol (by centrifuging, rinsing with distilled water, and keeping in a vacuum oven) and five spectra were collected from randomly selected areas of the sample.

Reproducibility Check

Due to the crucial importance of reproducibility in spectroscopic studies [12], some other potentially important factors were also considered. According to Schroeder et al. [13], key parameters needed to be carefully considered for a reproducible Raman experimental setup including: (1) the adjustment of laser power levels and monitoring of laser output stability; (2) the duration of data collection as it effects (a) the number of scans that can be obtained to improve signal to noise, and (b) the amount of thermo-luminescence; and (3) sample preparation methods, which can be accomplished in different forms. It is also believed that the repeatability of Raman measurements could be evaluated by increasing the number of the spectra collected from every sample [12]. Behrens et al. used a different approach for checking the reproducibility of their research [14]. In addition to concentrating on the instrumental reproducibility aspect of the data (which has also been considered by some other research groups [13,15]), they limited their experiments to only one objective, one laser power, and one aperture size. Sadezky et al. took other factors into account to create a unique pattern for their experiments [15] including the focus of the laser beam and the exposure time. The spectral resolution and spectral region covered could be other important factors in establishing a reproducible pattern [16]. It is believed that storage conditions would minimise the changes in spectral characteristics [17]. Kateinen et al. also focused on instrumental reproducibility, and followed a constant timing pattern for all experiments [18]. The reproducibility factors that were considered by Niemela et al. [19] are as follows: (a) laser power, (b) Raman range, (c) resolution, (d) size of spot, (e) proper focus, (f) constant number of spectra from every sample, and (g) constant measurement time.

It is important to establish a set of sample-specific parameters to produce reproducible results. These parameters for cell line works are tabulated in Table 6.1 and have been considered to establish a reproducible method for all of the experiments conducted in this research work.

Spectral Analysis

A total of 60 spectra were collected from six pairs of 833K samples, including six sensitive and six resistant ones. Figure 6.1 illustrates the spectra collected from sensitive and resistant samples, respectively.

In order to have a clear understanding of the chemical bands and functional groups existing in 833K samples, a brief explanation of the definitions of the observed peaks is presented in Table 6.2. This information has been derived from our published review works [20,21].

TABLE 6.1

Important Parameters for Reproducibility

Reproducibility Factor	Achieving Methods
Laser power	Fixed on 100%
Number of spectra collected from each sample	5 spectra from each sample (intra-/intersample reproducibility check)
Instrumental reproducibility	Using the same instrument for all studies
Objective	Doing all spectra collections with 10X
Aperture size	100-μm pinhole
Focus	Achieved visually
Exposure number and time	64 × 2 seconds
Spectral resolution	Low
Spectral region	96–3430 cm^{-1}
Sample storage conditions	Freezer (until a max of 4 hours before experiments)
Thermal stability	Air-conditioner at 20–22°C
Number of cells in each sample	3 × 10^6/5 mL
Sample preparation	A unique method for all

Raman spectra of the cell lines (see Figure 6.2) reveal interesting and important differences that help indistinguishing between R and S spectra. Details of these differences are summarised as follows:

- The 483 cm^{-1} peak does not exist in R samples. This peak is attributed to phosphate vibration in nucleic acids and phospholipids.
- The 862 cm^{-1} shoulder cannot be seen in R samples. This shoulder belongs to stretching C–C and bending CCH vibrations in proteins and polysaccharides.
- The CH region shows significant differences. While the shoulders of the CH region in R samples have almost turned to a separate peak, they illustrate weaker intensities in S ones. This can demonstrate the difference existing in CH chains between samples in nucleic acids, proteins, and lipids.
- The ratio of I880 cm^{-1}/I937 cm^{-1} is much greater in R samples in comparison to S samples. These peaks are due to vibrations in CH$_2$ (rocking), C–C (stretching), and in CH$_3$ (rocking) and C–C (stretching) in proteins, respectively.

Quantitative Analysis of the Spectra of 833k Samples

In this section, quantitative analysis of the spectral differences between the R and S samples is presented to demonstrate the ability of Raman microspectroscopy to precisely and statistically analyse cells.

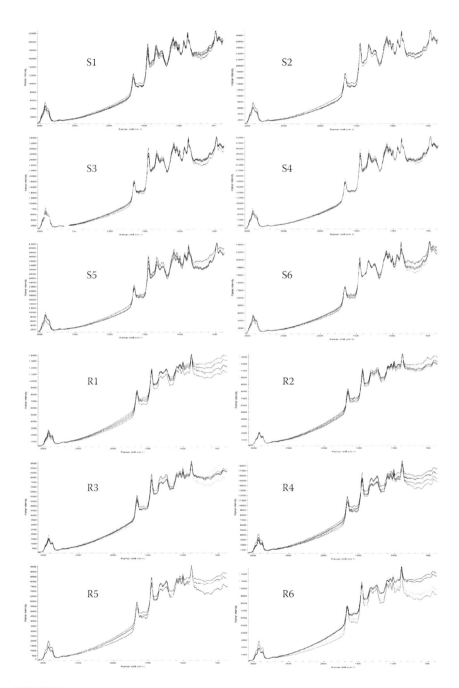

FIGURE 6.1
Spectra of 833K samples.

TABLE 6.2

Peak Definitions of 833K Samples

Peak Position in S (cm⁻¹)	Peak Position in R (cm⁻¹)	Definition	Assignment
483	n/a	Phosphate	Nucleic acids + phospholipids
862 (shoulder)	n/a	ν(C–C), δ(CCH)	Proteins + polysaccharides
880	884	ρ(CH₂) + ν(C–C)	Proteins (proline, valine, hydroxypro, tryptophan)
936	941	ρ(CH₃) terminal + ν(C–C)	Proteins (α-helix, proline, valine, keratin)
1004	1005	ν(CO), ν(CC), δ(OCH)	Proteins (phenylalanine) + polysaccharides
1049	1053	C–O stretching + bending	Carbohydrates
1089	1097	ν_sPO₂⁻	Nucleic acids + phospholipids
1249	1246	ν(CN), δ(NH)	Proteins (amide III, α-helix, collagen, tryptophan)
1333	1319	CH₃CH₂ wagging	Nucleic acids + proteins
1453	1454	δ(CH₂) δ(CH₃)	Proteins + phospholipids
1663	1671	ν(C=O)	Proteins (amide I, α-helix, collagen, elastin)
2928	2930	CH region	Nucleic acids + proteins + lipids

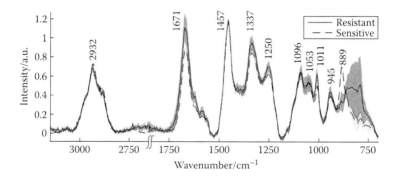

FIGURE 6.2

Raman spectra of resistant and sensitive testicular cell lines.

Kateinen et al. [18] chose one fifth of the spectra of their study for quantitative and statistical analysis of their results. However, the authors carried out the process of quantification on two thirds of the spectra. This included a total of 40 randomly chosen spectra, consisting of 20 R and 20 S spectra. The peak heights and peak areas of the selected spectra were calculated using Omnic Software version 7.3a.

TABLE 6.3

Comparative Analysis of Quantitative Measurements

Type of Samples	Peak 1 Position	Peak 2 Position	Ratio of Heights	Mathematical Diagnosis
S	878–882	934–939	1.04–2.42	<2
R	881–886	936–945	5.94–13.69	>6
			Ratio of Areas	
S	878–882	934–939	1.72–3.21	<3
R	881–886	936–945	6.21–20.59	>6

The results of peak height ratio calculations show that there are significant quantitative differences between R and S samples. In spite of minor differences in the position of the peaks, the main dissimilarities are in the ratios. Whilst almost all of the height ratios of S samples are less than 2, this number is more than 6 for R samples. Another difference is the method of distribution of the quantities. The ratios of R spectra are very widely distributed, ranging from 5.94 to 13.69. On the other hand, S numbers are all in a relatively narrow range of 1.04 to 2.42. Peak areas, in a very similar way, illustrate the quantitative dissimilarities. Whilst almost all the ratios of S samples are less than 3, this number is more than 6 for R samples. The method of distribution of the quantities is also different. The ratios of R spectra are very widely distributed, ranging from 6.21 to 20.59. On the other hand, S numbers are all in a relatively narrow range of 1.72 to 3.21. These results demonstrate that in proteins of resistant cell lines, there are much more $\rho(CH_2)$ and $v(C–C)$ vibrations in comparison with $\rho(CH_3)$ and $v(C–C)$ vibrations. Table 6.3 summarizes these findings. Similar results can be obtained if peak areas are measured instead of peak height.

Statistical Analysis

The results of cellular studies might not be particularly useful if there is no statistical data to support them. As a result, statistical analysis is used to provide supportive data to spectral findings.

There has been much interest recently in using vibrational spectroscopy and computer-based pattern recognition techniques to discriminate between different types of tissue and tumours [22,23]. These techniques could potentially complement traditional histopathological diagnosis by providing a rapid and objective analysis of tissue samples. Here, we demonstrate the ability of PLS-DA, a comparatively simple classification algorithm, to distinguish between the R and S cell lines.

PLS-DA involves two steps. In the first step, a PLS decomposition of the spectral and class data is performed, reducing the large number of wavelength variables to a small number of latent variables called *scores* [24]. In the second, these scores are used as descriptor variables in a standard linear discriminant analysis, or LDA [25]. This approach is generally believed to be superior to alternatives such as manual wavelength selection or principal components analysis because the PLS decomposition automatically finds a linear combination of the variables that is optimal for LDA [26].

The number of latent variables was chosen by a cross-validation [24] in which all the spectra from one sample of each type were left out of each iteration. The obtained misclassification rate was 0% with two latent variables.

After cross-validation, the PLS-DA model was calculated from the full training set and applied to both the training and test sets; the results are shown in Figure 6.3.

The high-dimensional spectral space has been reduced to a two-dimensional space of PLS scores (T1 and T2). All the spectra are correctly classified, and the test samples fit closely with the training samples.

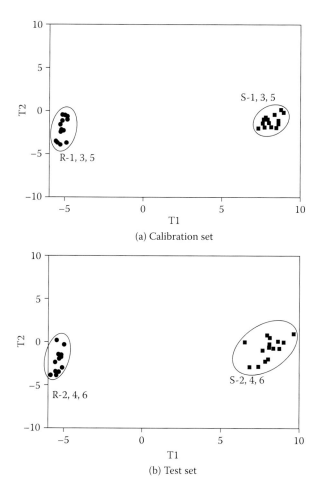

(a) Calibration set

(b) Test set

FIGURE 6.3
PLS-DA scores for (a) the calibration set and (b) the test set.

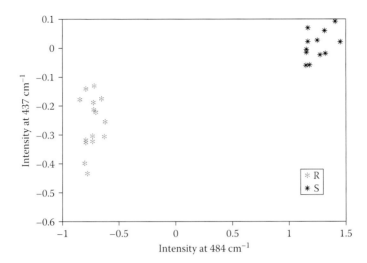

FIGURE 6.4
The effectiveness of I484/I437 for distinguishing cell types.

Ratio Classifier

While the PLS-DA classifier works very well, the quantitative analysis section of this research work suggests that a classifier using a wavenumber ratio could also work well. As a result, it was decided to focus on single peaks and their ratios. This was aimed at investigating the effectiveness of presenting the spectra in the form of numbers and mathematical diagnosis of the available spectral information. In this case, in contrast to what was done in the previous part, all twelve pairs of spectra were analysed and the ratio between the bands at 437 and 484 cm^{-1} was chosen (Figures 6.1 and 6.2).

Figure 6.4 illustrates a 2D plot of the resultant information showing similar findings to those of spectral (Figures 6.1 and 6.2) and quantitative investigations. R and S samples are clearly positioned separately. This result is in complete agreement with the other findings of this research that were reported previously, as R spectra are spread in a wider area in comparison with S spectra.

Treated Cell Lines

In this section, spectra of human ovarian A2780-R and S samples, which were treated with different concentrations of cisplatin, were collected and analysed. Samples were prepared in a manner similar to that used in the previous part of this research. However, they were treated with a chemotherapeutic medication (Cis-diamminedichloroplatinum, CDDP) for 24 hours

TABLE 6.4

Analysis of Cell Lines Treated with Different Concentrations of cisplatin

A2780 R Samples	A2780 S Samples
Control samples (A2780R Control)	Control samples (A2780S Control)
A2789RCis-10 µm CDDP	A2789SCis-0.3 µm CDDP
A2789RCis-20 µm CDDP	A2789SCis-1 µm CDDP
A2789RCis-30 µm CDDP	A2789SCis-3 µm CDDP

after preparation of normal cells. A total of eight samples were analysed. Table 6.4, illustrates the details of the analysed samples. Each sample was treated with the medication for 24 hours.

Five spectra were collected from each sample, except for A2789SCis-3 µm CDDP, with three spectra (two of the spectra were not of appropriate quality and were not useable).

Spectral Analysis

To begin with, the spectra of control cell lines (not treated) are shown in Figure 6.5. This figure also contains a summary of peak definitions and assignments. The two separate images show the distinct spectral differences in both the CH and fingerprint areas. Peak definitions are tabulated in Table 6.5.

Having a general look at the spectra, it can be concluded that the treated S samples do not show the desirable reproducibility that is seen in the R samples, particularly those treated with 1- and 3-µm drugs. This means that the reactions of the S samples that have been treated with cisplatin are not as constant as those of the R samples. This can be justified by the fact that these cell lines are *sensitive* to chemotherapy, and their reactions to treatment are not predictable. A similar finding can be seen only in 30-µm spectra of the R samples. In other words, in low concentration of cisplatin, R cell lines undergo a relatively constant modification as a result of the drug, but as the concentration of the medication increases, the modifications do not follow the previous constant pattern, and show some inconsistencies.

Figures 6.6 and 6.7 illustrate the spectra of all R and S samples in one image, respectively. It seems that spectral changes of the R samples follow a gradual pattern, consistent with the gradual increase in the drug strength (Figure 6.8). However, in the S samples, while a relatively significant dissimilarity can be observed between the control and 0.3-µm sample, the differences between the latter and the 1-µm and 3-µm samples are not that strong (Figure 6.9). This means that due to the sensitive nature of the S samples, the lowest concentration of the medication affects them considerably and higher concentrations do not increase this effect in a constant way.

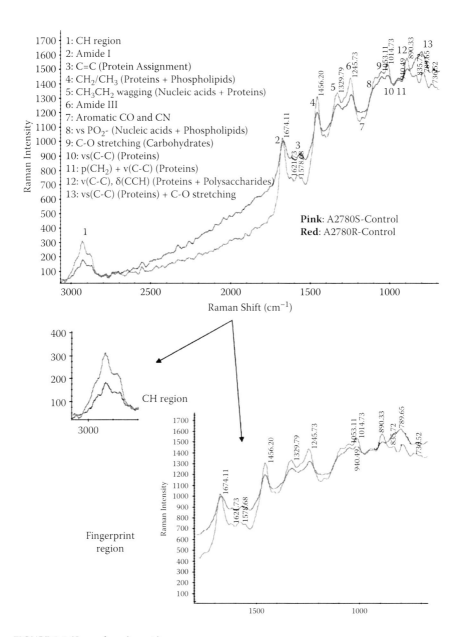

FIGURE 6.5 (See colour insert.)
Spectra of A2780 R and S (control samples).

TABLE 6.5

Peak Assignments and Definitions of Treated Cell Lines

Peak Position (cm⁻¹)	Assignment
483	Phosphate (Nucleic acids + phospholipids)
780	O–P–O of DNA
862	ν(C–C), δ(CCH) (Proteins + polysaccharides)
890	ρ(CH$_2$) + ν(C–C) (Proteins (proline, valine, tryptophan))
940	ρ(CH$_3$) + ν(C–C) (Proteins (proline, valine, keratin))
1013	ν_s(C–C) (Proteins (phenylalanine))
1050	C–O stretching (Carbohydrates)
1090	ν_sPO$_2^-$ (Nucleic acids + phospholipids)
1200	Aromatic CO and CN
1245	ν(CN), δ(NH) (Proteins (amide III, collagen, tryptophan))
1325	CH$_3$CH$_2$ wagging (Nucleic acids + proteins)
1450	δ(CH$_2$) δ(CH$_3$) (Proteins + phospholipids)
1550	ν(C=C) (Tryptophan and porphyrin, protein assignment)
1670	ν(C=O) (Proteins (amide I, collagen, elastin))
2930	CH region (Nucleic acids + proteins + lipids)

Apart from these general findings, there are some equally interesting and more specific findings, which are described in the following text.

Spectra of R Samples

The fingerprint region: In the fingerprint region of spectra of the treated R samples, the peaks flatten gradually with an increase in the drug concentration. This finding can be seen in Figure 6.10. However, a relatively different condition is seen in two areas:

a. In the amide I region, a shoulder is being added to the main peak along with the increase in the strength of the medication.
b. A similar observation can be seen in the C=C peak (~1550 cm⁻¹).

Both of these findings indicate creation of new chemical groups in the cells as a result of the medication.

The CH region: The decrease in the peak intensities can obviously be seen (Figure 6.11).

Spectra of S Samples

The fingerprint region: The sudden fall of the peak intensities and the collapse of the overall structure of the spectra can be seen in Figure 6.12. In other words, while in the R spectra, application of stronger medications will result in a step-by-step fall in the peaks, in the S spectra, after applying the lowest strength

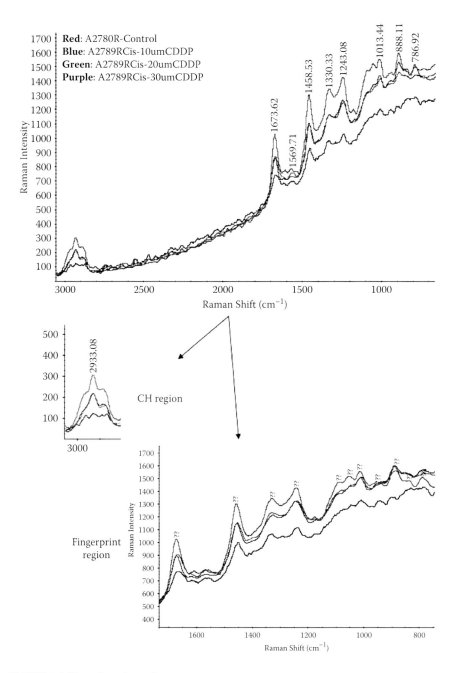

FIGURE 6.6 (See colour insert.)
Spectra of R samples treated with different concentrations of cisplatin.

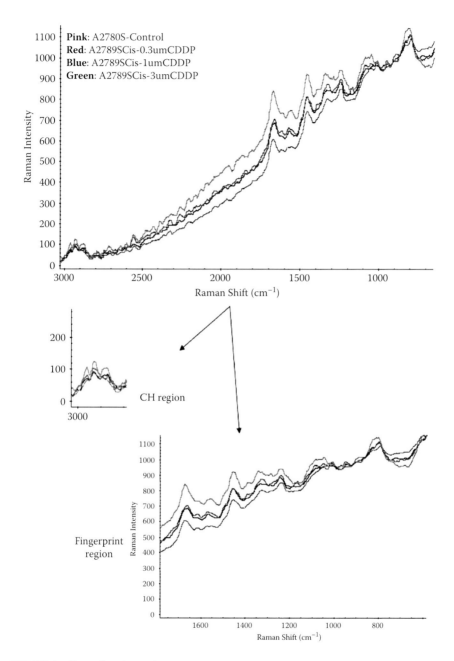

FIGURE 6.7 (See colour insert.)
Spectra of S samples treated with different concentrations of cisplatin.

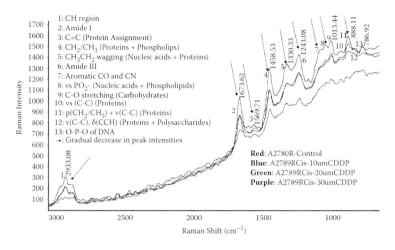

FIGURE 6.8 (See colour insert.)
Changes in spectra of treated R samples.

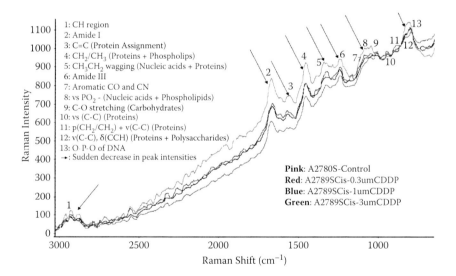

FIGURE 6.9 (See colour insert.)
Changes in spectra of treated S samples.

of cisplatin and the resulted spectral changes, increasing the concentration of the medication does not increase the modification dramatically. This condition can be observed in the ~780, 1000–1100, and 1200–1400 cm⁻¹ spectral regions.

The CH region: Figure 6.13 clearly shows the significant change in the peak intensities of CH bonds resulting from the application of cisplatin, and

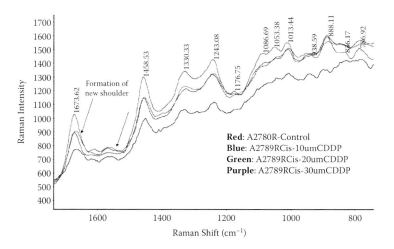

FIGURE 6.10 (See colour insert.)
Treated R spectra, fingerprint region.

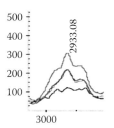

FIGURE 6.11 (See colour insert.)
Treated R spectra, CH region.

similar to other spectral areas, applying stronger medications does not result in further change in the peak intensities.

The results clearly indicate that Raman spectroscopy can be objectively applied on the treated cell lines. The technique can open new horizons in oncologic and cytologic information.

A summary of the findings are given here:

- The sensitive cell lines undergo significant changes under the influence of chemotherapeutic drugs. This effect is observed in the cell lines treated with the lowest concentrations of the medication, and subsequent increases in the strength of the medication do not lead to dramatic changes.

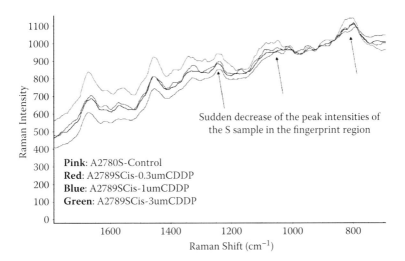

FIGURE 6.12 (See colour insert.)
Treated S spectra, fingerprint region.

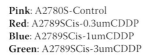

FIGURE 6.13 (See colour insert.)
Treated S spectra, CH region.

- Resistant cell lines illustrate a totally different process: Those treated with 10 and 20 μm of the medication show moderately severe changes in comparison with the control samples. However, more pronounced changes can be observed for cells treated with 30 μm of the drugs. In other words, a relatively step-by-step method of modification is seen for these cell lines.

- These findings can be attributed to the sensitive and resistant nature of the cells. In other words, sensitive cell lines illustrate stronger reactions to the medications, and undergo sudden and relatively sharp changes, when treated with lower strengths of the medication. On the other hand, resistant cells are stronger, and can withstand the effect of the drug comparably more than sensitive cell lines.

- The spectra of the treated S samples do not show the desirable reproducibility. This can be attributed to the sensitive nature of the cells, and the fact that they undergo severe and unpredictable changes when treated with chemotherapeutic medications. Similar findings can be observed only for the R samples treated with the strongest agent used in this study (30 µm). Once again, this can be attributed to the resistant nature of the cells.

Other Testicular Cell Lines

Similar experiments conducted for 833k samples were employed on two other types of testicular cancer cell lines, namely, GCT27 and Susa; and similar results were obtained. Spectra for GCT27 and Susa samples are given in Figures 6.14 and 6.15.

Lymphoma Cell Lines

At this part, a total of three DHL4 samples were analysed and compared to testicular samples. Figures 6.16 and 6.17 illustrate the comparative images of DHL4, and Susa R and S, respectively, with the responsible cellular compounds for each peak. Table 6.6 presents the peak definitions of the spectra.

As predicted, the difference between DHL-4 and Susa samples is more outstanding than that of cell lines belonging to the same tissues. However, lymphoma cells are more different from the S type Susa cell line than the R type. This could be an interesting topic for further analysis. It seems that proteins and carbohydrates in the spectra range of 1000–1200 cm^{-1} illustrate

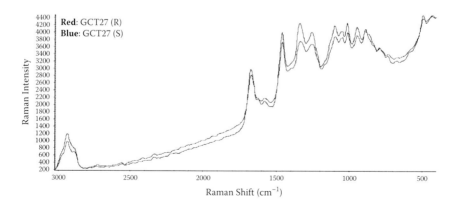

FIGURE 6.14 (See colour insert.)
Comparative image of GCT27 spectra.

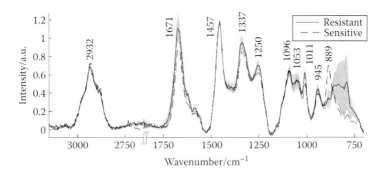

FIGURE 6.15
Mean spectra of Susa R and S samples.

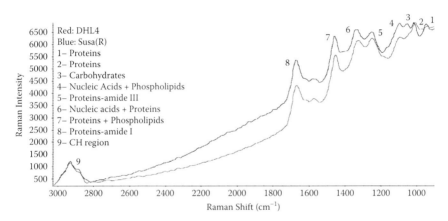

FIGURE 6.16 (See colour insert.)
Peak definitions of spectra of DHL4 and Susa (R).

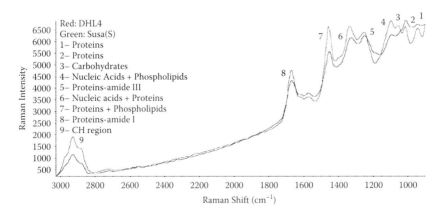

FIGURE 6.17 (See colour insert.)
Peak definitions of spectra of DHL4 and Susa (S).

TABLE 6.6

Peak Definitions of DHL4 Samples

Peak Position (cm^{-1})	Definition	Assignment
889	$\rho(CH_2) + \nu(C-C)$	Proteins (proline, valine, hydroxypro, tryptophan)
1011	$\nu(CO), \nu(CC), \delta(OCH)$	Proteins (phenylalanine) + polysaccharides
1053	C–O stretching + bending	Carbohydrates
1096	$\nu_s PO_2^-$	Nucleic acids + phospholipids
1250	$\nu(CN), \delta(NH)$	Proteins (amide III, α-helix, collagen, tryptophan)
1336	CH_3CH_2 wagging	Nucleic acids + proteins
1457	$\delta(CH_2) \, \delta(CH_3)$	Proteins + phospholipids
1671	$\nu(C=O)$	Proteins (amide I, α-helix, collagen, elastin)
2931	CH region	Nucleic acids + proteins + lipids

the strongest spectral differences in both images, resulting mainly from vibrations of CC, CO, and phosphate. Other spectra differences can be observed in areas related to amide I, amide III, and CH vibrations.

The results of this part of the project prove that one could expect relatively significant spectral differences between spectra of cell lines belonging to different tissues.

Analysis of cancer cell lines by spectroscopic means is increasing rapidly, as the technique offers a huge potential to determine the spectral and chemical differences between resistant and sensitive subtypes, which will help to develop a defined chemical pathway of cancer progression. Analysis of cell lines by spectroscopic means is a valuable way forward to be added to the armoury of both biologists and clinicians. These samples are easy to handle, and due to the existence of a single type of cell, it is possible to collect pure and more pronounced Raman spectra. Raman microspectroscopy can be applied to sensitive and resistant subtypes of testicular cancer cell lines, and differentiate them from one another successfully. Furthermore, this study shows that the technique can effectively elucidate the chemical groups and possible cellular chemical compounds that can contribute to different reactions of the cells to the chemotherapeutic agents.

The PLS-DA model seems to be effective in supporting spectroscopic findings. With this model, the overall misclassification rate in the test set is 0%. The cross-validation misclassification rate and the training misclassification rate are also 0. In addition, a simple classification technique (using specific spectral bands and their ratios as classification tools) is effective in distinguishing between resistant and sensitive cell lines. The method is reproducible and should be valuable in studying other types of cell lines and human tissue samples. Ultimately, the work will allow precise monitoring of patients before starting the actual course of treatment. The technique is also capable of analysing the effects of chemotherapeutic medications on cancer cell lines, as well as picking spectral differences between cell lines of different organs.

Taking all of the information presented in the statistical analysis section, the following conclusions are offered:

- Possibility of classification of R and S testicular cancer cell lines through Raman spectroscopy
- Reproducibility of the technique
- Potential for recognising the chemical differences between the two types
- Effective quantitative approach
- Good classification performance with PLS-DA
- The fingerprint region (300–1830 cm^{-1}) is more important for classification, but the C–H stretch region (2650–3130 cm^{-1}) is also useful and improves results in some cases.
- It is possible to choose single peaks or areas to create calibration models. However, as PLS does this automatically and includes a wide spectral range, it is better to use this facility.

The spectral database developed in this study suggests enormous potential to offer preoperative diagnosis of sensitive and resistant samples, in order to offer patients better management resulting in better outcomes. However, diagnosing cell lines is only a first step, and to be clinically useful, the method must be applicable to tissue samples. It may also be possible in the future to relate the chemical differences found in this study to sensitivity of tissues to cisplatin and assess the extent of sensitivity of the tissues preoperatively and noninvasively in vivo by Raman spectroscopy to guide treatment decisions.

As recently reported on the BBC by science reporter Katia Moskvitch ("Painless Laser Device Could Spot Early Signs of Disease") [27]: "The method, called Raman spectroscopy, could help spot the early signs of breast cancer, tooth decay and osteoporosis. Scientists believe that the technology would make the diagnosis of illnesses faster, cheaper and more accurate."

There have been a number of research groups working on the chemical structural characterisation of biological molecules to detect and monitor disease processes. The authors were amongst the first to report infrared and Raman spectroscopy of natural bones, including, human, bovine, and sheep bone [28–31], where it was predicted that spectroscopic means would be the way forward for routine diagnosis and to monitor the chemical structural properties of biological tissue, which changes with aging and disease processes.

This also encouraged the authors to monitor chemical changes in cancerous tissues such as breast cancer tissue, both by Fourier transform infrared (FTIR) and Raman spectroscopic means. Again, the authors first reported how FTIR and Raman spectroscopy can be used not only to distinguish between the types of breast cancer tissue, but the grades of the cancer tissues can also be precisely classified [32–33]. The authors also reported on how dietary flavonoids inhibit anticancer effects of a proteasome inhibitor, bortezomib, and on how Raman spectroscopy can be used for monitoring the chemical reactions between quercetin and bortezomib (Fi), quercetin and MG-262 (Fii), and quercetin and boric acid (Fiii,Fiv) [34]. There is a growing need to understand spectral data more precisely and accurately, as more and more researchers across the disciplines are working on spectroscopic techniques applied to biological molecules. As the biological molecules have complex structures, it becomes even more difficult to accurately interpret the spectra of biological molecules. To try and better understand the problem, the authors published two comprehensive review papers where nearly every single spectral peak that has been reported for a biological molecule in infrared and Raman spectra was listed [1,35].

Further studies can be carried out by using infrared and Raman spectroscopy to analyse cell lines, investigating differences in testicular cell lines, stem cells, bone cancers, and other cell lines. The authors firmly believe that in the near future, both the Raman and infrared spectroscopy will become the techniques of choice to detect and monitor the progression of cancers and other diseases and concur with the words of Professor Michael Morris (University of Michigan, USA) [36]. "You can replace a lot of procedures, a lot of diagnostics that are out there right now. The big advantage is that it's non-invasive, pretty fast—much faster than classical procedures—and more accurate."

To accelerate this concept in becoming a reality, there is a need to define a clear way forward and form cross-discipline partnerships, ultimately for the benefit of patients.

Summary

Both FTIR and Raman spectroscopic techniques are capable of determining functional groups and chemical bands of the cells that undergo changes due to the development processes of different diseases, evaluating the changes in the proteins, nucleic acids, lipids, and carbohydrates due to the disease process. Cancer cell lines can objectively be analysed using Raman spectroscopy. Statistical methods employed seem to be quite effective in distinguishing between resistant and sensitive cell lines (96–100% accuracy). Once a spectral database is established and there is good understanding of chemical structural characterisation of biological tissues, then yes, we again agree with Professor Morris, "In principle, it will take a couple of seconds to interpret the results." [36].

Acknowledgements

The authors would like to thank the technical and academic staffs of the Barry Reed Oncology Laboratory, St. Bartholomew's Hospital, London, United Kingdom for providing the samples.

References

1. Rehman, S. Movasaghi, Z, Tucker, A.T., and Rehman I. U. 2007. Raman spectroscopic analysis of breast cancer tissues: Identifying differences between normal, invasive ductal carcinoma and ductal carcinoma in situ of the breast tissue. *J. Raman Spectroscopy* 38, 1345–1351.
2. Dukor, R. K. 2002. Vibrational spectroscopy in the detection of cancer. *Biomedical Applications* 3335–3359.
3. Moreira L. M., Silveira L., Santos F. V., et al. 2008. Raman spectroscopy: A powerful technique for biochemical analysis and diagnosis. *Spectroscopy* 22(1), 1–19.
4. Jess, P. R. T., Smith, D. D. W., Mazilu, M., et al. 2007. Early detection of cervical neoplasia by Raman spectroscopy by Raman spectroscopy. *International Journal of Cancer* 121, 2723–2728.
5. Krishna, C. M., Sockalingum, G. D., Kegelaer, G., et al. 2005. Micro-Raman spectroscopy of mixed cancer cell populations. *Vibrational Spectroscopy* 38, 95–100.
6. Kuhnert, N., and Thumser, A. 2004. An investigation into the use of Raman microscopy for the detection of labelled compounds in living human cells. *J. Label. Comp. Radiopharm.* 47, 493–500.
7. Chan, J. W., Taylor, D. S., Zwerdling, T., et al. 2006. Micro-Raman spectroscopy detects individual neoplastic and normal hematopoietic cells. *Biophysical Journal* 90, 648–656.
8. http://cancerhelp.cancerresearchuk.org/about-cancer/what-is-cancer/statistics/incidence-survival-and-mortality.
9. American Cancer Society. *Cancer Facts and Figures*. 2007. Atlanta, GA: American Cancer Society.
10. Forman, D., Pike, M. C., Davey, G., et al. 1994. Aetiology of testicular cancer: Association with congenital abnormalities, age at puberty, infertility, and exercise. *BMJ* 308(6941), 1393–1399.
11. Swerdlow, A. J., Huttly, S. R., and Smith, P. G. 1987. Testicular cancer and antecedent diseases. *British Journal of Cancer* 55 (1), 97–103.

12. Hwang, M.-S., Cho, S., Chung, H., and Woo, Y.-A. 2005. Nondestructive determination of the ambroxol content in tablets by Raman spectroscopy. *Journal of Pharmaceutical and Biomedical Analysis* 38, 210–215.
13. Schroeder, P. A., Melear, N. D., and Pruett, R. J. 2003. Quantitative analysis of anatase in Georgia kaolins using Raman spectroscopy. *Applied Clay Science* 23, 299–308.
14. Behrens, H., Roux, J., Neuville, D. R., and Siemann, M. 2006. Quantification of dissolved H_2O in silicate glasses using confocal micro Raman spectroscopy. *Chemical Geology* 229, 96–112.
15. Sadezky, A., Muckenhuber, H., Grothe, H., et al. 2005. Raman microspectroscopy of soot and related carbonaceous materials: Apectral analysis and structural information. *Carbon* 43, 1731–1742.
16. Krafft, C., Neudert, L., Simat, T., and Salzer, R. 2005. Near-infrared Raman spectra of human brain lipids. *Spectrochimia Acta*, Part A 61, 1529–1535.
17. Lakshmi, R. J., Kartha, V. B., Krishna, C. M., et al. 2002. Tissue Raman spectroscopy for the study of radiation damage: Brain irradiation of mice. *Radiation Research* 157, 175–182.
18. Katainen, E., Elomaa, M., Laakkonen, U-M., et al. 2007. Qualification of the amphetamine content in seized street samples by Raman spectroscopy. *Journal of Forensic Science* 52(!), 88–92.
19. Niemela, P., Paallysaho, M., Harjunen, P., et al. 2005. Quantitative analysis of amorphous content of lactose using CCD-Raman spectroscopy. *Journal of Pharmaceutical and Biomedical Analysis* 37, 907–911.
20. Movasaghi, Z., Rehman, S., and Rehman, I. U. 2007. Raman spectroscopy of biological tissues. *Applied Spectroscopy Reviews* 42, 493–541.
21. Movasaghi, Z., Rehman, S., and Rehman, I. U. 2008. Fourier transform infrared (FTIR) spectroscopy of biological tissues. *Applied Spectroscopy Reviews* 43, 134–179.
22. Taleb, A., et al. 2006. Raman microscopy for the chemometric analysis of tumor cells. *J. Phys. Chem. B* 110(39), 19625–19631.
23. Barnes, R. J., Dhanoa, M. S., and Lister, S. J. 1989. Standard normal variate transformation and de-trending of near-infrared diffuse reflectance spectra. *Applied Spectroscopy* 43, 772–777.
24. Brereton, R. G. 2003. *Chemometrics: Data Analysis for the Laboratory and Chemical Plant*. Chichester: John Wiley & Sons.
25. Martens, H. A., and Dardenne, P. 1998. Validation and verification of regression in small data sets. *Chemometrics and Intelligent Laboratory Systems* 44, 99–121.
26. Barker, M., and Rayens, W. 2003. Partial least squares for discrimination. *Journal of Chemometrics* 17, 166–173.
27. http://www.bbc.co.uk/news/science-environment-11390951.
28. Rehman, I. U., Smith, R., Hench, L. L., and Bonfield, W. 1995. Structural evaluation of human and sheep bone and comparison with synthetic hydroxyapatite by FT-Raman spectroscopy. *Journal of Biomedical Materials Research* 29, 1287–1294.
29. Rehman, I. U., Smith, R., Hench, L. L., and Bonfield, W. 1994. FT-Raman spectroscopic analysis of natural bones and their comparison with bioactive glasses and hydroxyapatite. *Bioceramics* 7, edited by Ö. H. Andersson and A. Yli-Urpo 80–84.
30. Rehman, I. U. 1995. Characterisation of synthetic and natural materials by Fourier transform Raman spectroscopy. *Journal of Advanced Materials* 27(1), 59–63.

31. Rehman, I. U., and Bonfield, W. 1995. Structural characterisation of natural and synthetic bioceramics by photoacoustic-FTIR spectroscopy. *Bioceramics*, vol. 8, Edited by L. L. Hench, J. Wilson and D. Greenspan, 163–168.

32. Rehman, S., Movasaghi, Z., Joel, S. P., Darr, J. A., Ruban, A. V., and Rehman, I. U. 2007. Raman spectroscopic analysis of breast cancer tissues: Identifying differences between normal, invasive ductal carcinoma and ductal carcinoma in situ of the breast tissue. *Journal of Raman Spectroscopy* 38(10), 1345–1351.

33. Rehman, S., Movasaghi, Z., Tucker, A., Darr, J. A., and Rehman, I. U. 2010. Fourier transform infrared spectroscopic analysis of breast cancer tissues: Identifying differences between normal breast, invasive ductal carcinoma and ductal carcinoma in situ of the breast. *Applied Spectroscopy Reviews*, doi: 10.1080/05704928.2010.483674.

34. Liu, Feng-Ting, Agrawal, Samir G., Movasaghi, Zanyar, Wyatt, Peter, B., Rehman, Ihtesham, U, Gribben, John G., Newland, Adrian C. and Li, Jia. 2008. Dietary flavonoids inhibit the anti-cancer effects of the proteasome inhibitor bortezomib. *Blood* 112(9), 3540–3541.

35. Rehman, S., Movasaghi, Z, Tucker, A. T., and Rehman, I. U. 2008. Fourier transform infrared spectroscopy of biological tissues. *Applied Spectroscopy Reviews* 43, 134–179.

36. http://www.bbc.co.uk/news/science-environment-11390951.

7

Chemical Structural Analysis of Bone by Spectroscopy

Introduction

In this chapter, Raman and infrared spectroscopy of natural bones is described. These two studies on natural bones are presented as case studies. In addition, we explain how both of the vibrational techniques have been crucial in providing a baseline for the modification of synthetic apatite powders that are now routinely used as bone replacement materials. Prior to outlining the procedural details, it is important to understand the chemical structural properties of natural bone.

Cortical bone consists of two primary components: an inorganic or mineral phase, which is mainly a carbonated form of a nanoscale crystalline calcium phosphate, closely resembling hydroxyapatite (HA) [HA: $Ca_{10} (PO_4)_6(OH)_2$], and an organic phase, which is composed largely of type I collagen fibres [1–4]. Additional calcium phosphate phases have been proposed as minority constituents of the mineral phase of the bone tissue, such as octacalcium phosphate and amorphous calcium phosphate [5]. Other constituents of bone tissue include water and other organic molecules such as glycosaminoglycans, glycoproteins, lipids, and peptides. Ions such as sodium, magnesium, fluoride, and citrate are also present [4], as well as hydrogenophosphate [6]. Hence, the mineral phase in bone may be characterised essentially as nonstoichiometric substituted apatite. Such a distinction is important in the development of synthetic calcium phosphates for application as skeletal implants. As commercial calcium phosphates are generally based on hydroxyapatite, rather than a biological apatite, it is not surprising that the biological activity of such implants is less than that of Bioglass™ and AW™ glass ceramic, which develop oxyhydroxy carbonate (OHC) surfaces [7–8].

An understanding of bone function and its interfacial relationship to an implant clearly depends on the associated structure and composition. Traditional techniques used for structural and compositional analysis include light microscopy, electron microscopy, X-ray diffraction, and chemical analysis [9]. However, in preparation for such analysis, the tissue can be subjected to processes that may alter structure and/or composition. The ideal situation would be one in which minimal tissue preparation is required and physiologic conditions are maintained as closely as possible, and this can be achieved by employing both the Raman and infrared spectroscopic techniques [10].

Raman Spectroscopy of Bone

Raman spectroscopy is increasingly becoming recognized as a significant analytical method for biomedical applications and has the following advantages:

1. The need for tissue preparation is minimal. The measurement itself is nondestructive to the tissue and only very small amounts of material (micrograms to nanograms) are required.

2. "Wet" samples at atmospheric pressure can also be analysed, which is representative of a physiologic condition.

3. Molecular (ultrastructural-level) information is available from both infrared and Raman techniques, allowing investigation of functional groups, bonding types, and molecular conformations. The use of interferometers and Fourier transformation has allowed the realization of several spectrochemical advantages over conventional, dispersive spectrophotometers, such as increased signal throughput, greater frequency precision, and a decrease in the time required to obtain a spectrum [11].

However, by the incorporation of the Nd:YVO$_4$ near-infrared laser into Raman instrument design, the most significant feature of such technology is the virtual elimination of fluorescence from most biological samples. The problem of fluorescence previously restricted the Raman spectrochemical analysis of bone, and it was suggested that the fluorescing species lie within the organic phase of the tissue [12]. The fluorescence effect obscured much of the Raman signal originating from the analyte and restricted the applicability of the technique.

The problem of fluorescence was overcome previously by the deproteination of bone samples to remove the organic phase. Walters et al. [12] obtained conventional Raman and FTIR spectra of cortical bone from an ox femur. A conventional Raman investigation was also performed by Gevorkyan et al. [13] on the cortical bone of the human femur and on hydroxyapatite samples. The values quoted for the various phosphate peaks were in broad agreement with those reported by others [6,14–16], and the values given for the corresponding peaks arising from the bone tissue agreed well with those of the hydroxyapatites.

The two studies of bone tissue described earlier [12,13], although revealing some detail concerning the mineral phase of the tissue, were affected by sample fluorescence, particularly when attempts were made to characterise the organic phase of the tissue. Because of the presence of high fluorescence, very noisy data were obtained and useful information was lost in the background.

In a study of synthetic carbonated apatites, Nelson and Featherstone [14] employed chemical analysis, X-ray diffraction, and infrared and Raman spectroscopy to characterise samples, which were prepared using a variety of methods. Other studies on hydroxyapatite, bone, and related compounds [15–20] broadly agreed with the spectral findings of these workers.

Composition of cortical bone tissue has been determined in the physiologic state (i.e., without deproteination), as well as in the deproteinated state (which isolates the mineral phase of the tissue). The spectra obtained are compared with those reported by other workers [12,13]. Raman spectra obtained from commercial hydroxyapatite have been compared with the inorganic matrix (the natural apatite), which were considered in the context of previous studies [14–23] making it possible to compare and contrast in detail the spectra from natural and synthetic calcium phosphate and to establish a correlation between the spectral lines.

In this reported study, cortical bone was taken from the midshaft region of human (56 years old, male) and sheep (mature bone) femoral tissue. The tissue was obtained in a frozen state, at 0°C. Commercial synthetic hydroxyapatite powder (P120 type) was obtained from Plasma Biota1 (UK). Calcium carbonate, calcium phosphate, calcium hydroxide, and sodium hydroxide samples were obtained from BDH (UK) for use as controls.

Preparation of Bone Samples

Sample preparation in cases is extremely important, and it becomes even more crucial when biological samples are involved. Cubes of cortical bone tissues used in that study were prepared, measuring 4 mm along each axis. The size was selected to fit into the sample holders used and provided a sufficiently large surface area-to-volume ratio to ensure effective penetration of the deproteinating agent into the tissue. At the same time, the samples were large enough to be easily handled. Four samples were prepared: two from the human tissue and two from the sheep tissue; one from each of these groups was then deproteinated. The method of preparation involved cutting the tissue using a small hacksaw and a diamond band saw (EXAKT cutting-grinding system) with a recirculating water system. The time during which the bone was exposed to room temperature (during the preparation process) was kept to a minimum (10 min). The samples were refrozen as soon as cutting was completed and then allowed to thaw for 30 min at room temperature prior to spectrochemical analysis. The transverse faces of the cubes (with respect to the long axis of the femur) were analysed.

The deproteination procedure used was based on that of Termine and coworkers [16]. The deproteinating agent used (hydrazine) was anhydrous to prevent changes to the mineral of the tissue. A prepared bone sample was immersed in 5 mL of 99% anhydrous hydrazine in a glass-stoppered flask and treated as follows: at room temperature for 1 h; hydrazine reagent changed; incubated at 60°C for 1 h; hydrazine reagent changed; incubated at 60°C for 15 h; hydrazine reagent changed; and incubated at 60°C for 24 h.

The bone samples were then immersed in a graded series of aqueous alcohol solutions (5 mL, 30 min each) of 50, 75, 87.5, 100, 100, and 100% concentrations. The samples were then immersed in 5 mL of 95% acetone for 2 h. Finally, the samples were placed in a desiccator and left to dry at room temperature for 48 h before analysis was carried out.

Raman Spectral Data Acquisition

Raman spectra of the bones and hydroxyapatite samples were recorded using a Nicolet spectrophotometer, equipped with a Nd:YVO$_4$ near-infrared laser, which eliminated the problems of sample fluorescence and photodecomposition. The low-power near-infrared laser in the Raman spectrophotometer makes it possible to analyse fresh tissues without loss of sensitivity due to the fluorescence of organics. The analysed bone samples were attached to a metal post (sample holder). Hydroxyapatite, calcium carbonate, calcium phosphate, calcium hydroxide powder, and sodium hydroxide pellet samples were analysed in a glass tube 5 mm in diameter, similar to that used in nuclear magnetic resonance (NMR) spectroscopic analysis. Scattered radiation was collected and collimated by 180° reflective optical geometry. The spectra were obtained by using 450 mW of laser power, 100–600 scans (depending on the samples), and 4 cm^{-1} resolution.

Spectral Findings

The most striking feature of the spectra of bone tissues, (without deproteination) obtained from this work was the absence of fluorescence from the samples (Figures 7.1 and 7.2), allowing well-defined peaks to be resolved. Raman spectroscopy of the whole bone samples revealed numerous peaks associated with an organic phase, i.e., the peaks in the 3000–2800 cm^{-1} region were due to C=H and the peaks at 1650–1600 cm^{-1} were due to the amide band of collagen (a combination of C=O stretching, C=N stretching, and N=H bending modes).

The human, bovine, baboon, and sheep tissue provided almost identical spectra, both in the fresh and deproteinated conditions (Figures 7.1 and 7.2 and Table 7.1). The deproteination procedure clearly resulted in a considerable decrease in the scattering intensity of peaks arising from the organic phase. The peaks in the region of 3000–2900 cm^{-1} wavenumber were due to the C–H stretching mode and 1450 cm^{-1} to the C–H, bending mode, while other peaks, which were present due to the inorganic phase of the bone, remained substantially unaffected.

The peaks for the phosphate moiety centred at 952 cm^{-1} (PO:- stretching mode) and 584 cm^{-1} (PO:- bending mode) remained unaffected. Similarly,

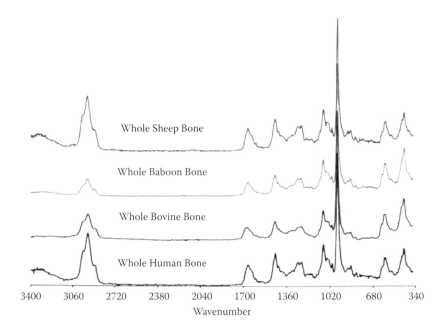

FIGURE 7.1
Raman spectra of whole human, bovine, baboon, and sheep bones. (With permission.)

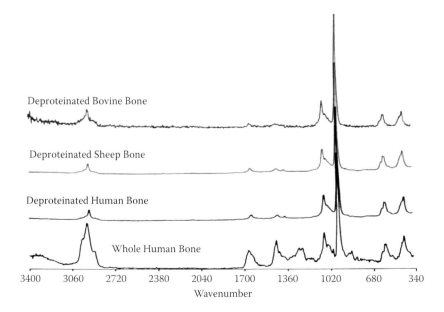

FIGURE 7.2
Raman spectra of whole human, deproteinated human, and sheep bones. (With permission.)

TABLE 7.1

Raman Peak Positions (cm^{-1}) for Normal (Untreated) and Deproteinated Bone Tissue, Synthetic Hydroxyapatite, Calcium Carbonate, Calcium Phosphate, Calcium Hydroxide, and Sodium Hydroxide with Assignments for Peaks (With permission).

Peak Assignment	Normal Bone		Deproteinated Bone				Synthetic Hydroxylapatite (HA) P120	Calcium Carbonate	Standard Spectra		
	Human Femur	Sheep Femur	Human Femur	Sheep Femur	Bovine Bone	Baboon Bone			Calcium Phosphate	Calcium Hydroxide	Sodium Hydroxide
CH stretch	3000–2900	3000–2900	2915	2915	2916	2916					
Amide I band	1660	1660	1636	1637	1639	1639					
CO$_3^{2-}$								1747			
CO$_3^{2-}$								1434			
C–H (bend)	1450	1450	1433	1434	1435	1435					
Amide II band	1262	1262	1261	1261	1261	1262					
C–O	1242	1242	1240	1240	1241	1241					
CO$_3^{2-}$								1084			
OH$^-$										931	1079
CO$_3^{2-}$ (stretch)	1062	1062	1062	1062	1062	1062					
CO$_3^{2-}$ (bend)	1036	1036	1036	1036	1036	1036			1045		
PO$_4^{3-}$ (symmetric stretch)	952	952	952	952	952	952	952		952		
OH$^-$							774		773	801	373
PO$_4^{3-}$ (asymmetric stretch)							690			721	
PO$_4^{3-}$ (bend)	584	585	584	584	585	585					
CO$_3^{2-}$ (bend)								710	706		
PO$_4^{3-}$ (bend)	429	429	429	429	429	429	436		431		
OH$^-$										355	293
										249	207

the peaks present due to the carbonate group of the in cm^{-1} (CO:- stretching mode) and 1036 cm^{-1} (CO:- bending mode) also remained unchanged. The difference in the spectra from the fresh and deproteinated bone can be seen in Figures 7.1 and 7.2.

Hydroxyapatite is an intense Raman scatterer, which is indicated by the exceptional signal-to-noise ratio. The values of the observed frequencies for hydroxyapatite are tabulated in Table 7.1 and the spectrum is given in Figure 7.3. A sharp but weak band at 952 cm^{-1} was attributed to a symmetric P=O stretching mode on the basis of previous conventional Raman studies of solid and aqueous phosphate [19–21]. The broad bands present at 774 and 690 cm^{-1} were assigned to an asymmetric P=O stretching mode on the basis of comparison with the spectrum of calcium phosphate. The correlation between the spectra of bone tissue and that of synthetic hydroxyapatite was only present in the shared peak at 952 cm^{-1} of symmetric P=O (Figure 7.3 and Table 7.1). The hydroxyapatite spectrum was dominated by the asymmetric P=O bands, whereas the spectra of bone samples were dominated by the intense symmetric P=O bands.

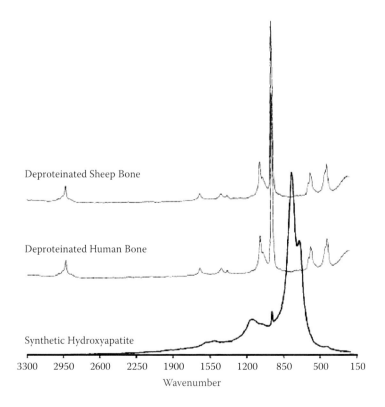

FIGURE 7.3
Raman spectra demonstrating the differences between the synthetic hydroxyapatite, deproteinated human, and sheep bones. (With permission.)

The Raman spectra of calcium carbonate, calcium phosphate, calcium hydroxide, and sodium hydroxide (standard samples) are given in Figure 7.4, and their peak positioning is tabulated in Table 7.1. Similarly, the Raman spectra of deproteinated human bone, hydroxyapatite, synthetic hydroxyapatite, calcium carbonate are given in Figure 7.5 for comparison. All the standard samples were Raman scatterers and the resultant spectra showed very good signal-to-noise ratios, with very little fluorescence.

The results showed that the samples from human and sheep bone were very closely matched in terms of their composition (Figures 7.1, 7.2, and 7.3 and Table 7.1). The results of Walters et al. [12] on an ox femur and Gevorkyan et al. [13] on a human femur support this finding, with all of the spectra showing many shared peaks.

Isolation of the mineral phase of the bone tissue was achieved using anhydrous hydrazine. Termine et a1. [16] reported an 85% decrease in protein

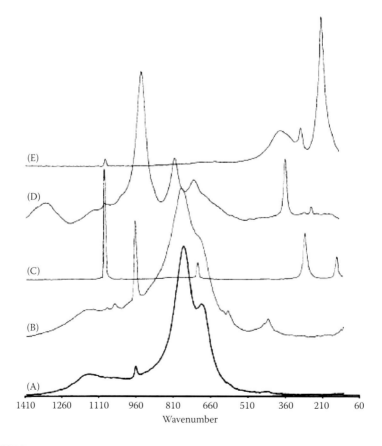

FIGURE 7.4

Raman spectra of commercial synthetic hydroxyapatite (A), calcium phosphate (B), calcium carbonate (C), calcium hydroxide (D), and sodium hydroxide (E). (With permission.)

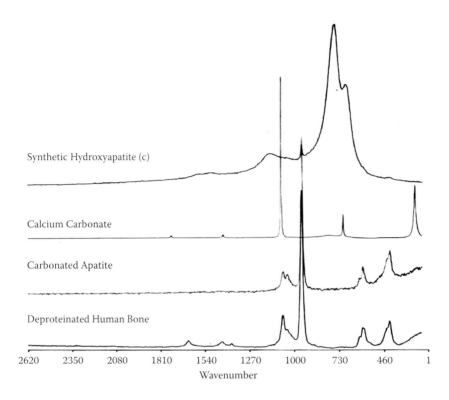

FIGURE 7.5
Raman spectra of commercial synthetic hydroxyapatite (A), calcium phosphate (B), calcium carbonate (C), calcium hydroxide (D), and sodium hydroxide (E). (With permission.)

content, using 0.6-g samples, as a result of this procedure. Walters et al. [12] proposed that when larger samples are used, as was done in this study, the decrease in the protein content of the sample surface is still equal to about 85%. Figure 7.2 shows a considerable reduction in the scattering intensities of peaks that arose from the organic phase of the tissue indicating the deproteination of human and sheep bones. The peaks at 1660, 1262, and 1242 cm^{-1} were likely due to the amide bands of collagen (a combination of C=O stretching, C–N stretching, and N–H bending modes), which disappeared (with an 80–90% decrease, as very small peaks were still present) upon deproteination. This percentage decrease was calculated by measuring the peak areas of the C–H peaks (organic phase) present in the region of 3000–2800 cm^{-1} and taking a ratio with the unchanged peak of symmetric phosphate peak centred at 952 cm^{-1}. Walters et al. [12] concluded that the chemical and morphologic changes in the bone mineral as a result of hydrazine treatment are minimal. The present results support this view, as there were no significant changes in the wavenumber and scattering intensity upon deproteination, whereas the associated changes in the organic peaks were noticeable (Figures 7.1 and 7.2).

A significant finding of this work was the appearance of peaks in the bone tissue (organic phase) spectra in the region 3000–2880 cm^{-1}. The peaks at 2934 and 2935 cm^{-1} (symmetric C–H) stretch mode in the spectra of normal bone tissue were due to an organic substance, judging by the virtual elimination of these peaks upon deproteination (Figure 7.2). The peak at 2915 cm^{-1} (symmetric C–H) in the deproteinated spectra was due to the organic bone tissue after the deproteination procedure. The current results confirm that complete deproteination of bone tissue by hydrazine was not achieved.

The only shared peak between the spectra for synthetic hydroxyapatite spectrum and bone tissue was that at 952 cm^{-1} (Figures 7.3, 7.4, and 7.5, and Table 7.1), which was associated with the phosphate (symmetric) stretch mode. Other regions of these spectra did not match each other as closely. The P120 HAP sample used in this study produced considerable signals at 774 and 690 cm^{-1}, which were assigned as phosphate vibrational modes. Previously, it was thought that there were strong chemical and structural similarities between hydroxyapatite with natural bone [22]. It was reported that there are significant structural differences between the spectral details of synthetic hydroxyapatite and the inorganic matrix of the natural bones. The only shared peak at 952 cm^{-1} had a higher relative intensity for the bone samples, which indicated more of a symmetric P–O stretching mode than that associated with synthetic hydroxyapatite. No other peaks in all of the spectra matched.

The broad band in the region of 900 and 600 cm^{-1}, which dominated the spectrum of synthetic hydroxyapatite, was a combination of two peaks, 774 and 690 cm$^{-1,}$ respectively (Figure 7.3). Tudor et al. [23] suggested two possibilities for the breadth of this band, one being the overlapping of two hydroxide bands and other a combination of carbonate and hydroxide bands. To evaluate these possibilities, the spectrum of hydroxyapatite was compared with the spectra of calcium hydroxide, calcium carbonate, and calcium phosphate. It was established that the broad peaks at 774 and 690 cm^{-1} were not due to calcium hydroxide or calcium carbonate, as the peaks were at different positions (Figure 7.4 and Table 7.1). Calcium hydroxide had two broad peaks centred at 801 and 721 cm^{-1}, which were completely absent in the spectrum of hydroxyapatite.

A comparison of the spectra of hydroxyapatite with calcium phosphate indicated that in both cases the broad bands matched very well in the region of 900 and 600 cm^{-1} (Figure 7.4). Both of these peaks were due to asymmetric P–O modes. Hence, it may be concluded that there are two possibilities for the broadening of the peaks: (1) the presence of multicrystalline phases of the phosphate moiety in the hydroxyapatite, and (2) phosphate bands that are subject to interference from the hydroxide group of the hydroxyapatite. The second possibility is unlikely, as a broad band was also observed in the case of the calcium phosphate spectrum (Figure 7.4). However, the first possibility was confirmed by comparing the hydroxyapatite spectra with the spectrum of sodium hydroxide, which indicated that sharp, well-defined hydroxide bands also appeared at a much lower wavenumber, centred at 293 and 207 cm^{-1}, which were completely absent in the case of synthetic hydroxyapatite. Similarly, two well-defined peaks present at 931 cm^{-1} for calcium hydroxide and 1079 cm^{-1} (O–H symmetric) were completely absent in the spectrum of synthetic hydroxyapatite [24].

Comparison of Natural and Synthetic Apatites

Comparing the spectra of natural bones with that of synthetic hydroxyapatite, it is apparent that there are significant structural differences. The spectra of bone tissue provide a baseline from which to improve the existing synthetic hydroxyl/carbonate/substituted apatite materials.

The spectrum of calcium carbonate (Figures 7.4 and 7.5 and Table 7.1) showed that the peaks at 1433, 1062, and 710 cm^{-1} (carbonate) were also present in the spectrum of bone, indicating that the inorganic phase of the bone contains a substantial amount of carbonate. In contrast, commercial synthetic hydroxyapatite material has only a trace of carbonate moiety.

Calcined bone has also been considered in the past as an implantable and has been of interest to the spectroscopic community. The spectrum of calcined bone is given in Figure 7.6, where the absence of a carbonate group is evident. The carbonate group is less stable compared to the phosphate group and is destroyed during the calcination process.

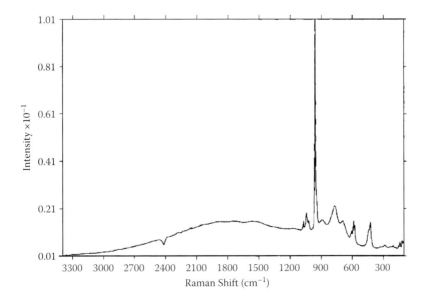

FIGURE 7.6
Raman spectra of calcined bone.

Infrared Spectroscopy of Bone

Fourier transform infrared (FTIR) spectroscopy is a widely used analytical technique that is routinely applied to the characterization of biomaterials. However, preparing samples of biomaterials for infrared spectroscopy is often a tedious process. The main sampling problem in FTIR characterisation of biomaterials is that nearly all solid materials are too opaque in their normal forms for direct transmission analysis in the mid-infrared region. This problem can be solved by reducing the optical density of samples to a suitable level by employing various sampling techniques [25,26]. These procedures, however, can alter the nature of the sample and are time-consuming.

Various alternative techniques, such as diffuse reflectance (DRIFT) and attenuated total reflectance (ATR) can be employed, but are limited in applications due to stringent surface requirements. Another approach is to avoid totally the opacity of the mid-infrared spectral region and work within the near-infrared region by using overtone and combination absorbance bands for analysis, where absorption coefficients are relatively lower and samples are less opaque [27]. Unfortunately, a limited amount of information is available within the near-infrared spectral region, whereas, the mid-infrared region provides most spectral bands for the required characterization.

Photo-acoustic sampling (PAS) provides a solution to these problems. A photo-acoustic signal is generated when infrared radiation absorbed by a sample is converted into heat within the sample. This heat diffuses to the sample surface and into the adjacent atmosphere. Thermal expansion of this gas produces the PAS signal. The signal generation process isolates a layer extending beneath the sample's surface, which has a suitable optical density for analysis, without altering the sample. The technique directly measures the absorbance spectrum of this layer.

The PAS technique in conjunction with FTIR can be employed to characterise natural bones and apatites, which helps to identify differences between the carbonated and standard hydroxyl-apatites and allows comparison of commercial hydroxyapatite powders from different sources. The contribution of the PAS-FTIR technique in the spectrochemical analysis of biomaterials proves to be an ideal technique, where neat samples (without the need of sample preparation) are analysed.

In addition to the human and sheep bone samples, commercial hydroxyapatite (HA) powders (P88, P120, P141, P149) from Plasma Biotal, HA powder from Merck, and carbonated apatite synthesized within our laboratories were also analysed to develop a relationship between the natural and synthetic apatites. The natural apatite was from human and sheep bone.

FTIR Spectral Data Acquisition and Analysis

FTIR spectra were obtained using a Nicolet spectrometer in conjunction with an MTech PAS cell. Good-quality spectra were obtained at 4 cm^{-1} resolution averaging 128 scans. It is deemed necessary to purge the sample with helium gas in combination with a drying agent by employing magnesium perchlorate. This ensured that the system was kept completely dry and free of atmospheric moisture.

The spectra of synthetic and natural apatites provided a number of spectral details indicating some similarities. Spectra of the apatites are given in Figures 7.7–7.12 and their peak positions are tabulated in Table 7.2. FTIR

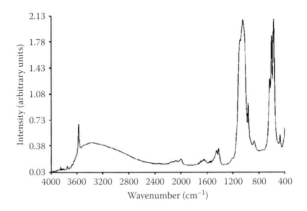

FIGURE 7.7
FTIR spectrum of commercial-grade hydroxyapatite Plasma Biotal (P88).

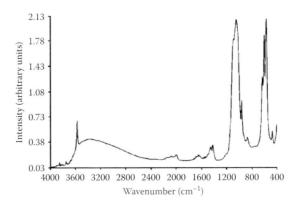

FIGURE 7.8
FTIR spectrum of commercial-grade hydroxyapatite Plasma Biotal (P120).

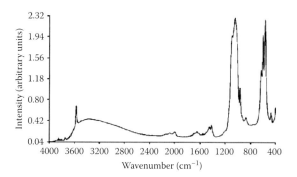

FIGURE 7.9
FTIR spectrum of commercial-grade hydroxyapatite Plasma Biotal (P141).

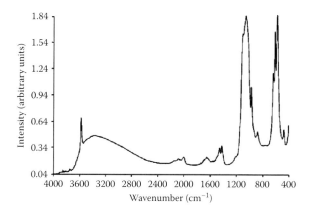

FIGURE 7.10
FTIR spectrum of commercial-grade hydroxyapatite Plasma Biotal (P149).

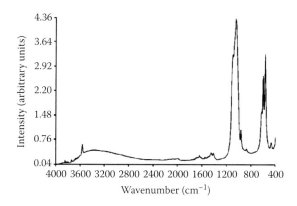

FIGURE 7.11
FTIR spectrum of commercial-grade hydroxyapatite from Merck.

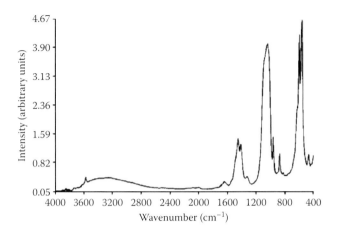

FIGURE 7.12
FTIR spectrum of carbonated hydroxyapatite.

spectra of both the synthetic and natural apatites revealed interesting findings (Figures 7.13 and 7.14), confirming the presence of hydroxyl, carbonate, and phosphate groups, details of which are provided in the following section.

Spectral Bands

Hydroxyl

Hydroxyl stretch is observed at 3569 cm^{-1} in the spectra of synthetic commercial hydroxyapatite and carbonated apatite powders. The hydroxyl band at 3569 cm^{-1}, in the spectrum of carbonated apatite has a lower intensity compared to that from commercial samples, which is also masked by broad H_2O absorptions (Figure 7.15). The large decrease in the hydroxyl band intensities of the carbonated apatite powder may be attributed to the increase in the carbonate substitution. This particular band is extremely valuable in calculating the presence of carbonate and hydroxyl group ratios both in the natural and synthetic apatites (Table 7.2).

The peak areas of hydroxyl stretch in the case of carbonated and P120 hydroxyapatite powders are 1.92 and 7.4, respectively, indicating that the hydroxyl group decreases with an increase in the carbonate substitution. The results obtained agree with the data reported by Elliott et al., who suggested the replacement of carbonate ions with hydroxyl ions [28].

Carbonate Bands

Theoretically, carbonate ions have four vibrational modes, three of which are observed in the infrared spectrum and two of which are observed in the Raman spectrum [24]. The carbonate m4 bands have very low intensity and are

TABLE 7.2

Infrared Spectral Band Positions for Natural Bones, Carbonated, and Hydroxylapatite Powders

Peak Assignment Wavenumber, cm^{-1}	Human Bone	Sheep Bone	Carbonated Apatite (CA)	Commercial Hydroxylapatite Powders				
				Merck (HA)	Plasma Biotal (P88)	Plasma Biotal (P120)	Plasma Biotal (P141)	Plasma Biotal (P149)
Hydroxyl stretch	—	—	3571	3570	3570	3568	3570	3569
Carbonate v_3	1650–1300	1650–1300	1650–1300	1650–1300	1650–1300	1650–1300	1650–1300	1650–1300
– (m)	1650	1650	1650	1648	1648	1648	1650	1651
– (m)	1473	1471	1454	1455	1454	1454	1454	1454
– (m)	1420	1421	1417	1417	1419	1419	1419	1419
– (w)	1327	1328	1321	—	—	—	—	—
Phosphate v_3	1190–976	1190–976	1190–976	1190–976	1190–976	1190–976	1190–976	1190–976
– (m)	1096	1096	—	1091	1088	1092	1089	1089
– (v_s)	1085	1085	1041	1042	1045	1042	1043	1043
– (s)	1056	1056	—	—	—	—	—	—
Phosphate v_1 (m)	961	961	961	962	962	962	961	961
Carbonate v_2 (ms)	873	873	873	877	875	874	873	875
Phosphate v_4	660–520	660–520	660–520	660–520	660–520	660–520	660–520	660–520
– (m)	—	—	629	632	633	633	633	633
– (v_s)	605	601	603	602	602	602	602	602
– (v_s)	565	579	567	566	567	566	566	565
Phosphate v_2 (w)	467	465	469	472	473	472	472	472
	445	445	—	—	—	—	—	—

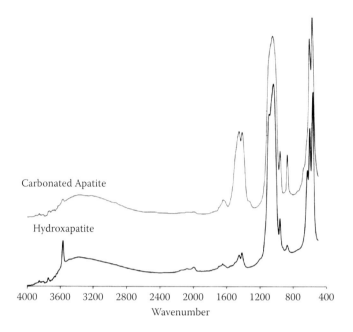

FIGURE 7.13
FTIR spectra of carbonated and hydroxyl apatites. (With permission.)

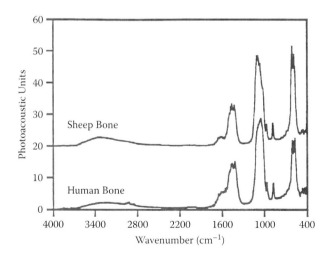

FIGURE 7.14
FTIR spectra of human and sheep bone.

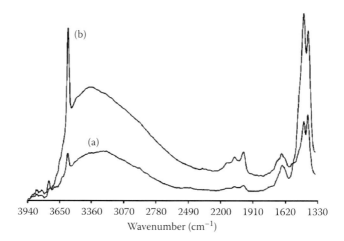

FIGURE 7.15
FTIR spectra of carbonated (a) and hydroxyl (b) apatites, carbonate and hydroxyl bands. (With permission.)

seldom seen in the infrared spectrum [20,21]. Usually, v_1 and v_4 have strong vibrational bands in the Raman spectra of bone and carbonated apatite [22,24], and v_2 and v_3 vibrational modes are observed in the infrared spectra. Carbonate ions occupy two different sites in carbonated apatite: peaks in the region of 1650 to 1300 cm^{-1} are due to the v_3 vibrational mode carbonate ion and the peak at 873 cm^{-1} is due to the v_2 vibrational mode [29]. These carbonate bands in the region of 1650 to 1300 cm^{-1} are assigned to surface carbonate ions, rather than to carbonate ions in the lattice of phosphate ions. The v_3 has a peak split in two peaks centred at 1649 and 1470 cm^{-1}, respectively, with the distribution of the carbonate v_3 sites depending on the maturation and formation of apatite crystals. Occupancy of the v_2 sites is considered to occur competitively between the OH$^-$ and carbonate groups at the interface of growing crystal, whereas occupancy of the v_3 sites depends on competition between the phosphate and carbonate ions [20,30]. This presence of v_2 and v_3 vibrational modes of carbonates in carbonated apatite may contribute to the decrease of hydroxyl ions in the spectrum, as the hydroxyl band present in the spectrum of carbonated apatite is weaker in intensity than that in commercial HA powders.

The spectra of synthetic hydroxyapatites also have peaks in the region of 1650 and 1300 cm^{-1}. Carbonated apatite has two well-defined peaks for the v_3 sites centred at 1649 and 1470 cm^{-1}, whereas synthetic commercial hydroxyapatite has three sites for v_3 vibrational mode centred at 1648, 1454, and 1419 cm^{-1}.

Results obtained by peak area calculation of the carbonate v_3 band indicate that carbonated apatite has more carbonate moiety than hydroxyapatite. The peak areas of carbonated apatite and hydroxyapatite powders are 165 and 13, respectively. The spectra of the carbonated bands of carbonated apatite and hydroxyapatite powders are compared in Figure 7.15.

Phosphate

Theoretically, there are four vibrational modes present for phosphate ions, v_1, v_2, v_3, and v_4. All these modes are Raman and infrared active and are observed for all the spectra of carbonated apatite and hydroxyapatite powders.

In the carbonated spectrum, a single intense v_3 band is present at 1046 cm^{-1}, whereas, in the hydroxyapatite spectra, the v_3 band has three different sites present at 1096, 1085, and 1056 cm^{-1} (Figure 7.16 and Table 7.1). The intense v_3 band is thought to be responsible for totally obscuring the v_1 carbonate bands [20,24,29,30].

The peak area of the v_3 band was calculated and used to determine the phosphate-to-carbonate ions ratio. The peak areas of the v_3 band of carbonated and hydroxyl apatite powders are given in Table 7.2. This ratio increases with an increase in the phosphate content.

Phosphate v_1 band is present at 961 cm^{-1} and can be observed in all the spectra of hydroxyapatite and carbonated apatites (Figures 7.7–7.12 and Tables 7.1 and 7.2).

The phosphate v_4 band is present in the region of 660 and 520 cm^{-1} and is a well-defined and sharp band, observed in the carbonated and hydroxylapatites. It has two sites in the case of carbonated apatite, centred at 603 and 567 cm^{-1}, and hydroxyapatite spectra have three sites observed at 633, 602, and 566 cm^{-1} (see Figure 7.16 and Table 7.1). This splitting of the v_4 vibrational band indicates the low site symmetry of molecules, as two and three observed bands confirm the presence of more than one distinction site for the phosphate group [31].

The phosphate v_1 band is observed in the region of 475 and 440 cm^{-1} and has two sites. These are weak bands, not as strong as the v_3 and v_4 bands.

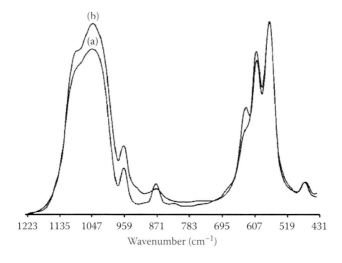

FIGURE 7.16
FTIR spectra of carbonated (a) and hydroxyl (b) apatites, phosphate (v_3) and (v_4) bands. (With permission.)

In carbonated apatite, these sites are at 467 and 450 cm^{-1}, whereas in commercial hydroxylapatite there is only one site, at 472 cm^{-1} [32].

FTIR spectra of synthetic commercial hydroxyapatite and synthetic carbonated apatite powder indicate that there are a number of differences between the two samples. The most obvious change in the spectrum of carbonated apatite is the large decrease of the hydroxyl peak centred at 3568 cm^{-1}, compared to the commercial sample's spectra, which have a well-defined, sharp peak at the same position (see Figure 7.7 and Table 7.2). Infrared spectra of commercial hydroxyapatites have a hydroxyl band at 624 cm^{-1}, which is absent in carbonated apatite [33].

There is a hydroxyl stretch identified at 3568 cm^{-1} and 3570 cm^{-1} in the spectra of the commercial hydroxyapatite powders P120 and P88, respectively, but it is not observed in the spectra of bones. It can be observed that the intensity of the hydroxyl bands for the commercial hydroxyapatite powders are approximately the same. The amount of carbonate substitution is calculated by measuring the peak area of the spectra of the carbonate and hydroxyl bands. The peak area of hydroxyl stretch can be calculated for the commercial hydroxyapatite powders since the bone does not have hydroxyl stretch. In the case of the P120 and P88 hydroxyapatite powders, the peak area is 6.62 and 6.78, respectively. It is apparent, even though the difference may be small, that a decrease in the hydroxyl groups will, in turn, increase the carbonate substitution. Obtained results are tabulated in Table 7.3.

Comparison of Natural and Synthetic Apatite

The spectra of the human and sheep bone are almost identical and the same can be said for the apatite powders.

While comparing the spectra of both the natural and inorganic matrix of bone, it is observed that both the human and sheep bones have three different sites present on the phosphate v_3 band at 1096 cm^{-1}, 1085 cm^{-1}, and 1056 cm^{-1} (see Figure 7.14 and Table 7.3). The intensity of the v_3 is thought to account for the obscurity of the v_1 carbonate bands. The phosphate-to-carbonate ions ratio can be calculated by using the already calculated carbonate v_3 peak area to calculate the peak area of the phosphate v_3 band. The phosphate-to-carbonate ions ratio

TABLE 7.3

Peak Area Calculation for Hydroxyl, Carbonate, and Phosphate Bands, Bone and Hydroxyapatite Spectra

Peak Assignments	Human Bone	Sheep Bone	HA (P120)	HA (P88)	HA (P141)	HA (P149)	HA (Merck)	CA
Hydroxyl stretch			6.62	6.78	6.59	6.37	5.06	1.92
Carbonate (v_3)	188.7	198.9	12.7	11.2	11.1	12.1	11.2	164.8
Phosphate (v_3)	314.9	341.8	225.8	165.7	182.0	138.1	351.6	314.5
PO_4^{3-}/CO_3^{2-}	1.67	1.72	1/17.8	14.8	1/16.4	1/11.4	1/31.4	1.9/1

is calculated and it is discovered that the phosphate-to-carbonate ions ratio for human and sheep bone is 1.67 and 1.72, respectively.

Similarly, the ratio was calculated for the commercial hydroxyapatite powders and the phosphate-to-carbonate ions ratios are tabulated in Table 7.3.

The phosphate v_1 band can be observed at 961 cm^{-1} for both the human and sheep bone. It is also present in the synthetic hydroxyapatite powders. The phosphate v_4 band exists between the 660 cm^{-1} and 500 cm^{-1} and produces a sharp, well-defined band. The phosphate v_4 band can be visible in both the bone and synthetic hydroxyapatite spectra. The phosphate v_2 band has two or three spectral sites. In human and sheep bones, the spectral bands are centred at 605 cm^{-1} and 565 cm^{-1} for human bone, and at 601 cm^{-1} and 579 cm^{-1}, for sheep bone. For the synthetic hydroxyapatite, the sites are centred at 633 cm^{-1}, 602 cm^{-1}, and 566 cm^{-1}. The splitting of the phosphate v_4 band into two and three observed sites proves the presence of more than one distinct site for the phosphate group.

The phosphate v_2 band has two distinct sites for bones and is not as strong as the v_3 and v_4 bands. In human and sheep bone, these sites are 467 cm^{-1} and 445 cm^{-1}, respectively. The synthetic hydroxyapatite has only one site and it is centred on 472 cm^{-1} for the synthetic hydroxyapatite powders.

Spectral information on the synthetic hydroxyapatite spectra gave numerous peaks parallel to that of human and sheep bone. Although this may be the case, the spectra still indicate that there are a number of differences between synthetic hydroxyapatite and bone. The most obvious difference is the presence of the hydroxyl stretch on the hydroxyapatite powders and the lack of hydroxyl stretch on the bone spectra. Also, there is a large (or larger) amount of carbonate present in bone compared to the synthetic hydroxyapatite; the two proofs being that the peak area is much larger in bone and the phosphate-to-carbonate ion ratio is far larger in the synthetic hydroxyapatite.

Summary

The quality of the structural and compositional details that can be revealed by both the FTIR (combined with minimal PAS-FTIR spectroscopy) and Raman spectroscopic techniques makes these two very attractive techniques for the analysis of natural bone tissues and synthetic analogue biomaterials. For analysing neat materials, bone, or apatite powders, PAS-FTIR offers an attractive choice for characterisation without the need to reduce the particle size or dilute with KBr, allowing the analysis of biomaterial in a physiological condition. Quantitative analysis is also possible, which confirms the presence of more carbonate moiety in the natural bone compared to commercial hydroxyapatite and substituted carbonated apatite powders.

Similarly, Raman spectroscopy proves to be a powerful technique to characterise both natural and synthetic materials. Both the FTIR and Raman spectroscopic are useful techniques to get a complete picture of the chemical structural characterisation of natural bone and its replacement materials.

References

1. Carter, D. R., and Spengler, D. M. 1978. Mechanical properties and composition of cortical bone. *Clin. Orthop. Rel. Res.* 135, 192–217.
2. Campo, R. D., and Tourtellotte, C. D. 1967. The composition of bovine cartilage and bone. *Biochim. Biophys. Acta,* 141, 614.
3. Pugliarello, M. C., Vittur, F., Bernard, B., deBonucci, E., and Ascenzi, A 1973. Analysis of bone composition at the microscopic level. *Calcif. Tissue Res.* 12, 209.
4. Riggs, B. L., Wahner, H. W., Dunn, W. L., Mazess, R. B., Offord, K. P., and Melton, L. J. 1981. Differential changes in bone mineral density of the appendicular and axial skeleton with ageing: Relationship to spinal osteoporosis. *J. Clin. Inves.* 67, 328–335.
5. Black, J. 1988. *Orthopaedic Biomaterials in Research and Practice,* 23–74. New York: Churchill Livingstone.
6. Legros, R., Balmain, N., and Bonel, G. 1987. Age-related changes in mineral of rat and bovine cortical bone. *Calcif. Tissue. Int.* 41, 137–144.
7. Rehman, I., Hench, L. L., and Bonfield, W. 1993. Comparison of hydroxycarbonate layers on bioactive glasses with human bone. *Bioceramics* 7, 123–128.
8. Rehman, I., Hench, L. L., Bonfield, W., and Smith, R. 1994. Analysis of surface layers on bioactive glasses. *Biomaterials* 15, 865–870.
9. Martin, R. B, and Burr, D. B. 1989. *Structure, Function and Adaptation of Compact Bone.* New York: Raven Press.
10. Gendreau, R. M. 1986. Fourier transform infrared spectroscopy as a biomedical research tool. *Trends Anal. Chem.* 5, 68–71.
11. Chase, B. 1987. Fourier transform Raman spectroscopy. *Anal. Chem.* 59, 881A–889A.
12. Walters, M. A., Leung, Y. C., Blumenthal, N. C., Legeros, R. Z., and Konsker, K. A. 1990. A Raman and infrared spectroscopic investigation of biological hydroxylapatite. *J. Inorg. Biochem.* 39, 19S200.
13. Gevorkyan, B. Z., Arnotskaya, N. Y., and Fedorova, Y. N. 1984. Study of the structure of bone tissue using the polarized spectra of combination scatter. *Biofizika* 29, 1140–1148.
14. Nelson, D. G. A., and Featherstone, J. D. B. 1982. Preparation, analysis and characterization of carbonated apatites. *Calcif. Tissue Int.* 34, S69–S81.
15. Blumenthal, N. C., and Posner, A. S. 1973. Hydroxyapatite: Mechanism of formation and properties. *Calcif. Tissue Res.* 13, 235–243.
16. Termine, J. D., Eanes, E. D., Greenfield, D. J., Nylen, M. U., and Harper, R. A. 1973. Hydrazine-deproteinated bone mineral, physical and chemical properties. *Calcified Tissue Res.* 12, 73–90.

17. Pleshko, N., Boskey, A., and Mendelsohn, R. 1991. A novel infrared spectroscopic method for the determination of crystallinity of hydroxyapatite minerals. *Biophys.* 60, 786–793.

18. Doyle, B. B., Bendit, E. G., and Blout, E. R. 1975. Infrared spectroscopy of collagen and collagen-like polypeptides. *Biopolymers* 14, 935–957.

19. Chapman, A. C., and Thirlwell, P. 1964. Spectra of phosphorus compounds-I. The infrared spectra of orthophosphates. *Spectrochem. Acta* 20, 937–945.

20. Rehman, I. U., and Bonfield, W. 1997. Characterisation of hydroxyapatite and carbonated apatite by photo-acoustic FTIR spectroscopy. *Journal of Materials Science; Materials in Medicine* 8, 1–4.

21. Smith, R., and Rehman, I. U. 1994. Fourier transform Raman spectroscopic studies of human bone. *Journal of Material. Science; Materials in Medicine* 5 (9–10), 775–778.

22. Cook, S. D., Thomas, K. A., Key, J. K., and Jarcho, M. 1988. Hydroxyapatite-coated porous titanium for use as an orthopedic biologic attachment system. *Clin. Orthop. Xelat. Res.* 230, 303.

23. Tudor, A. M., Melia, C. D., Davies, M. C., Anderson, D., Hasting, G., Morrey, S., Sandoz, J. D., and Barbosa, M. 1993. The analysis of biomedical hydroxyapatite powders and hydroxyapatite coatings on metallic medical implants by near-infrared Fourier transform Raman spectroscopy. *Spectrochem. Acta* 49A, 675–680.

24. Rehman, I. U., Smith, R., Hench, L. L., and Bonfield, W. 1995. Structural evaluation of human and sheep bone and comparison with synthetic hydroxyapatite by FT-Raman spectroscopy. *Journal of Biomedical Materials Research* 29, 1287–1294.

25. Rockley, M. G. 1980. Fourier-transformed infrared photoacoustic spectroscopy of solids. *Spectroscopy* 34, 405–406.

26. Colthup, N. B., Dally, L. H., and Wiberley, S. E. 1990. *Introduction to Infrared and Raman Spectroscopy*, 3rd ed. Boston: Academic Press.

27. Osborne, B. G., and Fearrn, T. 1986. *Near-Infrared Spectroscopy in Food Analysis*. Essex. UK: Longman Scientific and Technical.

28. Elliott, J. C., Holcomb, D. W., and Young, R. A. 1985. Infrared determination of the degree of substitution of hydroxyl by carbonate ions in human dental enamel. *Calcified Tissue International* 37(4), 372–375.

29. Paschalis, E. P., DiCarlo, E., Betts, F., Sherman, P., Mendelsohn, R., and Boskey, A. L. 1996. FTIR micro-spectroscopic analysis of human osteonal bone. *Calcified Tissue International* 59(6), 480–487.

30. Paschalis, E. P., Betts, F., DiCarlo, E., Mendelsohn, R., and Boskey, A. L. 1997. FTIR micro-spectroscopic analysis of normal human cortical and trabecular bone. *Calcified Tissue International* 61(6), 480–486.

31. Kimura-Suda, H., Kajiwara, M., Matsumoto, N., Murayama, H., and Yamato, H. 2009. Characterization of apatite and collagen in bone with FTIR imaging. *Molecular Crystals and Liquid Crystals* 505(1), 64/[302]–69/[307].

32. Paschalis, E. P., Betts, F., DiCarlo, E., Mendelsohn, R., and Boskey, A. L. 1997. FTIR microspectroscopic analysis of human iliac crest biopsies from untreated osteoporotic bone. *Calcified Tissue International* 61(6), 487–492.

33. Elkin, S. L., Vedi, S., Bord, S., Garrahan, N. J., Hodson, M. E., and Compston, J. E. 2002. Histomorphometric analysis of bone biopsies from the iliac crest of adults with cystic fibrosis. *American Journal of Respiratory and Critical Care Medicine* 166(11), 1470–1474.

8

FTIR and Raman Characteristic Peak Frequencies in Biological Studies

Introduction

This chapter describes some of the most widely used peak frequencies of the two spectroscopic methods, Fourier transform infrared (FTIR) and Raman spectroscopy, and their assignments. It is designed to prepare a database of molecular fingerprints for defining the chemical compounds and introducing the most important characteristic peaks.

In spectroscopic studies, accurate peak definitions can have a great influence on the reliability of the results. Combing through the literature, it becomes apparent that most scientists have mainly used the previously published articles to define the data acquired from the collected spectra. However, without having a reliable and detailed database that can cover most of the famous peaks in the spectral range, it would be a time-consuming task to find the meaning of different peak intensities in the related articles. This chapter tries to alleviate that shortcoming.

In spite of applying different methods, there seems to be a considerable similarity in defining the peaks of identical spectral areas. As a result, it is believed that preparing a unique collection of the frequencies encountered in spectroscopic studies can lead to significant improvements in both the quantity and quality of spectroscopic studies of biological molecules. As a result, this chapter presents a precise database (Table 8.1) on the major characteristic peak frequencies reported in the literature on studies of natural tissues by spectroscopic means and will surely be of considerable assistance to scientists focusing in this subject area.

TABLE 8.1

Major Characteristic Peak Frequencies Reported in the Literature

Peak	Assignment	FT-IR	Raman	Reference Number
200–70 cm^{-1}	CH$_3$ torsion, usually not detected experimentally and theoretically		*	1
250 cm^{-1}	(CH$_3$)$_{ester}$ modes: deformations torsion (with very low Raman activity)		*	1
415 cm^{-1}	Phosphatidylinositol		*	2
418 cm^{-1}	Cholesterol		*	3
428 cm^{-1}	Symmetric stretching vibration of ν_2PO$_4$$^{3-}$(phosphate of HA)		*	3
429 cm^{-1}	Cholesterol, cholesterol ester		*	2
436 cm^{-1}	Thymine (ring breathing modes of the DNA/RNA bases)		*	4
445 cm^{-1}	N–C–S stretch (one of three thiocyanate peaks, with 2095 and 735 cm^{-1})		*	5
447/54 cm^{-1}	Ring torsion of phenyl(2)		*	6
472/5 cm^{-1}	C$_\alpha$=C$_\alpha$, torsion and C–OH$_3$ torsion of methoxy group(4)	*		6
477 cm^{-1}	Polysaccharides, amylose		*	7
	Polysaccharides, amylopectin		*	7
481 cm^{-1}	DNA		*	8
484–90 cm^{-1}	Glycogen		*	9
490 cm^{-1}	Glycogen		*	10
505/8 cm^{-1}	C–OH$_3$ torsion of methoxy group(1)		*	6
509 cm^{-1}	S–S disulfide stretching band of collagen		*	3
	ν(S–S) *gauche-gauche-gauche* (amino acid cysteine)		*	7
519 cm^{-1}	Phosphatidylinositol		*	2
521 cm^{-1}	C$_\alpha$=C$_\alpha$, torsion and ring torsion of phenyl(1)	*		6
524 cm^{-1}	S–S disulfide stretching in proteins		*	9, 10
	Phosphatidylserine		*	2
	ν(S–S) *gauche-gauche-trans* (amino acid cysteine)		*	7
538 cm^{-1}	Cholesterol ester		*	2
	Disulfide bridges in cysteine (this peak seems to be the main factor for the FT-Raman determination of the boundaries between healthy and pathological tissue)		*	11

TABLE 8.1 (*Continued*)

Major Characteristic Peak Frequencies Reported in the Literature

Peak	Assignment	FT-IR	Raman	Reference Number
540 cm^{-1}	ν(S–S) *trans-gauche-trans* (amino acid cysteine)		*	7
548 cm^{-1}	Cholesterol		*	2
540 cm^{-1}	Glucose-saccharide band (overlaps with acyl band)		*	2
573 cm^{-1}	Tryptophan/cytosine, guanine		*	9
576 cm^{-1}	Phosphatidylinositol		*	2
583/6 cm^{-1}	OH out-of-plane bending (free)		*	6
589 cm^{-1}	Symmetric stretching vibration of $v_4PO_4^{3-}$ (phosphate of HA)		*	3
	Glycerol		*	2
596 cm^{-1}	Phosphatidylinositol		*	2
600–800 cm^{-1}	Nucleotide conformation		*	12
600–900 cm^{-1}	CH out-of-plane bending vibrations	*		13
606 cm^{-1}	Ring deformation of phenyl(1)	*		6
607 cm^{-1}	Glycerol		*	2
608 cm^{-1}	Ring deformation of phenyl(1)	*		6
	Cholesterol		*	2
	Has no identifiable theoretical match; it may be assigned to in-plane N–C=O bend		*	14
614 cm^{-1}	Cholesterol ester		*	2
617 cm^{-1}	Thymine (ring breathing modes of the DNA/RNA bases)		*	4
618 cm^{-1}	C–C twisting (protein)		*	12
620 cm^{-1}	C–C twist aromatic ring (one of C–C vibrations to be expected in aromatic structure of xylene)		*	15
621 cm^{-1}	C–C twisting mode of phenylalanine		*	9, 10, 16
622 cm^{-1}	Protein peak		*	17
	C–C twisting mode of phenylalanine in tumours		*	17
630 cm^{-1}	Glycerol		*	2
630–70 cm^{-1}	ν(C–S) *gauche* (amino acid methionine)		*	7
635 cm^{-1}	OH out-of-plane bend (associated)	*		5
640 cm^{-1}	C–S stretching and C–C twisting of proteins-tyrosine		*	12
642 cm^{-1}	Protein peak		*	17

(Continued)

TABLE 8.1 (*Continued*)

Major Characteristic Peak Frequencies Reported in the Literature

Peak	Assignment	FT-IR	Raman	Reference Number
	C–C twist of tyrosine in tumours		*	17
643 cm^{-1}	C–C twisting mode of tyrosine		*	3, 9
645 cm^{-1}	C–C twisting mode of tyrosine		*	16
	Ring, cyclic deformation		*	14
646 cm^{-1}	C–C twisting mode of tyrosine		*	3
649 cm^{-1}	Ring, cyclic deformation		*	14
657 cm^{-1}	CH$_{2, 6, 6'}$ out-of-plane bend and ring puckering	*		6
659 cm^{-1}	Ring deformation, cyclic rock		*	14
660 cm^{-1}	CH$_{2, 6, 6'}$ out-of-plane bend and ring puckering	*		6
662 cm^{-1}	C–S stretching mode of cystine (collagen type I)		*	3, 10
664/6 cm^{-1}	T, G, and ν(C–S) of cysteine (DNA/RNA)		*	17
666 cm^{-1}	G, T (ring breathing modes in the DNA bases)-tyrosine-G backbone in RNA		*	12
667/9 cm^{-1}	C–S stretching mode of cystine (collagen type I)		*	3, 10
669 cm^{-1}	C–S stretching mode of cytosine		*	9
669 cm^{-1}	ν_7(δ: Porphyrin deformation), observed in the spectra of single human RBC		*	18
671 cm^{-1}	C–S stretching mode of cystine		*	16
672 cm^{-1}	Guanine (ring breathing modes of DNA/RNA bases)		*	4
676 cm^{-1}	Ring deformation		*	14
678 cm^{-1}	Ring breathing modes in the DNA bases		*	12
	G (ring breathing modes in the DNA bases)/C-2′-endo-anti			
686 cm^{-1}	Ring deformation		*	14
700–45 cm^{-1}	ν(C–S) *trans* (amino acid methionine)		*	7
700–1000 cm^{-1}	Out-of-plane bending vibrations	*		6
702 cm^{-1}	Cholesterol, cholesterol ester		*	2
702/3 cm^{-1}	Ring, C–NR$_2$, cyclic stretch with C–H rock		*	14
702/04/17 cm^{-1}	CH$_{5',2'}$ out-of-plane bend and ring puckering	*	*	6
710–50 cm^{-1}	A weak signal due to in-plane chain deformation modes (C–O–C) of the ester moiety		*	1

TABLE 8.1 (*Continued*)

Major Characteristic Peak Frequencies Reported in the Literature

Peak	Assignment	FT-IR	Raman	Reference Number
717 cm^{-1}	CH$_{5',2'}$ out-of-plane bend and ring puckering	*	*	6
717 cm^{-1}	C–N (membrane phospholipids head)/adenine		*	9
718 cm^{-1}	Choline group		*	2
	C–N (membrane phospholipid head)/nucleotide peak symmetric choline stretch, phospholipids, (H3C)N+ stretch band		*	14
719 cm^{-1}	C–N (membrane phospholipid head)/nucleotide peak		*	10
	Symmetric stretch vibration of choline group N$^+$(CH$_3$)$_3$, characteristic for phospholipids		*	2
	Phosphatidylcholine, sphingomyelin		*	2
720 cm^{-1}	DNA		*	8
	C–N stretching in lipids/adenine		*	16
	C–N (membrane phospholipid head)/nucleotide peak symmetric choline stretch, phospholipids, (H3C)N+ stretch band		*	17
722 cm^{-1}	DNA		*	8
724 cm^{-1}	C–N (membrane phospholipid head)/nucleotide peak symmetric choline stretch, phospholipids, (H3C)N+ stretch band		*	17
725 cm^{-1}	A (ring breathing mode of DNA/RNA bases)		*	12
726 cm^{-1}	C–S (protein), CH$_2$ rocking, adenine		*	9
727 cm^{-1}	Adenine (ring breathing modes of DNA/RNA bases)		*	4
727/8 cm^{-1}	C–C stretching, proline (collagen assignment)		*	19
733 cm^{-1}	Phosphatidylserine		*	2
735 cm^{-1}	C–S stretch (one of three thiocyanate peaks, with 2095 and 445 cm^{-1})		*	5
737 cm^{-1}	C–O$_{4'}$ bend	*	*	6

(*Continued*)

TABLE 8.1 (*Continued*)

Major Characteristic Peak Frequencies Reported in the Literature

Peak	Assignment	FT-IR	Raman	Reference Number
740 cm^{-1}	Ring deformation		*	14
741/46 cm^{-1}	C–O$_{4'}$ bend	*	*	6
746 cm^{-1}	T (ring breathing mode of DNA/RNA bases)		*	12
748 cm^{-1}	DNA		*	8
749 cm^{-1}	Symmetric breathing of tryptophan (protein assignment)		*	3, 10, 20
750 cm^{-1}	CH$_{2,6}$ out-of-plane bending, observed in the spectra of single human RBC		*	6
	Lactic acid		*	8
752 cm^{-1}	ν_{15} (porphyrin breathing mode), (the most informative about the status of RBC and a direct measure of the heme groups of the hemoglobins)		*	18
	DNA		*	8
	Symmetric breathing of tryptophan		*	17
752/3/4/5 cm^{-1}	Symmetric breathing of tryptophan (protein assignment)		*	3, 10, 20
755 cm^{-1}	Symmetric breathing of tryptophan		*	9, 16
757 cm^{-1}	CH$_{2,6}$ out-of-plane bending	*		6
758 cm^{-1}	Symmetric ring breathing of tryptophan, ν_s(O–P–O)		*	17
759 cm^{-1}	Tryptophan		*	9
	Ethanolamine group		*	2
	Phosphatidylethanolamine		*	2
	CH$_{2,6}$ out-of-plane bending	*		6
760 cm^{-1}	Tryptophan, δ(ring)		*	7
762 cm^{-1}	Symmetric ring breathing of tryptophan, ν_s(O–P–O)		*	17
763 cm^{-1}	In-phase ring CH wag		*	14
766 cm^{-1}	Pyrimidine ring breathing mode		*	5
770 cm^{-1}	CH$_{\alpha,\alpha'}$ wagging and CH$_{6',6}$ out-of-plane bending and ring puckering	*		6
776 cm^{-1}	Phosphatidylinositol		*	2
777 cm^{-1}	Ring deformation, cyclic deformation, C–H twist		*	14
780 cm^{-1}	Uracil-based ring breathing mode		*	5
781 cm^{-1}	Cytosine/uracil ring breathing (nucleotide)		*	9, 10

TABLE 8.1 (*Continued*)

Major Characteristic Peak Frequencies Reported in the Literature

Peak	Assignment	FT-IR	Raman	Reference Number
782 cm⁻¹	DNA		*	8
	Ring deformation, cyclic deformation, C–H twist		*	14
	U, C, T ring breathing, DNA		*	17
784 cm⁻¹	Phosphodiester; cytosine		*	21
	U, C, T ring breathing, DNA		*	17
785 cm⁻¹	U, T, C (ring breathing modes in the DNA/RNA bases)		*	12
	Backbone O–P–O			
	Phosphodiester; cytosine		*	21
786 cm⁻¹	DNA: O–P–O, cytosine, uracil, thymine			9
	Pyrimidine ring breathing mode		*	4
787 cm⁻¹	Can be taken as a measure of the relative quantity of nucleic acids present		*	22
	Phosphatidylserine		*	2
788 cm⁻¹	DNA:O–P–O backbone stretching/thymine/cytosine		*	16
789 cm⁻¹	CH$_{\alpha,\alpha'}$ wagging and CH$_{6',6}$ out-of-plane bending and ring puckering	*		6
792 cm⁻¹	CH$_{\alpha,\alpha'}$ wagging and CH$_{2',5',6'}$ out-of-plane bending and ring puckering	*		6
793 cm⁻¹	Guanine in a $C_3'endo/syn$ conformation in the Z conformation of DNA	*		23
794 cm⁻¹	CH$_{\alpha,\alpha'}$ wagging and CH$_{2',5',6'}$ out-of-plane bending and ring puckering	*		6
800–1200 cm⁻¹	Backbone geometry and phosphate ion interactions		*	12
802 cm⁻¹	Uracil-based ring breathing mode		*	5
802–5 cm⁻¹	Left-handed helix DNA (Z form)	*		23
805 cm⁻¹	$C_3endo/anti$ (A-form helix) conformation	*		23
810/2 cm⁻¹	Phosphodiester (Z-marker)		*	21
813 cm⁻¹	Ring CH deformation	*		6
	One of the two most distinct peaks for RNA (with 1240 cm⁻¹)		*	12
813 cm⁻¹	C–C stretching (collagen assignment)		*	19

(*Continued*)

TABLE 8.1 (*Continued*)

Major Characteristic Peak Frequencies Reported in the Literature

Peak	Assignment	FT-IR	Raman	Reference Number
815 cm^{-1}	Proline, hydroxyproline, tyrosine, $v_2PO_2^-$ stretch of nucleic acids		*	3
817 cm^{-1}	C–C stretching (collagen assignment)		*	19
820 cm^{-1}	Protein band		*	24, 25
	Structural protein modes of tumours		*	26
	Proteins, including collagen I		*	8
820–930 cm^{-1}	C–C stretch of proline and hydroxyproline		*	27
822 cm^{-1}	Phosphodiester		*	21
823 cm^{-1}	Out-of-plane ring breathing, tyrosine (protein assignment)		*	20
823/5/6 cm^{-1}	Phosphodiester		*	21
826 cm^{-1}	O–P–O stretch DNA		*	9
827 cm^{-1}	Proline, hydroxyproline, tyrosine, $v_2PO_2^-$ stretch of nucleic acids		*	3
828 cm^{-1}	Out-of-plane ring breathing, tyrosine/O–P–O stretch DNA		*	9, 10
	Phosphodiester		*	21
829 cm^{-1}	C_2' endo conformation of sugar	*		23
830 cm^{-1}	Proline, hydroxyproline, tyrosine, $v_2PO_2^-$ stretch of nucleic acids		*	3
	C–H out-of-plane bending in benzoid ring		*	28
	Protein peak		*	17
	Out-of-plane ring breathing tyrosine/v(O–P–O), DNA in tumours		*	17
830 cm^{-1}	Tyrosine (Fermi resonance of ring fundamental and overtone)		*	7
831 cm^{-1}	Asymmetric O–P–O stretching, tyrosine		*	12
	C_2' endo conformation of sugar	*		23
832/3 cm^{-1}	Ring deformation, C–C stretch		*	14
833 cm^{-1}	DNA:O–P–O backbone stretching/out-of-plane ring breathing in tyrosine		*	16
	C_2' endo conformation of sugar		*	23
835 cm^{-1}	C_2 *endo/anti* (B-form helix) conformation	*		23
835–40 cm^{-1}	Left-handed helix DNA (*Z* form)	*		23
835/6 cm^{-1}	CH$_{6',5',2'}$ ring out-of-plane bending	*		6

TABLE 8.1 (*Continued*)

Major Characteristic Peak Frequencies Reported in the Literature

Peak	Assignment	FT-IR	Raman	Reference Number
838 cm^{-1}	Deformative vibrations of amine groups		*	28
840 cm^{-1}	α-anomers		*	2
	Glucose-saccharide band (overlaps with acyl band)		*	2
	Saccharide (α)		*	2
840–60 cm^{-1}	Polysaccharide structure		*	29, 30
842 cm^{-1}	Glucose		*	9
847 cm^{-1}	Monosaccharides (α-glucose), (C–O–C) skeletal mode		*	7
	Disaccharide (maltose), (C–O–C) skeletal mode		*	7
850 cm^{-1}	Most probably due to single bond stretching vibrations for the amino acids and valine and polysaccharides		*	29
	Tyrosine (Fermi resonance of ring fundamental and overtone)		*	7
850–950 cm^{-1}	Signal free area of amphetamine		*	31
850–990 cm^{-1}	Ring stretch, C–N–R2 stretch and cyclic C–C–C bend		*	14
852 cm^{-1}	Proline, hydroxyproline, tyrosine		*	3
	Tyrosine ring breathing		*	12
	Glycogen		*	8
	Ring breathing of tyrosine and ν(C–C) hydroxyproline ring specific to collagen		*	17
853 cm^{-1}	Ring breathing mode of tyrosine and C–C stretch of proline ring		*	9, 10
	Glycogen		*	8
853/4 cm^{-1}	Structural proteins like collagen		*	32
854 cm^{-1}	CH$_{2,6}$ out-of-plane bending	*	*	6
	(C–O–C) skeletal mode of α-anomers (polysaccharides, pectin)		*	7
	Protein peak		*	17
	Ring breathing of tyrosine and ν(C–C) hydroxyproline ring specific to collagen		*	17
	Ring breathing in tyrosine/C–C stretching in proline		*	16
855 cm^{-1}	Proline, tyrosine		*	20

(Continued)

TABLE 8.1 (*Continued*)

Major Characteristic Peak Frequencies Reported in the Literature

Peak	Assignment	FT-IR	Raman	Reference Number
	ν(C–C), proline + δ(CCH) ring breathing, tyrosine (protein assignment and polysaccharide)		*	20
	δ(CCH) phenylalanine, olefinic (protein assignment and polysaccharide)		*	33
855/6 cm⁻¹	Proline, hydroxyproline, tyrosine		*	3
	C–C stretching, proline (collagen assignment)		*	19
856 cm⁻¹	Amino acid side chain vibrations of proline and hydroxyproline, as well as a (C–C) vibration of the collagen backbone hydroxyproline (collagen type I)		*	3
857 cm⁻¹	CH$_{2,6}$ out-of-plane bending	*		6
859 cm⁻¹	Tyrosine, collagen		*	9
860 cm⁻¹	C$_3$*endo/anti* (A-form helix) conformation	*		23
	Phosphate group		*	2
	Phosphatidic acid		*	2
864 cm⁻¹	Structural proteins like collagen		*	34
865 cm⁻¹	Ring deformation, OC–N and cyclic C–C–C stretch		*	14
867 cm⁻¹	Ribose vibration, one of the distinct RNA modes (with 915 and 974 cm⁻¹)		*	12
868 cm⁻¹	Left-handed helix DNA (Z form)	*	*	23
	C–C stretching, hypro (collagen assignment)		*	19
	Monosaccharides (β-fructose), (C–O–C) skeletal mode		*	7
	Disaccharide (sucrose), (C–O–C) skeletal mode		*	7
	Polysaccharides, amylose		*	7
	Polysaccharides, amylopectin		*	7
	Proline		*	17
869 cm⁻¹	Proline		*	9
870 cm⁻¹	Most probably due to single-bond stretching vibrations for the amino acids proline and valine and polysaccharides		*	29
	C–C stretching, hypro (collagen assignment)		*	19
	Proline		*	17

TABLE 8.1 (*Continued*)

Major Characteristic Peak Frequencies Reported in the Literature

Peak	Assignment	FT-IR	Raman	Reference Number
871 cm^{-1}	CH$_{2,2'}$ out-of-plane bending	*		6
873 cm^{-1}	Hydroxyproline, tryptophan		*	3
874 cm^{-1}	CH$_{2,2'}$ out-of-plane bending	*		6
	C–C stretching, hypro (collagen assignment)		*	19
875 cm^{-1}	Antisymmetric stretch vibration of choline group N$^+$(CH$_3$)$_3$, characteristic for phospholipids		*	2
	Phosphatidylcholine, sphingomyelin		*	2
	Hydroxyproline (of collagen)		*	35
876 cm^{-1}	ν(C–C), hydroxyproline (protein assignment)		*	20
	C–C stretching, hydroxyproline (collagen assignment)		*	19
	ν(C–C), hydroxyproline in tumours		*	17
878 cm^{-1}	C$_3$endo/anti (A-form helix) conformation	*		23
879 cm^{-1}	Hydroxyproline, tryptophan		*	3
880 cm^{-1}	Tryptophan, δ(ring)		*	7
883 cm^{-1}	ρ(CH$_2$) (protein assignment)		*	33
884 cm^{-1}	Proteins, including collagen I		*	8
885 cm^{-1}	Disaccharide (cellobiose), (C–O–C) skeletal mode		*	7
886 cm^{-1}	Ring deformation and symmetric C–N–C stretch		*	14
889 cm^{-1}	C–C, C–O deoxyribose	*		36
890 cm^{-1}	Protein bands		*	24, 25
	C$_3$endo/anti (A-form helix) conformation	*		23
	C$_2$endo/anti (B-form helix) conformation	*		23
	Structural protein modes of tumours		*	26
	β-anomers		*	2
	Ring deformation and symmetric C–N–C stretch		*	14
891 cm^{-1}	Saccharide band (overlaps with acyl band)		*	2
892 cm^{-1}	C–C, C–O deoxyribose	*		23
	Fatty acid, saccharide (β)			

(*Continued*)

TABLE 8.1 (*Continued*)

Major Characteristic Peak Frequencies Reported in the Literature

Peak	Assignment	FT-IR	Raman	Reference Number
893 cm⁻¹	CH₂' out-of-plane bending	*		6
	Backbone, C–C skeletal		*	12
893/5/6 cm⁻¹	Phosphodiester, deoxyribose		*	21
896 cm⁻¹	ν(C–C) protein backbone a-helix (proline/glycogen), collagen		*	17
898 cm⁻¹	Monosaccharides (β-glucose), (C–O–C) skeletal mode		*	7
	Disaccharide (maltose), (C–O–C) skeletal mode		*	7
899 cm⁻¹	CH₂' out-of-plane bending	*		6
900 cm⁻¹	C–C skeletal stretching in protein		*	16
900–50 cm⁻¹	Variations in the secondary structures of proteins		*	37
900–1300 cm⁻¹	Phosphodiester region	*		38
	The nucleic acids	*		39
900–1350 cm⁻¹	Phosphodiester stretching bands region (for absorbances due to collagen and glycogen)	*		40
900–1450 cm⁻¹	CH₂/CH₃ deformations (scissoring, wagging, twisting, and rocking)		*	1
907 cm⁻¹	Formalin contamination peak on fixed tissues		*	41
913 cm⁻¹	Glucose		*	2
915 cm⁻¹	Ribose vibration, one of the distinct RNA modes (with 915 and 974 cm⁻¹)		*	12
918 cm⁻¹	Proline, hydroxyproline		*	3
	Glycogen and lactic acid		*	8
920 cm⁻¹	C–C stretch of proline ring/glucose/lactic acid		*	9, 10
	C–C, praline ring (collagen assignment)		*	19
924 cm⁻¹	Ring CH wag, cyclic C–C stretch and C–N–C bend		*	14
	Ring CH wag		*	14
925–9 cm⁻¹	Left-handed helix DNA (Z form)	*		23
928–40 cm⁻¹	ν(C–C), stretching; probably in amino acids proline and valine (protein band)		*	30, 42
931 cm⁻¹	Carbohydrates peak for solutions and solids		*	24
	Carbohydrates		*	30

TABLE 8.1 (*Continued*)

Major Characteristic Peak Frequencies Reported in the Literature

Peak	Assignment	FT-IR	Raman	Reference Number
932 cm^{-1}	Skeletal C–C, α-helix		*	9
933 cm^{-1}	Proline, hydroxyproline, ν(C–C) skeletal of collagen backbone		*	3
934 cm^{-1}	C–C backbone (collagen assignment)		*	19
935 cm^{-1}	C–C stretching mode of proline, valine, and protein backbone		*	10, 20
	(α-helix conformation)/glycogen (protein assignment)		*	9
	P(CH$_3$) terminal, proline, valine + ν(CC) α-helix keratin (protein assignment)		*	33
936 cm^{-1}	Protein peak		*	17
937 cm^{-1}	Proline (collagen type I)		*	3
	Amino acid side chain vibrations of proline and hydroxyproline, as well as a		*	3
	(C–C) vibration of the collagen backbone			
	C–C backbone (collagen assignment)		*	19
	Glycogen		*	8
	ν(C–C) residues (α-helix)		*	43
	Out-of-phase ring CH wag		*	14
937/8 cm^{-1}	Proline, hydroxyproline, ν(C–C) skeletal of collagen backbone		*	3
938 cm^{-1}	Unassigned	*		38
	C–C stretch backbone		*	9
939 cm^{-1}	Structural proteins like collagen		*	32
	C–C skeletal stretching in protein		*	16
940 cm^{-1}	Carotenoid	*		25
	CH$_2$ rocking mode		*	44
941 cm^{-1}	Skeletal modes (polysaccharides, amylose)		*	7
	Skeletal modes (polysaccharides, amylopectin)		*	7
943 cm^{-1}	Cyclic C–C, ring and C–NR2 stretch		*	14
	Out-of-phase ring CH wag		*	14
945 cm^{-1}	CH$_{6',5'}$ out of plane	*		6
948 cm^{-1}	Structural proteins like collagen		*	34

(*Continued*)

TABLE 8.1 (*Continued*)

Major Characteristic Peak Frequencies Reported in the Literature

Peak	Assignment	FT-IR	Raman	Reference Number
950 cm^{-1}	Most probably due to single-bond stretching vibrations for the amino acids proline and valine and polysaccharides		*	29
	CH$_{6',5'}$ out of plane		*	6
950–1050 cm^{-1}	Amphetamine has a group of three bands at this region		*	31
950–1250 cm^{-1}	Symmetric and asymmetric stretching vibrations from the phosphate group	*		30
951 cm^{-1}	ν_s(CH$_3$) of proteins (α-helix)		*	43
956 cm^{-1}	Carotenoids (absent in normal tissues)		*	45
957 cm^{-1}	Hydroxyapatite, carotenoid, cholesterol		*	9
959 cm^{-1}	CH$_2$ rock		*	46
	Ring, C–N–R2 stretch, cyclic C–C–C bend		*	14
960 cm^{-1}	Symmetric stretching vibration of ν_1PO$_4^{3-}$ (phosphate of HA)		*	3
	CH$_{2,6'}$ out-of-plane bending	*		6
	Calcium—Phosphate stretch band (high quantities of cholesterol)		*	13, 24
	Quinoid ring in-plane deformation		*	28
	Deoxyribose-phosphate	*		47
961 cm^{-1}	C–O deoxyribose, C–C	*		48
962 cm^{-1}	Phosphate symmetric stretching vibration of calcium hydroxyapatite		*	49
963 cm^{-1}	Unassigned in protein assignments		*	20
	CH$_{2,6'}$ out-of-plane bending		*	6
	δ(C=O) (polysaccharides, pectin)	*		7
963/4 cm^{-1}	C–C, C–O deoxyribose	*		23
964 cm^{-1}	C–C and C–O in deoxyribose of DNA of tumour cells	*		23
	C–O deoxyribose, C–C	*		48
965 cm^{-1}	C–O stretching of the phosphodiester and the ribose	*		50
966 cm^{-1}	C–O deoxyribose, C–C	*		48
	DNA	*		51

TABLE 8.1 (*Continued*)

Major Characteristic Peak Frequencies Reported in the Literature

Peak	Assignment	FT-IR	Raman	Reference Number
	Hydroxyapatite		*	9
968 cm⁻¹	Lipids		*	24
	δ(=CH) wagging		*	17
970 cm⁻¹	Phosphate monoester groups of phosphorylated proteins and cellular nucleic acids		*	24
	Symmetric stretching mode of dianionic phosphate monoesters of phosphorylated proteins or cellular nucleic acids	*		52
	DNA	*		53
	δ(=CH) wagging		*	17
971 cm⁻¹	νPO₄= of nucleic acids and proteins	*		51
	Symmetric stretching mode of dianionic phosphate monoester in phosphorylated proteins, such as phosvitin	*		52
		*		51
	ν(C–C) wagging		*	43
972 cm⁻¹	OCH₃ (polysaccharides, pectin)	*		7
972/3 cm⁻¹	C–C backbone (collagen assignment)		*	19
973 cm⁻¹	ρ(CH₃), δ(CCH) olefinic (protein assignment)		*	33
974 cm⁻¹	Ribose vibration, one of the distinct RNA modes (with 874 and 915 cm⁻¹)		*	12
978 cm⁻¹	CH₂,₆′ out-of-plane bending	*		6
985 cm⁻¹	OCH₃ (polysaccharides-cellulose)	*		7
990–1010 cm⁻¹	Ring stretch modes of benzene derivatives		*	14
994 cm⁻¹	C–O ribose, C–C	*		48
995 cm⁻¹	Ring breathing	*		6
996 cm⁻¹	Carbohydrates peak for solutions and solids		*	24
	C–O ribose, C–C	*		48
	Carbohydrates (weak shoulder)		*	30
999 cm⁻¹	ν₄₅(CC), observed in the spectra of single human RBC		*	18
1000 cm⁻¹	Phenylalanine		*	54
	Bound and free NADH		*	54

(Continued)

TABLE 8.1 (*Continued*)

Major Characteristic Peak Frequencies Reported in the Literature

Peak	Assignment	FT-IR	Raman	Reference Number
	Aromatic stretch of a substituted benzene ring		*	14
	Ring stretch		*	14
1006 cm^{-1}	Symmetric ring breathing mode of phenylalanine		*	16
1000–50 cm^{-1}	Ring stretching vibrations mixed strongly with CH in-plane bending	*		13
1000–140 cm^{-1}	Protein amide I absorption	*		55
1000–150 cm^{-1}	A reasonably specific area for nucleic acids in the absence of glycogen	*		50
1000–200 cm^{-1}	C–OH bonds in oligosaccharides such as mannose and galactose	*		56
	Mainly from phosphate or oligosaccharide groups	*		56
1000–300 cm^{-1}	CH in-plane bending vibrations	*		13
	The aromatic CH bending and rocking vibrations	*		6
1000–350 cm^{-1}	Region of the phosphate vibration	*		48
1000–650 cm^{-1}	Porphyrin ring of heme proteins	*		57
1001 cm^{-1}	Symmetric ring breathing mode of phenylalanine		*	9, 10
1002 cm^{-1}	C–C aromatic ring stretching		*	15
	Phenylalanine			54
	Phenylalanine and tryptophan (ring breathing modes of DNA/RNA bases)		*	4
1002 cm^{-1}	Phenylalanine		*	3
	ν_s Symmetric ring breathing mode, phenylalanine		*	17
	Phenylalanine (collagen assignment)		*	19
1003 cm^{-1}	Phenylalanine, C–C skeletal		*	12
	Phenylalanine residues		*	58, 59
1004 cm^{-1}	Phenylalanine (of collagen)		*	3
	ν_s(C–C), Symmetric ring breathing, phenylalanine (protein assignment)		*	20, 33
	Phenylalanine (collagen assignment)			19
	Phenyl breathing mode		*	33

TABLE 8.1 (*Continued*)

Major Characteristic Peak Frequencies Reported in the Literature

Peak	Assignment	FT-IR	Raman	Reference Number
	ν(C–C) phenylalanine		*	43
	Protein peak		*	17
	Phenylalanine		*	60
	Phenylalanine ring breathing mode		*	35
	ν_s Symmetric ring breathing mode, phenylalanine		*	17
1005 cm^{-1}	Aromatic stretch of a substituted benzene ring		*	14
	Ring stretch		*	14
1006 cm^{-1}	Carotenoids (absent in normal tissues)		*	45
	Phenylalanine, δ(ring)		*	7
1008 cm^{-1}	CH$_{\alpha,\alpha'}$ out-of-plane bending and C$_\alpha$=C$_{\alpha'}$ torsion	*		6
	Phenylalanine		*	22
	ν(CO), ν(CC), δ(OCH), ring (polysaccharides, pectin)		*	7
1009/10/1 cm^{-1}	Stretching C–O deoxyribose	*		48
1011 cm^{-1}	CH$_{\alpha,\alpha'}$ out-of-plane bending and C$_\alpha$=C$_{\alpha'}$ torsion	*		6
1015 cm^{-1}	νC–O of DNA ribose	*		61
1016 cm^{-1}	Carbohydrates peak for solids		*	24
1018 cm^{-1}	Stretching C–O ribose		*	48
	ν(CO), ν(CC), δ(OCH), ring (polysaccharides, pectin)	*		7
1019 cm^{-1}	Ring, C–N–R2 stretch, in-plane CH bend		*	14
1020 cm^{-1}	DNA	*		40
1020–50 cm^{-1}	Glycogen	*		38
1020–230 cm^{-1}	Phosphates of DNA	*		39
1021 cm^{-1}	Deoxyribose-phosphate (shoulder)	*		47
1022 cm^{-1}	Glycogen	*		38, 52
1022 cm^{-1}	Glycogen		*	24
1023 cm^{-1}	Glycogen		*	24
1024 cm^{-1}	Glycogen (C–O stretch associated with glycogen)	*		38, 52
1025 cm^{-1}	Carbohydrates peak for solutions	*	*	24
	Vibrational frequency of –CH$_2$OH groups of carbohydrates (including glucose, fructose, glycogen, etc.)			62
	Glycogen			24, 40

(*Continued*)

TABLE 8.1 (*Continued*)

Major Characteristic Peak Frequencies Reported in the Literature

Peak	Assignment	FT-IR	Raman	Reference Number
	–CH$_2$OH groups and the C–O stretching vibration coupled with C–O bending of the C–OH groups of carbohydrates (including glucose, fructose, glycogen, etc.)	*		63
	O–CH$_3$ stretching mode (with a very low Raman activity)		*	1
	Carbohydrates		*	30
	Ring, C–N–R2 stretch, in-plane CH bend		*	14
1028 cm^{-1}	Glycogen absorption due to C–O and C–C stretching and C–O–H deformation motions	*		62
	O–H stretching coupled with C–O bending of C–OH groups of carbohydrates	*		61
1029/30 cm^{-1}	O–CH$_3$ stretching of methoxy groups	*	*	6
1030 cm^{-1}	Glycogen vibration	*		55, 64
	CH$_2$OH vibration	*		64
	v_sC–O	*		65
	Collagen and phosphodiester groups of nucleic acids	*		63
1030 cm^{-1}	Stretching C–O ribose	*		48
1030 cm^{-1}	Phenylalanine of collagen		*	3
	Collagen	*		66
	v(CC) skeletal, keratin (protein assignment)		*	33
1031 cm^{-1}	δ(C–H), phenylalanine (protein assignment)		*	20
	C–H in-plane bending mode of phenylalanine			10
	Carbohydrate residues of collagen			24
	v(CC) skeletal *cis* conformation, v(CH$_2$OH), v(CO) stretching coupled with C–O bending	*		66, 67
	Collagen	*		40, 51
	One of the triad peaks of nucleic acids (along with 1060 and 1081)	*		50
	Phenylalanine, C–N stretching of proteins		*	12
	C–H in-plane bending mode of phenylalanine		*	9

TABLE 8.1 (*Continued*)

Major Characteristic Peak Frequencies Reported in the Literature

Peak	Assignment	FT-IR	Raman	Reference Number
1032 cm⁻¹	CH₂CH₃ bending modes of collagen and phospholipids		*	20
	C–C skeletal stretch (one of C–C ring vibration to be expected in aromatic structure of xylene)		*	15
	Phenylalanine of collagen		*	3
	O–CH₃ stretching of methoxy groups	*		6
	Proline (collagen assignment)		*	19
	Proline		*	17
	δ(C–H), phenylalanine		*	60
1033 cm⁻¹	Differences in collagen content		*	24
	ν(CC) skeletal *cis* conformation, ν(CH₂OH), ν(CO) stretching coupled with C–O bending	*		66, 68
	Phenylalanine mode		*	12
	ν(CO), ν(CC), ν(CCO) (polysaccharides, pectin)		*	7
1034 cm⁻¹	Phenylalanine of collagen		*	3
1035 cm⁻¹	Skeletal *trans* conformation (CC) of DNA	*		62
	ν(CC) skeletal *cis* conformation, ν(CH₂OH), ν(CO) stretching coupled with C–O bending	*		66
	Glycogen	*		52
	Collagen		*	69
	ν(CO), ν(CC), ν(CCO), (polysaccharides-cellulose)	*		7
1036 cm⁻¹	C–H in plane bending mode of phenylalanine		*	16
1037 cm⁻¹	ν(CC) skeletal *cis* conformation, ν(CH₂OH), ν(CO) stretching coupled with C–O bending	*		66
1039/40 cm⁻¹	Stretching C–O ribose	*		48
1040 cm⁻¹	ω(CH₂)		*	1
1040/41 cm⁻¹	Formalin peaks appearing in the fixed normal and tumour tissues		*	50
1043 cm⁻¹	Carbohydrates peak for solutions and solids		*	24
	Proline (collagen assignment)		*	19
	Carbohydrates		*	30

(Continued)

TABLE 8.1 (*Continued*)

Major Characteristic Peak Frequencies Reported in the Literature

Peak	Assignment	FT-IR	Raman	Reference Number
1044 cm^{-1}	$v_3PO_4^{3-}$ (symmetric stretching vibration of $v_3PO_4^{3-}$ of HA)		*	3
1045 cm^{-1}	Glycogen band (due to OH stretching coupled with bending)	*		70, 71
	C–O stretching frequencies coupled with C–O bending frequencies of the C–OH groups of carbohydrates (including glucose, fructose, glycogen, etc.	*		62
	–CH$_2$OH groups and the C–O stretching vibration coupled with C–O bending of the C–OH groups of carbohydrates (including glucose, fructose, glycogen, etc.)	*		63
1045/545 cm^{-1}	Gives an estimate carbohydrate concentrations (lower in malignant cells)	* *		62 63
1046 cm^{-1}	C–O stretching of carbohydrates		*	17
1048 cm^{-1}	Glycogen		*	8
1050 cm^{-1}	$v_sCO-O-C$	*		65
	C–O stretching coupled with C–O bending of the C–OH of carbohydrates	*		64
	Glycogen	*		38
1050–80 cm^{-1}	Indicates a degree of oxidative damage to DNA	*		40
1050–100 cm^{-1}	Phosphate and oligosaccharides	*		56
	PO$_2^-$ stretching modes, P–O–C antisymmetric stretching mode of phosphate ester, and C–OH stretching of oligosaccharides	*		56
1052 cm^{-1}	Phosphate I band for two different C–O vibrations of deoxyribose in DNA in *A* and *B* forms of helix or ordering structure	*		23
	O–H stretching coupled with C–O bending of C–OH groups of carbohydrates	*		61
1053 cm^{-1}	vC–O and δC–O of carbohydrates	*		51
	C–O stretching, C–N stretching (protein)		*	12
	Shoulder of 1121 cm^{-1} band, due to DNA	*		72

TABLE 8.1 (*Continued*)

Major Characteristic Peak Frequencies Reported in the Literature

Peak	Assignment	FT-IR	Raman	Reference Number
1055 cm^{-1}	In the solid, the most significant difference between the two nucleic acids is the ratio intensity of the bands in this area.		*	24
	Oligosaccharide C–O bond in hydroxyl group that might interact with some other membrane components	*		56
	Mainly from phospholipid phosphate and partly from oligosaccharide C–OH bonds	*		56 56
	Phosphate ester	*		56
	PO$_2^-$ stretching and C–OH stretching of oligosaccharides	*		56
	Phosphate residues	*		56
	Membrane-bound oligosaccharide C–OH bond (a part of it may originate from the hydroxyl group of the sugar residue)	*		56
	ν(CO), ν(CC), δ(OCH) (polysaccharides, pectin)	*		7
1056/7 cm^{-1}	Stretching C–O deoxyribose	*		48
1057 cm^{-1}	Lipids		*	24
1059 cm^{-1}	2-methylmannoside	*		56
	Oligosaccharide C–OH stretching band	*		56
	Mannose and mannose-6-phosphate	*		56
1059–1192 cm^{-1}	Phosphodiester-deoxyribose (to detect structural disorders in tumour DNA)	*		73
1060 cm^{-1}	Stretching C–O deoxyribose	*		48
	One of the triad peaks of nucleic acids (along with 1031 and 1081 cm^{-1})	*		50
	ν(CO), ν(CC), δ(OCH) (polysaccharides-cellulose)	*		7
	ω(CH$_2$)		*	1
1061 cm^{-1}	C–C in-plane bending (one of C–C ring vibration to be expected in aromatic structure of xylene)		*	15

(*Continued*)

TABLE 8.1 (*Continued*)

Major Characteristic Peak Frequencies Reported in the Literature

Peak	Assignment	FT-IR	Raman	Reference Number
1063 cm⁻¹	C–C skeletal stretch random conformation	*		15
	Ring deformation, cyclic C–C stretch	*		14
1064 cm⁻¹	Stretching C–O ribose	*		48
	Skeletal C–C stretch of lipids		*	9
	Acyl chains		*	2
	ν(C–C) *trans*		*	43
1065 cm⁻¹	C–O stretching of the phosphodiester and the ribose	*		50
	Nucleic acids, in the absence of glycogen	*		50
	Palmitic acid		*	2
	Fatty acid		*	2
1066/7 cm⁻¹	Proline (collagen assignment)		*	19
1068 cm⁻¹	Stretching C–O ribose	*		48
1069 cm⁻¹	PO_2^- stretching mode of the DNA backbone		*	4
1070 cm⁻¹	Triglycerides (fatty acids)		*	49
	Mannose and mannose-6-phosphate	*		56
1070–80 cm⁻¹	Nucleic acid band	*		74
1070–90 cm⁻¹	Symmetric PO_2^- stretching of DNA (represents more DNA in cell)		*	75
1071 cm⁻¹	Phosphate I band for two different C–O vibrations of deoxyribose in DNA in disordering structure	*		23
	Glucose		*	2
	Skeletal C–C stretch in lipids		*	16
1073 cm⁻¹	Carbonate symmetric stretching vibration of calcium carbonate apatite		*	49
	Triglycerides (fatty acids)		*	49
1075 cm⁻¹	Symmetric phosphate stretching modes or $\nu(PO_2^-)$ sym. (phosphate stretching modes originate from the phosphodiester groups in nucleic acids and suggest an increase in the nucleic acids in the malignant tissues)	*		66
	$\nu(PO_2^-)$ symmetric stretching of phosphodiesters	*		66

TABLE 8.1 (*Continued*)

Major Characteristic Peak Frequencies Reported in the Literature

Peak	Assignment	FT-IR	Raman	Reference Number
1076 cm⁻¹	C–C (lipid in normal tissues)		*	54
	Symmetric stretching vibration of $v_3PO_4^{3-}$ (phosphate of HA)	*		
	Skeletal *cis* conformation (CC) of DNA	*		62
1076 cm⁻¹	Symmetric phosphate [PO_2^- (sym)] stretching	*		62
1078 cm⁻¹	v(C–C) or v(C–O), phospholipids (lipid assignment)		*	20
	Pronounced symmetric phosphate stretch		*	20
	Phospholipids		*	20
	C–C or C–O stretching mode of phospholipids		*	20
	Carbohydrate peak for solids		*	24
	$v_sPO_2^-$	*		65
	Phosphate I in RNA	*		23
	Symmetric phosphate	*		38, 52
	Glycogen absorption due to C–O and C–C stretching and C–O–H deformation motions	*		50
	DNA in healthy samples, in the absence of glycogen indicating the role of phosphates during diseases	*		50 53
	C–C or C–O stretch (lipid), C–C or PO_2 stretch (nucleic acid)		*	9
	C–OH stretching band of oligosaccharide residue	*		56
	v(CC) skeletal		*	33
	v(CC) vitor$_s$ (PO_2^-) nucleic acid		*	33
1079 cm⁻¹	$v_sPO_2^-$	*		38, 52
1080 cm⁻¹	Amide II (?)		*	24
	Typical phospholipids		*	54
	vPO_2^-	*		38, 76, 77
	Phosphate vibration	*		55
	Phosphate vibrations (phosphodiester groups in nucleic acids)		*	55
	Symmetric phosphate [PO_2^- (sym)] stretching	*		64
	Collagen		*	69

(Continued)

TABLE 8.1 (*Continued*)

Major Characteristic Peak Frequencies Reported in the Literature

Peak	Assignment	FT-IR	Raman	Reference Number
	Collagen and phosphodiester groups of nucleic acids	*		63
1081 cm^{-1}	$v_1CO_3^{2-}$, $v_3PO_4^{3-}$, $v(C–C)$ skeletal of acyl backbone in lipid (*gauche* conformation)		*	3
	Symmetric phosphate stretching modes or $v(PO_2^-)$ sym. (Phosphate stretching modes originate from the phosphodiester groups in nucleic acids and suggest an increase in the nucleic acids in the malignant tissues.)	*		66
	$v(PO_2^-)$ symmetric stretching of phosphodiesters	*		77
	Phosphate I in RNA	*		23
	One of the triad peaks of nucleic acids (along with 1031 and 1060)	*		50
1082 cm^{-1}	Carbohydrate residues of collagen		*	24
	Carbohydrates peak for solutions		*	24
	Nucleic acids		*	24
	PO_2^- symmetric	*		78
	Phosphate band	*		79
	Collagen	*		40
	Symmetric phosphate stretching band of the normal cells	*		51
	Carbohydrates		*	30
	Phospholipids O–P–O and C–C		*	46
1083 cm^{-1}	C–N stretching mode of proteins (and lipid mode to a lesser degree)		*	10
	PO_2^- symmetric			
	C–N stretching mode of proteins (and lipid mode to a lesser degree)	*		78
	$v_sPO_2^-$ of phosphodiesters of nucleic acids		*	9
		*		61
1084 cm^{-1}	Phosphodiester groups in nucleic acids		*	24
	DNA (band due to PO_2^- vibrations)			71

TABLE 8.1 (*Continued*)

Major Characteristic Peak Frequencies Reported in the Literature

Peak	Assignment	FT-IR	Raman	Reference Number
	Symmetric phosphate [PO$_2^-$ (sym)] stretching	*		40, 64
	PO$_2^-$ symmetric	*		78
	Stretching PO$_2^-$ symmetric	*		23
	Absorbance by the phosphodiester bonds of the phosphate/sugar backbone of nucleic acids	*		40
	Nucleic acid region	*		40
	Nucleic acid–phosphate band	*		72
	C–N stretching of proteins and lipids, ν(C–C) and ν(C–O) of phospholipids		*	17
1084–6 cm^{-1}	ν_s(PO$_2^-$) of nucleic acids	*		51
1085 cm^{-1}	PO$_2^-$ symmetric (phosphate II)	*		48
	PO$_2^-$ symmetric	*		78
	Mainly from absorption bands of the phosphodiester group of nucleic acids and membrane phospholipids, and partially protein (amide III). The band originating from sugar chains (C–OH band) overlaps.	*		78
	Mainly from phospholipid phosphate and partly from oligosaccharide C–OH bonds	*		56
	Phosphate ester	*		56
1086 cm^{-1}	Symmetric phosphate stretching modes or ν(PO$_2^-$) sym. (Phosphate stretching modes originate from the phosphodiester groups in nucleic acids and suggest an increase in the nucleic acids in the malignant tissues.)	*		66
	PO$_2^-$ symmetric	*		78
	ν(PO$_2^-$) symmetric stretching of phosphodiesters	*		77, 80
	ν(C–C) *gauche*		*	43
1087 cm^{-1}	PO$_2^-$ symmetric (phosphate II)	*		48
	ν_1CO$_3^{2-}$, ν_3PO$_4^{3-}$, ν(C–C) skeletal of acyl backbone in lipid (*gauche* conformation)		*	3

(Continued)

TABLE 8.1 (*Continued*)

Major Characteristic Peak Frequencies Reported in the Literature

Peak	Assignment	FT-IR	Raman	Reference Number
1087–90 cm^{-1}	C–C stretch		*	9
	PO$_2^-$ stretch			
1088 cm^{-1}	C–N stretching of proteins and lipids, ν(C–C) and ν(C–O) of phospholipids	*		17
1088–90 cm^{-1}	Phosphate I (stretching PO$_2^-$ symmetric vibration) in *B*-form DNA	*		23
1089 cm^{-1}	Stretching PO$_2^-$ symmetric in RNA	*		23
1090 cm^{-1}	Symmetric phosphate stretching vibrations		*	24, 30
	Mannose and mannose-6-phosphate	*		56
	Phosphate ester (C–O–P) band	*		56
1090–100 cm^{-1}	Phosphate II (stretching PO$_2^-$ asymmetric vibration) in *A*-form RNA	*		23
1092/3 cm^{-1}	Phosphodioxy groups		*	21
1093 cm^{-1}	Symmetric PO$_2^-$ stretching vibration of the DNA backbone–phosphate backbone vibration as a marker mode for the DNA concentration –C–N of proteins		*	12
1094 cm^{-1}	DNA		*	54
	Stretching PO$_2^-$ symmetric (phosphate II)	*		48
	ν_{asym}(C–O–C) (polysaccharides-cellulose)	*		7
	Ring, C–N–R2 stretch		*	14
1095 cm^{-1}	Lipid		*	38
	Stretching PO$_2^-$ symmetric	*		50
	ν(C–N)		*	3
	DNA: O–P–O backbone stretching		*	16
1096 cm^{-1}	Phosphodioxy (PO$_2^-$) groups		*	2
	Ring, C–N–R2 stretch	*	*	14
1099 cm^{-1}	ν(C–N)		*	3
1099/100 cm^{-1}	Stretching PO$_2^-$ symmetric (phosphate II)	*		48
1100 cm^{-1}	Phospholipids (stretching C–C skeletal vibrations in the *gauche* conformation)		*	35

TABLE 8.1 (*Continued*)

Major Characteristic Peak Frequencies Reported in the Literature

Peak	Assignment	FT-IR	Raman	Reference Number
	Symmetric stretching vibrations of the PO_2 and the C–O stretching vibrations of the deoxyribose	*		47
1100–200 cm^{-1}	Vibrational modes of benzene-derived molecules		*	14
1100–300 cm^{-1}	Due to antisymmetric stretching vibrations of the PO_2 group (~1240 cm^{-1}) and the symmetric stretching vibrations of the PO_2 and the C–O stretching vibrations of the deoxyribose (~1100 cm^{-1}).	*		47
1100–375 cm^{-1}	Several bands of moderate intensity, belonging to amide III and other groups (proteins)		*	43
1100–450 cm^{-1}	OH vibrations (deformations [bendings] modes)		*	1
1100–800 cm^{-1}	ν(C–C): Lipids, fatty acids		*	7
1104 cm^{-1}	Symmetric stretching P–O–C	*		48
	Phenylalanine (proteins)		*	43
1105 cm^{-1}	Carbohydrates peak for solutions		*	24, 30
	Carbohydrates	*		72
1107 cm^{-1}	ν(C–O), ν(CC), ring (polysaccharides, pectin)	*		7
1110 cm^{-1}	ν(C–O) ν(CC) ring (polysaccharides, cellulose)	*		7
1111 cm^{-1}	Benzoid ring deformation		*	28
1112 cm^{-1}	Saccharide band (overlaps with acyl band)		*	2
1115/9/13 cm^{-1}	Symmetric stretching P–O–C	*		48
1115/6 cm^{-1}	$CH_{2,6}$ in-plane bend and C_1–C_α–H_α bend		*	6
1117 cm^{-1}	Glucose		*	2
1117–9 cm^{-1}	C–C stretch (breast lipid)		*	9
1119 cm^{-1}	C–O stretching mode	*		52
1120 cm^{-1}	The strong C–O band of ribose (serves as a marker band for RNA in solutions)		*	24, 30
	Carotene		*	54
	Mannose-6-phosphate	*		56
	Phosphorylated saccharide residue	*		56
	ν(C–C) of lipids from *trans* segments and ν(C–N) of proteins		*	17

(*Continued*)

TABLE 8.1 (*Continued*)

Major Characteristic Peak Frequencies Reported in the Literature

Peak	Assignment	FT-IR	Raman	Reference Number
1121 cm^{-1}	Symmetric phosphodiester stretching band	*		40
	RNA	*		40
	Shoulder of 1121 cm^{-1} band, due to RNA	*		72
	νC–O of RNA ribose	*		61
1122 cm^{-1}	νC–O of carbohydrates	*		51
	ν_{22} (porphyrin half ring), observed in the spectra of single human RBC		*	17
	ν_s(CC) skeletal			
	ν_{sym}(C–O–C) (polysaccharides, cellulose)		*	33
	ν(C–C) *trans*		*	7
			*	43
1123 cm^{-1}	(C–N), proteins (protein assignment)		*	20
	C–C stretching mode of lipids and protein, C–N stretch		*	9, 10
	Glucose		*	9
1124 cm^{-1}	ν(C–C) skeletal of acyl backbone in lipid (*trans* conformation)		*	3
1125 cm^{-1}	CH$_{2,6}$ in-plane bend and C$_1$–C$_\alpha$–H$_\alpha$ bend	*		6
	ν(CO) ν(CC) ring (polysaccharides, cellulose)	*		7
1126 cm^{-1}	Paraffin		*	24
	ν(C–C) skeletal of acyl backbone in lipid (*trans* conformation)		*	3
	C–N stretching vibration (protein vibration)		*	12
	ν(C–O), Disaccharides, sucrose	*		7
	ν(C–O) + ν(C–C), Disaccharides, sucrose		*	7
	ν(C–C) of lipids from *trans* segments and ν(C–N) of proteins		*	17
1127 cm^{-1}	ν(C–N)		*	43
1128/9 cm^{-1}	ν(C–C) skeletal of acyl backbone in lipid (*trans* conformation)		*	3
1129 cm^{-1}	Skeletal C–C stretching in lipids		*	16
1130 cm^{-1}	C–C skeletal stretch *trans* conformation		*	15

TABLE 8.1 (*Continued*)

Major Characteristic Peak Frequencies Reported in the Literature

Peak	Assignment	FT-IR	Raman	Reference Number
	Phospholipid structural changes (*trans* versus *gauche* isomerism)			
	Acyl chains		*	40
	Haemoglobin		*	2
			*	60
1130–460 cm^{-1}	CH$_3$ deformations (both symmetric and asymmetric)		*	1
1131 cm^{-1}	Palmitic acid		*	2
	Fatty acid		*	2
1137 cm^{-1}	Oligosaccharide C–OH stretching band	*		56
	2-methylmannoside	*		56
1138 cm^{-1}	Adenine (ring breathing modes of the DNA/RNA bases)		*	4
1144 cm^{-1}	Ring stretch, in-plane CH bend		*	14
1145 cm^{-1}	Phosphate and oligosaccharides	*		56
	Oligosaccharide C–O bond in hydroxyl group that might interact with some other membrane components			
	Membrane-bound oligosaccharide C–OH bond			
1149 cm^{-1}	Carbohydrates peak for solids		*	24
1150 cm^{-1}	C–O stretching vibration			52
	Glycogen	*		24
	C–O stretching mode of the carbohydrates		*	51
	Carotenoid	*		25
	CH$_8$, CH″$_8$ deformations		*	6
	ν(C–O–C), ring (polysaccharides, pectin)	*		7
1150–200 cm^{-1}	Phosphodiester stretching bands (sym. and asym.)	*		40
1151 cm^{-1}	Glycogen absorption due to C–O and C–C stretching and C–O–H deformation motions	*		50
1152 cm^{-1}	ν(C–N), proteins (protein assignment)		*	20
	ν(C–C), carotenoid		*	20
	Carotenoid peaks due to C–C and conjugated C=C band stretch		*	20
	CH$_8$, CH″$_8$ deformations	*		6

(Continued)

TABLE 8.1 (*Continued*)

Major Characteristic Peak Frequencies Reported in the Literature

Peak	Assignment	FT-IR	Raman	Reference Number
1153 cm⁻¹	Carbohydrates peak for solutions		*	24, 30
	Stretching vibrations of hydrogen-bonding C–OH groups	*		64
	ν C–O absorbance of glycogen	*		61
1155 cm⁻¹	C–C (and C–N) stretching of proteins (also carotenoids)		*	9, 10
	Glycogen		*	24
	C–O stretching vibration	*		38
	ν(C–C): Diagnostic for the presence of a carotenoid structure, most likely a cellular pigment	*	*	22
1156 cm⁻¹	C–C, C–N stretching (protein)		*	12
	(–C–C–) stretching of carotenoid		*	60
1156/7 cm⁻¹	Carotenoids (absent in normal tissue)		*	45
1157 cm⁻¹	In-plane vibrations of the conjugated =C–C=		*	81
	β-carotene accumulation (C=C stretch mode)		*	49
1157/9 cm⁻¹	In-plane CH bend, ring, cyclic stretch		*	14
1159–74 cm⁻¹	νC–O of proteins and carbohydrates	*		51
1160 cm⁻¹	CO stretching	*		64
	C–C stretching in proteins		*	16
1161 cm⁻¹	Stretching vibrations of hydrogen bonding C–OH groups	*		64
	Deformative vibrations of quinoid ring		*	28
1161/2 cm⁻¹	Mainly from the C–O stretching mode of C–OH groups of serine, threosine, and tyrosine of proteins)	*		66
	ν(CC), δ(COH), ν(CO) stretching	*		77, 82
1162 cm⁻¹	Stretching modes of the C–OH groups of serine, threonine, and tyrosine residues of cellular proteins	*		51
	δ(C–O–C), ring (polysaccharides, cellulose)	*		7
1163 cm⁻¹	Tyrosine (collagen type I)	*		3
	Tyrosine	*		3
	CH′₉, CH₇, CH′₇ deformations	*		6

TABLE 8.1 (*Continued*)

Major Characteristic Peak Frequencies Reported in the Literature

Peak	Assignment	FT-IR	Raman	Reference Number
1163/4 cm⁻¹	C–O stretching band of collagen (type I)	*		78
1164 cm⁻¹	Mainly from the C–O stretching mode of C–OH groups of serine, threosine, and tyrosine of proteins)	*		77
	ν(CC), δ(COH), ν(CO) stretching	*		68, 77
	C–O stretching (in normal tissue)	*		79
	Hydrogen-bonded stretching mode of C–OH groups	*		51
1167 cm⁻¹	N=Quinoid ring=N stretching and C–H in-plane bending		*	28
1167/8 cm⁻¹	Ring, cyclic stretch, in-plane CH bend		*	14
1168 cm⁻¹	Lipids		*	24
	ν(C=C) δ(COH) (lipid assignment)		*	33
	ν(C–C), carotenoid		*	33
1169 cm⁻¹	Tyrosine (collagen type I)		*	3
	Tyrosine		*	3
1170 cm⁻¹	C–H in-plane bending mode of tyrosine		*	9, 10
	ν_{as}CO–O–C	*		65
	C–O bands from glycomaterials and proteins	*		78
1171 cm⁻¹	Tyrosine (collagen type I)		*	3
	Tyrosine	*		3
	(CH) phenylalanine, tyrosine	*		43
	C–OH groups of serine, threonine, and tyrosine amino acids of proteins	*		61
1172 cm⁻¹	δ(C–H), tyrosine (protein assignment)		*	20
	Stretching vibrations of nonhydrogen bonding C–OH groups	*		64
	CO stretching	*		64
	CO stretching of collagen (type I)	*		78
	Stretching modes of the C–OH groups of serine, threonine, and tyrosine residues of cellular proteins	*		52

(Continued)

TABLE 8.1 (*Continued*)

Major Characteristic Peak Frequencies Reported in the Literature

Peak	Assignment	FT-IR	Raman	Reference Number
1172/3 cm⁻¹	CO stretching of the C–OH groups of serine, threosine, and tyrosine in the cell proteins as well as carbohydrates	*		64
1173 cm⁻¹	C–O stretching (in malignant tissues)	*		79
1173 cm⁻¹	Nonhydrogen-bonded stretching mode of C–OH groups	*		51
	Cytosine, guanine		*	21
	CH'$_9$, CH$_7$, CH'$_7$ deformations	*		6
	Tyrosine (collagen type I)		*	3
1174 cm⁻¹	Tyrosine, phenylalanine, C–H bend (protein)		*	12
	Protein peak		*	17
	Haemoglobin		*	60
	δ(C–H), tyrosine in tumours		*	17
1175/6/7 cm⁻¹	Cytosine, guanine		*	21
1180 cm⁻¹	Cytosine, guanine		*	21
	(C–O)$_{ester}$ vibration		*	1
	O–CH$_3$ stretching vibration		*	1
	Cytosine, guanine, adenine		*	16
1180–84 cm⁻¹	Cytosine, guanine, adenine		*	9
1180–300 cm⁻¹	Amide III band region	*		83
1185/1/2 cm⁻¹	CH$_2$	*		48
1185–300 cm⁻¹	Antisymmetric phosphate vibrations		*	69
1187/8 cm⁻¹	CH$_{\alpha,2',8}$ in-plane bend	*		6
1188 cm⁻¹	Deoxyribose	*		48
1194 cm⁻¹	Ring stretch, in-plane CH bend, CH$_2$ twist		*	14
1200 cm⁻¹	Collagen	*		66
	Nucleic acids and phosphates		*	69
	Phosphate (P=O) band	*		56
	Aromatic C–O and C–N		*	84
1200–350 cm⁻¹	Amide III; due to C–N stretching and N–H bending		*	12, 54
1200–360 cm⁻¹	Electronic structure of nucleotides		*	12
1200–400 cm⁻¹	C–N bends of the peptide linkages		*	85
1200–500 cm⁻¹	Amide III of proteins		*	35
1201 cm⁻¹	PO$_2^-$ asymmetric (phosphate I)	*		48

TABLE 8.1 (*Continued*)

Major Characteristic Peak Frequencies Reported in the Literature

Peak	Assignment	FT-IR	Raman	Reference Number
1203 cm^{-1}	C–C$_6$H$_5$ stretch mode (one of C–C ring vibration to be expected in aromatic structure of xylene)		*	15
1204 cm^{-1}	Amide III and CH$_2$ wagging vibrations from glycine backbone and proline side chains		*	24
	Vibrational modes of collagen proteins; amide III	*		83
	C–O–C, C–O dominated by the ring vibrations of polysaccharides C–O–P, P–O–P	*		77, 86
	Collagen	*	*	40, 69
	Tyrosine, phenylalanine (IgG?)		*	43
	Ring stretch, in-plane CH bend, CH$_2$ twist			14
1205 cm^{-1}	Differences in collagen content		*	24
	C–O–C, C–O dominated by the ring vibrations of polysaccharides C–O–P, P–O–P	*		66, 77
1206 cm^{-1}	Amide III	*		50
	Collagen	*		51
	Hydroxyproline, tyrosine (collagen assignment)		*	19
	Hydroxyproline, tyrosine		*	9
1207 cm^{-1}	PO$_2^-$ asymmetric (phosphate I)	*		50
1208 cm^{-1}	ν(C–C$_6$H$_5$), tryptophan, phenylalanine (protein assignment)		*	20, 33
	Tryptophan		*	3
	A, T (ring breathing modes of the DNA/RNA bases); amide III (protein)		*	12
1209 cm^{-1}	PO$_2^-$ asymmetric (phosphate I)	*		48
	Tryptophan and phenylalanine ν(C–C$_6$H$_5$) mode		*	9, 10
1210 cm^{-1}	C–C$_6$H$_5$ stretching mode in tyrosine and phenylalanine		*	87
	ν$_{18}$(δ: C$_m$H), observed in the spectra of single human RBC		*	22
	Haemoglobin		*	60
	Tryptophan	*	*	35
1212 cm^{-1}	PO$_2^-$ asymmetric (phosphate I)	*		48

(Continued)

TABLE 8.1 (*Continued*)

Major Characteristic Peak Frequencies Reported in the Literature

Peak	Assignment	FT-IR	Raman	Reference Number
1216 cm^{-1}	Stretching of C–N		*	28
1217 cm^{-1}	PO$_2^-$ asymmetric (phosphate I)	*		48
1220 cm^{-1}	PO$_2^-$ asymmetric vibrations of nucleic acids when it is highly hydrogen-bonded	*		64
	Asymmetric hydrogen-bonded phosphate stretching mode	*		51
	C=N=C stretching		*	28
	Amide III (β-sheet)		*	16
1220–1 cm^{-1}	Amide III (β-sheet)		*	9
1220–4 cm^{-1}	Phosphate II (stretching PO$_2^-$ asymmetric vibration) in *B*-form DNA	*		23
1220–50 cm^{-1}	vPO$_2^-$	*		76
1220–300 cm^{-1}	Amide III (arising from coupling of C–N stretching and N–H bonding; can be mixed with vibrations of side chains) (protein band)		*	29
	Amide III			
			*	30, 60
1222 cm^{-1}	Phosphate stretching bands from phosphodiester groups of cellular nucleic acids	*		51
	CH$_{6,2',\alpha,\alpha'}$ rock	*		6
1222/3 cm^{-1}	PO$_2^-$ asymmetric (phosphate I)	*		20, 48
1223 cm^{-1}	v(PO$_2^-$), nucleic acids		*	20
	Cellular nucleic acids		*	20
	A concerted ring mode		*	5
	Proteins, including collagen I	*	*	8
1224 cm^{-1}	Amide III (β sheet structure)		*	43
1226 cm^{-1}	PO$_2^-$ asymmetric (phosphate I)	*		48
1230 cm^{-1}	Antisymmetric phosphate stretching vibration		*	24, 30
	Stretching PO$_2^-$ asymmetric	*		23, 50
	Overlapping of the protein amide III and the nucleic acid phosphate vibration	*		50
1230–300 cm^{-1}	Amide III (arising from coupling of C–N stretching and N–H bonding; can be mixed with vibrations of side chains)		*	24, 30
1234 cm^{-1}	A concerted ring mode		*	5

TABLE 8.1 (*Continued*)

Major Characteristic Peak Frequencies Reported in the Literature

Peak	Assignment	FT-IR	Raman	Reference Number
1235 cm^{-1}	Composed of amide III as well as phosphate vibration of nucleic acids	*		50
	CH$_{6,2',\alpha,\alpha'}$ rock	*		6
	Amide III		*	7
	In-phase ring, C–C, C–N–C stretch		*	14
1236 cm^{-1}	Amide III and asymmetric phosphodiester stretching mode ($v_{as}PO_2^-$), mainly from the nucleic acids	*		83
	$v_{as}PO_2^-$ of nucleic acids	*		51
1236–42 cm^{-1}	Relatively specific for collagen and nucleic acids	*		40
1236/7 cm^{-1}	Stretching PO$_2^-$ asymmetric (phosphate I)	*		48
1237 cm^{-1}	Amide III and CH$_2$ wagging vibrations from glycine backbone and proline side chains		*	24
	PO$_2^-$ asymmetric (phosphate I)	*		48
	PO$_2^-$ asymmetric	*		88
1238 cm^{-1}	Stretching PO$_2^-$ asymmetric (phosphate I)	*		48
	Asymmetric phosphate [PO$_2^-$ (asym.)] stretching modes	*		64
	Stretching PO$_2^-$ asymmetric	*		23
	Amide III	*		50
1238/9 cm^{-1}	Asymmetric PO$_2^-$ stretching	*		78
1239 cm^{-1}	Amide III		*	15
1240 cm^{-1}	One of the two most distinct peaks for RNA (with 813 cm^{-1})		*	12
	Differences in collagen content		*	24
	$v_{as}PO_2^-$	*		38, 66
	Collagen	*		51
	Asymmetric nonhydrogen-bonded phosphate stretching mode	*		51
	Asymmetric phosphate [PO$_2^-$ (asym.)] stretching modes (Phosphate stretching modes originate from the phosphodiester groups of nucleic acids and suggest an increase in the nucleic acids in the malignant tissues.)	*	*	10 66

(*Continued*)

TABLE 8.1 (*Continued*)

Major Characteristic Peak Frequencies Reported in the Literature

Peak	Assignment	FT-IR	Raman	Reference Number
	Mainly from absorption bands of the phosphodiester group of nucleic acids and membrane phospholipids, and partially protein (amide III).	*		78
	Amide III			
	PO_2^- asymmetric vibrations of nucleic acids when it is nonhydrogen-bonded			
	$v_{as}PO_2^-$	*		78
	Collagen	*		64
	Asymmetric phosphodiester stretching band			
	Amide III	*		65
	PO_2^- ionized asymmetric stretching	*		40
	$v(PO_2^-)$ asymmetric stretching of phosphodiesters	*		40
	Composed of amide III mode of collagen protein and the asymmetric stretching mode of the phosphodiester groups of nucleic acids	*		66 62, 77
	Asymmetric stretching mode of phosphodiester groups of nucleic acids	*		38, 66, 67, 70
	Collagen	*		51
	Antisymmetric stretching vibrations of the PO_2 group			
		*		51
			*	69
		*		47
1240–45 cm^{-1}	Phosphate I (stretching PO_2^- symmetric vibration) in *A*-form RNA	*		23
1240–65 cm^{-1}	Amide III(C–N stretching mode of proteins, indicating mainly α-helix conformation)	*		9, 83
1240–310 cm^{-1}	vC–N, amide III	*		76
1241 cm^{-1}	PO_2^- asymmetric (phosphate I)	*		48
	Phosphate band	*		79

TABLE 8.1 (*Continued*)

Major Characteristic Peak Frequencies Reported in the Literature

Peak	Assignment	FT-IR	Raman	Reference Number
	Asymmetric phosphate [PO_2^- (asym.)] stretching modes (Phosphate stretching modes originate from the phosphodiester groups of nucleic acids and suggest an increase in the nucleic acids in the malignant tissues. Generally, the PO_2^- groups of phospholipids do not contribute to these bands.)	*	*	3 64
	Phosphate stretching bands from phosphodiester groups of cellular nucleic acids	*		51
	v_{as} Phosphate	*		77
1242 cm^{-1}	PO_2^- asymmetric	*		78
	Collagen I and IV	*		78
	Amide III	*		40, 50
	Amide III collagen	*		51
	Amide III (β-sheet and random coils)		*	43
1243 cm^{-1}	Amide III		*	24
	$v(PO_2^-)$ asymmetric stretching of phosphodiesters	*		77, 80
	Asymmetric phosphate [PO_2^- (asym.)] stretching modes (Phosphate stretching modes originate from the phosphodiester groups of nucleic acids and suggest an increase in nucleic acids in the malignant tissues. Generally, the PO_2^- groups of phospholipids do not contribute to these bands.)	*	*	3 64
	Phosphate in RNA			
	C–O_4 aromatic stretch	*		23
	Amide III of collagen (CH_2 wagging, C–N stretching) and pyrimidine bases (C, T)		* *	6 9
1243/4 cm^{-1}	Collagen I	*		78
1244 cm^{-1}	Collagen I and IV	*		78
	Asymmetric phosphate stretching ($v_{as}PO_2^-$)	*		23, 38, 52

(Continued)

TABLE 8.1 (*Continued*)

Major Characteristic Peak Frequencies Reported in the Literature

Peak	Assignment	FT-IR	Raman	Reference Number
1244/5 cm⁻¹	PO₂⁻ asymmetric (phosphate I)	*		48
1245 cm⁻¹	PO₂⁻ asymmetric	*		52, 78
	Amide III		*	7
1246 cm⁻¹	Amide III (of collagen)		*	3
	PO₂⁻ asymmetric	*		38, 78
1247 cm⁻¹	PO₂⁻ asymmetric (phosphate I)	*		48
	Amide III (collagen assignment)		*	19
1247/8 cm⁻¹	Guanine, cytosine (NH₂)		*	21
1248 cm⁻¹	PO₂⁻ asymmetric	*		78
1250 cm⁻¹	Amide III	*	*	22
1250/2 cm⁻¹	Guanine, cytosine (NH₂)		*	21
1252 cm⁻¹	C–O₄ aromatic stretch		*	6
1254 cm⁻¹	Formalin contamination on fixed tissues		*	41
	C–N in-plane stretching		*	28
1255 cm⁻¹	Lipids		*	24
	Amide II (?)	*		50
1256 cm⁻¹	PO₂⁻ asymmetric (phosphate I)	*		48
1257 cm⁻¹	A, T (ring breathing modes of the DNA/RNA bases); Amide III (protein)		*	12
	Ring stretch		*	14
	Ring C–C stretch		*	14
1258 cm⁻¹	Amide III, adenine, cytosine		*	9
	Protein peak		*	17
	Amide III, β-sheet/adenine and cytosine in tumours		*	17
	Amide III (β-sheet), adenine, cytosine		*	16
1259 cm⁻¹	Guanine, cytosine (NH₂)		*	21
	Amide III		*	19
1260 cm⁻¹	Amide III (protein band)		*	24
	Protein band		*	25
	Second amide (?)		*	45
	Amide III (unordered)		*	9
	Structural protein modes of tumours		*	26
	Amide III vibration mode of structural proteins		*	49
1262 cm⁻¹	PO₂⁻ asymmetric (phosphate I)	*		48

TABLE 8.1 (*Continued*)

Major Characteristic Peak Frequencies Reported in the Literature

Peak	Assignment	FT-IR	Raman	Reference Number
1263 cm⁻¹	T, A (ring breathing modes of the DNA/RNA bases); =C–H bend (protein)		*	12
1264 cm⁻¹	Triglycerides (fatty acids)		*	49
	Ring stretch		*	14
	δ(=CH) of lipids		*	17
1265 cm⁻¹	PO_2^- asymmetric (phosphate I)	*		48
	$CH_{\alpha'}$ rocking	*		6
	Amide III of collagen		*	41
	Amide III (collagen assignment)		*	19
	Amide III		*	33, 35
	ν(CN),δ(NH) amide III, α-helix, collagen (protein assignment)		*	33
	Phenylalanine stretching (C–C6H5) mode		*	35
1266 cm⁻¹	Amide III (of proteins in the α-helix conformation)		*	3, 20
	ν(CN), δ(NH) amide III, α-helix, collagen, tryptophan (protein assignment)		*	20
	δ (=C–H) *cis*		*	7
	Amide III (α-helix)		*	43
1267 cm⁻¹	C–H (lipid in normal tissue)		*	54
	Amide III (collagen assignment)			
1268 cm⁻¹	δ (=C–H) (phospholipids)		*	43
	Ring stretch		*	14
1268/9 cm⁻¹	Amide III (collagen assignment)		*	19
1269 cm⁻¹	Structural proteins like collagen		*	32
	Cytosine (ring breathing modes of the DNA/RNA bases)		*	4
1270 cm⁻¹	Typical phospholipids		*	54
	Amide III band in proteins		*	89, 90
	Has traditionally been attributed to Amide III, a C–N stretch from alpha helix proteins		*	91
	C=C groups in unsaturated fatty acids		*	2
1272/3 cm⁻¹	$CH_{\alpha'}$ rocking	*	*	6
1275 cm⁻¹	Amide III		*	7
1276 cm⁻¹	N–H thymine	*		48

(*Continued*)

TABLE 8.1 (*Continued*)

Major Characteristic Peak Frequencies Reported in the Literature

Peak	Assignment	FT-IR	Raman	Reference Number
1278 cm⁻¹	Vibrational modes of collagen proteins; Amide III	*		83
	Proteins, including collagen I		*	8
	Structural proteins like collagen		*	58
1278/9 cm⁻¹	Deformation N–H	*		48
1279 cm⁻¹	Amide III (α-helix)		*	9
1280 cm⁻¹	Amide III and CH_2 wagging vibrations from glycine backbone and proline side chains		*	24
	Collagen	*		66
	Amide III	*		50
	Collagen		*	69
	Nucleic acids and phosphates		*	69
1282 cm⁻¹	Amide III band components of proteins	*		66, 77
	Collagen	*		40
1283 cm⁻¹	Differences in collagen content		*	24
	Collagen	*		51
1283–1339 cm⁻¹	Collagen	*		51
1284 cm⁻¹	Amide III band components of proteins	*		70
1287 cm⁻¹	Deformation N–H	*		48
1287/8 cm⁻¹	Cytosine		*	21
1288 cm⁻¹	Phosphodiester groups in nucleic acids		*	24
	N–H thymine	*		48
1290 cm⁻¹	Cytosine		*	21
1290–400 cm⁻¹	CH bending		*	90
1290/1 cm⁻¹	C–N–C and C–C stretch		*	14
1291/2 cm⁻¹	N–H thymine	*		48
	Cytosine		*	21
1294/5/6 cm⁻¹	Deformation N–H cytosine	*		48
1296 cm⁻¹	CH_2 deformation		*	15
1298 cm⁻¹	Palmitic acid		*	2
	Acyl chains		*	2
	Fatty acids		*	2
1299/300 cm⁻¹	CH_2 deformation (lipid)		*	9
1300 cm⁻¹	$-(CH_2)n-$ in-plane twist vibration (lipid band)		*	29
	Fatty acids		*	24
	Fatty acids		*	25

TABLE 8.1 (*Continued*)

Major Characteristic Peak Frequencies Reported in the Literature

Peak	Assignment	FT-IR	Raman	Reference Number
	$\delta(CH_2)$; lipids, fatty acids		*	7
	CH_2 twisting modes		*	33
	$\delta(CH)$, $\tau(CH_2)$ (α-helix)		*	43
	CH_2 twisting (lipids)		*	43
	CH twist		*	58
	$-(CH_2)n-$ in-phase twist vibration	*	*	30
1300–500 cm^{-1}	CH bending vibrations, weak NH vibrations, CH in-plane base deformations and vertical base stacking interactions	*		47
1301 cm^{-1}	Assign from Parker (lipid in normal tissue)		*	54
	Triglycerides (fatty acids)		*	49
	$\tau(CH_2)$, lipids		*	43
	Indicates the concentration of lipids (in-phase CH_2 twist mode)		*	38, 92
1302 cm^{-1}	$\delta(CH_2)$ twisting, wagging, collagen (protein assignment)		*	20, 33
	$\delta(CH_2)$ twisting, wagging, phospholipids (lipid assignment)		*	20, 33
	CH_3/CH_2 twisting or bending mode of lipid/collagen		*	3
	Amide III (protein)		*	12
	Methylene bending mode (a combination of proteins and phospholipids)		*	41
	(CH_2) twist, phospholipids and collagen		*	17
1303/4 cm^{-1}	CH_3, CH_2 twisting (collagen assignment)		*	19
1304 cm^{-1}	CH_2 deformation (lipid), adenine, cytosine		*	9
1305 cm^{-1}	Bending mode of CH_3CH_2 twisting of proteins		*	35
	C–N–C and cyclic C–C stretch		*	14
1306 cm^{-1}	Unassigned band	*		52
1307 cm^{-1}	Amide III	*		50
1307 cm^{-1}	CH_3/CH_2 twisting or bending mode of lipid/collagen		*	3
	CH_3/CH_2 twisting, wagging and/or bending mode of collagens and lipids		*	3

(Continued)

TABLE 8.1 (*Continued*)

Major Characteristic Peak Frequencies Reported in the Literature

Peak	Assignment	FT-IR	Raman	Reference Number
1308 cm⁻¹	C–N asymmetric stretching in asymmetric aromatic amines		*	28
	C–N–C and cyclic C–C stretch		*	14
	CH₂ deformation in lipids, adenine, guanine		*	16
1309 cm⁻¹	CH₃/CH₂ twisting or bending mode of lipid/collagen		*	3
	CH₃/CH₂ twisting, wagging and/or bending mode of collagens and lipids		*	3
1310 cm⁻¹	Amide III	*		76
1312 cm⁻¹	Amide III band components of proteins	*		70, 77
1313 cm⁻¹	CH₃CH₂ twisting mode of collagen/lipid		*	9, 10
1314 cm⁻¹	CH₃CH₂ twisting mode of collagen		*	42
	CH₂ twisting and wagging		*	60
1315 cm⁻¹	Guanine (B, Z-marker)		*	21
1317 cm⁻¹	Amide III band components of proteins	*		66, 77
	Collagen	*		51
	C–C stretch of Phenyl (2) CH$_{6',2',\alpha,\alpha'}$ in-plane bend	*		6
1317/8 cm⁻¹	Guanine (B, Z-marker)		*	21
1318 cm⁻¹	G (ring breathing modes of the DNA/RNA bases); C–H deformation (protein)		*	12
	Amide III (α-helix)		*	43
	Protein peak		*	17
	Guanine, CH₃CH₂ wagging nucleic acids, CH₃CH₂ wagging of collagen in tumours		*	17
1319 cm⁻¹	Guanine (B, Z-marker)		*	21
	CH₃CH₂ twisting (collagen assignment)		*	19
1321 cm⁻¹	Amide III (α-helix)		*	43
1322 cm⁻¹	CH₃CH₂ twisting, collagen			20
	CH₃CH₂ twisting and wagging in collagen			20
	CH₃CH₂ deforming modes of collagen and nucleic acids	*		41
1323 cm⁻¹	Guanine (B, Z marker)		*	21

TABLE 8.1 (*Continued*)

Major Characteristic Peak Frequencies Reported in the Literature

Peak	Assignment	FT-IR	Raman	Reference Number
1324 cm^{-1}	CH$_3$CH$_2$ wagging mode present in collagen and purine bases of DNA		*	87
1325–30 cm^{-1}	CH$_3$CH$_2$ wagging mode in purine bases of nucleic acids		*	91
1327/8 cm^{-1}	Stretching C–N thymine, adenine	*		48
1328 cm^{-1}	Benzene ring mixed with the CH in-plane bending from the phenyl ring and the ethylene bridge	*		6
	C–C stretch of phenyl (2) CH$_{6',2',\alpha,\alpha'}$ in-plane bend	*		6
1330 cm^{-1}	Typical phospholipids		*	54
	CH$_2$ wagging	*		77, 86
	Region associated with DNA and phospholipids		*	26
	Collagen		*	69
	Nucleic acids and phosphates		*	69
	Ring, C–NR2, cyclic C–C and amide C–N–H stretch		*	14
1332 cm^{-1}	–C stretch of phenyl (1) and C$_3$–C$_3$ stretch and C$_5$–O$_5$ stretch CH$_\alpha$ in-plane bend		*	6
1333 cm^{-1}	Guanine		*	21
	δ(OH) + δ(CH)			
1335 cm^{-1}	CH$_3$CH$_2$ wagging, collagen (protein assignment)		*	20
	CH$_3$CH$_2$ wagging, nucleic acid		*	20
	CH$_3$CH$_2$ wagging mode of collagen and polynucleotide chain (DNA purine bases)		*	10
	CH$_3$CH$_2$ twisting and wagging in collagen		*	20
	Cellular nucleic acids		*	20
	CH$_3$CH$_2$ deforming modes of collagen and nucleic acids		*	41
	An unassigned mode		*	5
	δ(CH), ring (polysaccharides, pectin)	*		7
	Amide III (C–N stretching mode of proteins, indicating mainly α-helix conformation)		*	12

(Continued)

TABLE 8.1 (*Continued*)

Major Characteristic Peak Frequencies Reported in the Literature

Peak	Assignment	FT-IR	Raman	Reference Number
1335–45 cm^{-1}	CH$_3$CH$_2$ wagging mode of collagen		*	9
1335 cm^{-1}	δ(CH), ring (polysaccharides, pectin)	*		7
	Adenine (ring breathing modes of the DNA/RNA bases)		*	4
	A mixture of biochemicals (nucleic acids and proteins due to extracellular matrix)		*	35
	Nucleic acids/proteins		*	35
	Ring, C–NR2, cyclic C–C, C–N–H stretch		*	14
1335/6 cm^{-1}	Guanine		*	21
1336 cm^{-1}	Polynucleotide chain (DNA purine bases)		*	9
	δ(CH$_3$) δ(CH$_2$) twisting, collagen (protein assignment)		*	33
	δ(CH), ring (polysaccharides, cellulose)	*		7
1337 cm^{-1}	Amide III and CH$_2$ wagging vibrations from glycine backbone and proline side chain		*	24
	Collagen	*		40
	A, G (ring breathing modes in the DNA bases), C–H deformation (protein)		*	12
1337 cm^{-1}	Tryptophan		*	3
	CH$_2$/CH$_3$ wagging, twisting and/or bending mode of collagens and lipids		*	3
	CH$_2$/CH$_3$ wagging and twisting mode in collagen, nucleic acid and tryptophan		*	3
1337/8 cm^{-1}	CH$_2$ wagging	*		68, 70, 77
1338 cm^{-1}	Protein peak		*	17
	Amide III, hydrated α-helix δ(N–H) and ν(C–N), desmosine/		*	17
	Isodesmosine (elastin) (A and G of DNA/RNA) in tumours			
1339 cm^{-1}	Tryptophan	*		3
	CH$_2$/CH$_3$ wagging, twisting and/or bending mode of collagens and lipids	*		3

TABLE 8.1 (*Continued*)

Major Characteristic Peak Frequencies Reported in the Literature

Peak	Assignment	FT-IR	Raman	Reference Number
	CH_2/CH_3 wagging and twisting mode in collagen, nucleic acid and tryptophan		*	3
1339 cm^{-1}	Collagen	*		51
	In-plane C–O stretching vibration combined with the ring stretch of phenyl	*		6
	C–C stretch of phenyl (1) and C_3–C_3 stretch and C_5–O_5 stretch CH_α in-plane bend		*	6
1340 cm^{-1}	Nucleic acid mode		*	45
	Differences in collagen content		*	24
	CH_2 wagging	*		66, 77
	Collagen	*		66
	Nucleic acid modes indicating the nucleic acid content in tissues		*	93
	DNA		*	
	Nucleic acid vibrational mode		*	37
	Polynucleotide chain (DNA bases)		*	4
			*	16
1342 cm^{-1}	CH_2 twisting and wagging		*	60
1343 cm^{-1}	CH_3, CH_2 wagging (collagen assignment)		*	19
	Glucose		*	2
1343/4 cm^{-1}	$\delta(CH)$, residual vibrations		*	43
1344 cm^{-1}	A and G of DNA/RNA and CH deformation of proteins		*	17
1347 cm^{-1}	An unassigned mode		*	5
1350 cm^{-1}	Carbon particle		*	94
1354 cm^{-1}	Ring and C–N stretch		*	14
1355/7 cm^{-1}	Guanine (N_7, B, Z-marker)		*	21
1358 cm^{-1}	Stretching C–O, deformation C–H, deformation N–H	*		48
1359 cm^{-1}	Tryptophan		*	3
1360 cm^{-1}	Tryptophan		*	7, 60
1361/2/3/5 cm^{-1}	Guanine (N_7, B, Z-marker)		*	21
1365 cm^{-1}	Tryptophan		*	3
	$\omega(CH_2) + \delta(OH)$		*	1
1366 cm^{-1}	Guanine, protein, lipids		*	17
1367 cm^{-1}	Stretching C–O, deformation C–H, deformation N–H	*		48

(*Continued*)

TABLE 8.1 (*Continued*)

Major Characteristic Peak Frequencies Reported in the Literature

Peak	Assignment	FT-IR	Raman	Reference Number
	$\nu_s(CH_3)$ (phospholipids)		*	43
	$\delta(OH) + \delta(CH_3)$		*	1
1368 cm^{-1}	$\delta(CH_2)$, $\nu(CC)$ (polysaccharides, pectin)	*		7
1369 cm^{-1}	Guanine, TRP (protein), porphyrins, lipids		*	9
1369/70 cm^{-1}	Stretching C–N cytosine, guanine	*		48
1370 cm^{-1}	The most pronounced saccharide band		*	2
1370/1 cm^{-1}	Stretching C–O, deformation C–H, deformation N–H	*		48
	Ring and C–N stretch		*	48
1370/1/3 cm^{-1}	Deformation N–H, C–H	*		23
1373 cm^{-1}	Stretching C–N cytosine, guanine	*		48
	T, A, G (ring breathing modes of the DNA/RNA bases)		*	12
	$\delta(OH) + \delta(CH_3)$		*	1
1376 cm^{-1}	Adenine (ring breathing modes of the DNA/RNA bases)		*	4
1378 cm^{-1}	Paraffin		*	24
1379 cm^{-1}	δCH_3 symmetric (lipid assignment)		*	33
1380 cm^{-1}	δCH_3	*		65, 76
	Stretching C–O, deformation C–H, deformation N–H	*		48
1384 cm^{-1}	Structural proteins like collagen		*	32
	$\delta(OH) + \delta(CH_3)$		*	1
1386 cm^{-1}	CH$_3$ band		*	9
	$\delta(OH) + \delta(CH_3)$		*	1
1388 cm^{-1}	Carbonyl and C–N–C stretch		*	14
1390 cm^{-1}	Carbon particle	*		50
1391 cm^{-1}	CH rocking		*	6
1392 cm^{-1}	C–N stretching, in Quinoid ring–Benzoid ring–Quinoid ring		*	28
1393 cm^{-1}	CH rocking		*	6
	$\delta(OH) + \delta(CH_3)$		*	1
1395 cm^{-1}	Less characteristic, due to aliphatic side groups of the amino acid residues	*		50
1396 cm^{-1}	Symmetric CH$_3$ bending of the methyl groups of proteins	*		64

TABLE 8.1 (*Continued*)

Major Characteristic Peak Frequencies Reported in the Literature

Peak	Assignment	FT-IR	Raman	Reference Number
1398 cm⁻¹	C=O symmetric stretch		*	15
	CH₂ deformation		*	3
	CH₃ symmetric deformation	*		95
1399 cm⁻¹	Extremely weak peaks of DNA and RNA; arises mainly from the vibrational modes of methyl and methylene groups of proteins and lipids and amide groups	*		64
	Symmetric CH₃ bending modes of the methyl groups of proteins	*		66
	δ[(CH₃)] sym.	*		80, 88
	δ[C(CH₃)₂] symmetric	*		77
1400 cm⁻¹	Symmetric stretching vibration of COO⁻ group of fatty acids and amino acids	*		52
	δₛCH₃ of proteins	*		51
	Symmetric bending modes of methyl groups in skeletal proteins	*		51
	Specific absorption of proteins	*		53
	Symmetric stretch of methyl groups in proteins	*		38
	NH in-plane deformation		*	5
	Haemoglobin		*	60
1400–30 cm⁻¹	ν(C=O)O⁻ (amino acids aspartic and glutamic acid)		*	7
1400–50 cm⁻¹	(CH₃)ₑₛₜₑᵣ modes: deformations		*	1
1400–500 cm⁻¹	Ring stretching vibrations mixed strongly with CH in-plane bending	*		13
1400/1/2 cm⁻¹	CH₃ symmetric deformation	*		95
1401 cm⁻¹	Symmetric CH₃ bending modes of the methyl groups of proteins	*		66
	δ[(CH₃)] sym.	*		80, 88
	Bending modes of methyl groups (one of the vibrational modes of collagen)		*	24
1401/2 cm⁻¹	Symmetric CH3 bending modes of the methyl groups of proteins	*		66
	δ[(CH₃)] sym.			

TABLE 8.1 (*Continued*)

Major Characteristic Peak Frequencies Reported in the Literature

Peak	Assignment	FT-IR	Raman	Reference Number
	Stretching C–N, deformation N–H, deformation C–H	*		66, 67, 88
	$\delta[C(CH_3)_2]$ symmetric	*		48
		*		77
1402 cm⁻¹	$CH_{\alpha,\alpha'}$ rocking	*		6
	$\delta_s(CH_3)\ \omega(CH_2)$		*	1
1403 cm⁻¹	Symmetric CH_3 bending modes of the methyl groups of proteins	*		66
	$\delta[(CH3)]$ sym.			
	$\delta_s CH_3$ of collagen	*		68
	$\delta[C(CH_3)_2]$ symmetric	*		51
		*		77
1404 cm⁻¹	Urea/triglycerides, which are abundant in the plasma	*		30
1404/5 cm⁻¹	CH_3 asymmetric deformation	*		95
1406 cm⁻¹	$\nu_s COO^-$ (IgG?)		*	43
1407 cm⁻¹	Sciss. (CH_2)		*	1
1409 cm⁻¹	$\nu_s COO^-$ (IgG?)		*	43
1412 cm⁻¹	Ring stretch		*	14
1412/4 cm⁻¹	Stretching C–N, deformation N–H, deformation C–H	*		48
	Ring stretch		*	14
1416 cm⁻¹	Deformation C–H, N–H, stretching C–N	*		48
1417 cm⁻¹	Stretching C–N, deformation N–H, deformation C–H	*		48
	C=C stretching in quinoid ring		*	28
1418 cm⁻¹	$CH_{2,6,7,\alpha}$ rock	*		6
1418/9 cm⁻¹	Deformation C–H	*		23
1419 cm⁻¹	$\nu_s(COO^-)$ (polysaccharides, pectin)	*		7
1420–50 cm⁻¹	CH_2 scissoring vibration (lipid band)		*	29, 30
1420–70 cm⁻¹	(prominent peak at 1445 cm⁻¹ is of diagnostic significance)		*	42
	CH_2 bending mode of proteins & lipids			
1420/1 cm⁻¹	Deoxyribose, (B, Z-marker)		*	21
1421 cm⁻¹	A, G (ring breathing modes of the DNA/RNA bases)		*	12
1422 cm⁻¹	$CH_{2,6,7,\alpha}$ rock	*		6
	Deoxyribose, (B, Z-marker)		*	21

TABLE 8.1 (*Continued*)

Major Characteristic Peak Frequencies Reported in the Literature

Peak	Assignment	FT-IR	Raman	Reference Number
1423 cm^{-1}	NH in-plane deformation		*	5
1424 cm^{-1}	Deoxyribose, (B, Z-marker)		*	21
1430 cm^{-1}	$\delta(CH_2)$ (polysaccharides, cellulose)	*		7
1431 cm^{-1}	CH$_2$ deformation		*	14
1436 cm^{-1}	CH$_2$ scissoring		*	15
1437 cm^{-1}	CH$_2$ (lipids in normal tissue)	*		54
	CH$_2$ deformation (lipid)	*		9
	Acyl chains	*		2
	$\delta_{as}(CH_3)$	*		1
1437–42 cm^{-1}	CH$_2$ deformation		*	2
1438 cm^{-1}	$\delta(CH_2)\,\delta_{as}(CH_3)$ in lipids		*	17
1439 cm^{-1}	CH$_2$ bending mode in normal tissue		*	96
	CH$_{\alpha,\alpha'}$ in-plane bend	*		6
	CH$_3$, CH$_2$ deformation (collagen assignment)		*	19
	CH$_2$ scissoring		*	19
	CH$_2$ deformation in normal breast tissue		*	19
	CH$_2$ deformation		*	19
	CH$_2$		*	54
	$\delta(CH_2)$ (lipids)		*	43
1440 cm^{-1}	CH$_2$ and CH$_3$ deformation vibrations		*	25
	CH deformation		*	2
	Cholesterol, fatty acid band		*	2
	$\delta(CH_2)$ (lipids)		*	43
	CH$_2$ bending (lipids)		*	43
	Indicates the concentration of lipids (C–H bend modes including CH$_2$ scissors and CH$_3$ degenerate deformation)		*	58, 92
1441 cm^{-1}	CH$_2$ scissoring and CH$_3$ bending in lipids		*	87
	Cholesterol and its esters		*	49
	C–H bending mode of accumulated lipids in the necrotic core of the atheromatous plaque		*	49
1442 cm^{-1}	Fatty acids		*	24, 25
	CH$_2$ bending mode		*	45

(*Continued*)

TABLE 8.1 (*Continued*)

Major Characteristic Peak Frequencies Reported in the Literature

Peak	Assignment	FT-IR	Raman	Reference Number
	Due to changes in chemical environment of the CH$_2$ bending mode		*	25
	CH$_{\alpha,\alpha'}$ in-plane bend	*		6
	CH$_3$, CH$_2$ deformation (collagen assignment)		*	19
	Triglycerides (fatty acids)		*	49
1443 cm^{-1}	CH$_2$ deformation (lipids and proteins)		*	9
	Triglycerides (fatty acids)		*	49
1444 cm^{-1}	Cholesterol band (associated to atherosclerotic spectrum)		*	49
	ν_{28}(C$_\alpha$C$_m$), observed in the spectra of single human RBC		*	18
	δ(CH$_2$), lipids, fatty acids	*	*	7
	δ(CH) (polysaccharides, pectin)	*		7
1445 cm^{-1}	δ(CH$_2$), δ(CH$_3$), collagen (protein assignment)		*	20, 33
	δ(CH$_2$), δ(CH$_3$), scissoring, phospholipids (lipid assignment)		*	20, 33
	CH$_2$CH$_3$ bending modes of collagen and phospholipids		*	20
	CH$_2$ scissoring		*	20
	CH$_2$ bending mode of proteins and lipids are of diagnostic significance		*	42
	CH$_2$ bending and scissoring modes of collagen and phospholipids		*	41
	Methylene bending mode (a combination of proteins and phospholipids)		*	41
	CH$_2$ bending modes		*	33
1446 cm^{-1}	CH$_2$ bending mode of proteins and lipids		*	9, 10
	Umbrella mode of methoxyl (2)	*		6
	CH$_2$ deformation		*	19
	Protein peak		*	17
1447 cm^{-1}	CH$_2$ bending mode of proteins and lipids		*	15
	CH$_2$ deformation (protein vibration); a marker for protein concentration		*	12
	δ_{as}(CH$_3$) δ(CH$_2$) of proteins		*	43

TABLE 8.1 (*Continued*)

Major Characteristic Peak Frequencies Reported in the Literature

Peak	Assignment	FT-IR	Raman	Reference Number
1448 cm^{-1}	CH$_2$CH$_3$ deformation		*	3
	Umbrella mode of methoxyl (2)	*		6
	CH$_2$ deformation		*	19
	CH$_2$		*	54
	Collagen		*	97
1449 cm^{-1}	Asymmetric CH$_3$ bending of the methyl groups of proteins	*		64
	δ_{as}(CH$_3$)		*	1
	δ(CH$_2$) δ_{as}(CH$_3$) in proteins		*	17
1450 cm^{-1}	CH$_2$ bending		*	54, 98, 99
	CH$_2$ bending mode in malignant tissues		*	96
	Bending modes of methyl groups (one of vibrational modes of collagen)			
	Methylene deformation in biomolecules			
	CH$_2$CH$_3$ deformation (collagen assignment)		*	24
	CH$_2$ deformation in IDC breast tissue	*		52
	C–H deformation bands (CH functional groups in lipids, amino acids side chains of the proteins and carbohydrates)		* *	19 22
	δ(C–H)			
	Polyethylene methylene deformation modes		*	22
	CH$_2$ bending (proteins)	*		50
	C–H bending deformation		*	43
	CH$_3$CH$_2$ twisting mode of proteins and nucleic acids and CH$_2$ bending mode of proteins and lipids		* *	85 35
	CH$_2$ proteins/lipids			
			*	35
1451 cm^{-1}	CH$_2$CH$_3$ deformation		*	3
	Asymmetric CH$_3$ bending modes of the methyl groups of proteins	*		66
	δ[(CH$_3$)] asym.	*		77, 88
	CH$_2$CH$_3$ deformation (collagen assignment)		*	19

(Continued)

TABLE 8.1 (*Continued*)

Major Characteristic Peak Frequencies Reported in the Literature

Peak	Assignment	FT-IR	Raman	Reference Number
1452 cm^{-1}	CH$_2$CH$_3$ deformation		*	60
	CH$_2$ deformation in lipids		*	16
1453 cm^{-1}	Protein bands		*	24, 25
	Umbrella mode of methoxyl (4)		*	6
	C–H bending mode of structural proteins		*	49
1453 cm^{-1}	Structural protein modes of tumours		*	26
	CH$_2$ deformation, amide (III) stretch		*	14
1454 cm^{-1}	CH$_2$ stretching/CH$_3$ assymetric deformation		*	15
	Overlapping asymmetric CH$_3$ bending and CH$_2$ scissoring (is associated with elastin, collagen, and phospholipids)		*	42
	Asymmetric methyl deformation	*		38
	Collagen and phospholipids		*	26
1455 cm^{-1}	C–O–H	*		48
	Less characteristic, due to aliphatic side groups of the amino acid residues	*		50
	δ_{as}CH$_3$ of proteins	*		51
	Symmetric bending modes of methyl groups in skeletal proteins	*		51
	Deoxyribose		*	21
	Umbrella mode of methoxyl (4)	*		6
	δ(CH$_2$)		*	43
	CH$_2$ scissoring of proteins and lipids		*	35
1455/6/7 cm^{-1}	Asymmetric CH$_3$ bending modes of the methyl groups of proteins	*		66
	δ[(CH$_3$)] asym.	*		66, 68, 77, 80
1457 cm^{-1}	Extremely weak peaks of DNA and RNA; arises mainly from the vibrational modes of methyl and methylene groups of proteins and lipids and amide groups	*		64
	Deoxyribose		*	21
	δ_{as}(CH$_3$)		*	1

TABLE 8.1 (*Continued*)

Major Characteristic Peak Frequencies Reported in the Literature

Peak	Assignment	FT-IR	Raman	Reference Number
1458 cm^{-1}	Nucleic acid modes		*	45
	$\delta_{as}CH_3$ of collagen	*		51
	Nucleic acid modes indicating the nucleic acid content in tissues		*	93
	Nucleic acid vibrational modes		*	4
	CH_2 deformation, amide (III) stretch		*	14
1459 cm^{-1}	Deoxyribose		*	21
	Umbrella mode of methoxyl (4)	*		6
	$\delta(CH_2)$		*	43
1460 cm^{-1}	CH_2/CH_3 deformation of lipids and collagen		*	3
	CH_2 wagging, CH2/CH3 deformation		*	3
	Deoxyribose		*	21
	Umbrella mode of methoxyl (1)	*		6
	$\delta_{as}(CH_3)$		*	1
1462 cm^{-1}	Paraffin	*		24
	δCH_2 disaccharides, sucrose		*	7
1463 cm^{-1}	Fermi interaction $\delta(CH_2)$, and $\gamma(CH_2)$		*	43
1464 cm^{-1}	Umbrella mode of methoxyl (1)	*		6
1465 cm^{-1}	Lipids		*	24
	CH_2 scissoring mode of the acyl chain of lipid	*		78
	$\delta_{as}(CH_3)$		*	1
1465/6 cm^{-1}	Umbrella mode of methoxyl (3)	*		6
1467 cm^{-1}	Cholesterol–methyl band	*		71
1468 cm^{-1}	δCH_2	*		76
	δCH_2 of lipids	*		51
1469 cm^{-1}	CH_2 bending of the acyl chains of lipids	*		64
1470 cm^{-1}	CH_2 bending of the methylene chains in lipids	*		51
	C=N stretching		*	28
	CH_2 deformation or C–N–H stretch, bend		*	14
1472 cm^{-1}	Paraffin		*	24
1476 cm^{-1}	δCH_2		*	65
1480 cm^{-1}	Polyethylene methylene deformation modes	*		50

(*Continued*)

TABLE 8.1 (*Continued*)

Major Characteristic Peak Frequencies Reported in the Literature

Peak	Assignment	FT-IR	Raman	Reference Number
	DNA		*	37
	CH_3 deformations		*	1
1480–543 cm^{-1}	Amide II	*		7
1480–575 cm^{-1}	Amide II (Largely due to a coupling of CN stretching and in-plane bending of the N–H group; is not often used for structural studies per se because it is less sensitive and is subject to interference from absorption bounds of amino acid side chain vibrations. This is extremely weak in Raman spectra.)		*	24, 30
1480–600 cm^{-1}	The region of the amide II band in tissue proteins. Amide II mainly stems from the C–N stretching and C–N–H bending vibrations weakly coupled to the C=O stretching mode.	*		83
1481 cm^{-1}	$\delta_{as}(CH_3)$		*	1
1482 cm^{-1}	Benzene	*		6
1482/3/5 cm^{-1}	C_8–H coupled with a ring vibration of guanine	*		48
1485 cm^{-1}	G, A (ring breathing modes in the DNA bases)		*	12
	Nucleotide acid purine bases (guanine and adenine)		*	9
1485–550 cm^{-1}	NH_3^+		*	15
1486 cm^{-1}	Deformation C–H	*		23
1487/8 cm^{-1}	Guanine (N_7)		*	21
	C=C, deformation C–H	*		48
1488 cm^{-1}	Collagen		*	97
1488/9 cm^{-1}	Deformation C–H	*		23
1489 cm^{-1}	In-plane CH bending vibration	*		6
1490 cm^{-1}	DNA		*	54
	C=C, deformation C–H	*		48
	In-plane CH bending vibration	*		6
	Formalin peak appearing in the fixed normal and tumour tissues		*	41
1491 cm^{-1}	C–N stretching vibration coupled with the in-plane C–H bending in amino radical cations		*	15
1492 cm^{-1}	Formalin peak appearing in the fixed normal and tumour tissues		*	41

TABLE 8.1 (*Continued*)

Major Characteristic Peak Frequencies Reported in the Literature

Peak	Assignment	FT-IR	Raman	Reference Number
1494 cm⁻¹	In-plane CH bending vibration	*		6
1495 cm⁻¹	Adenine, thymine (ring breathing modes of the DNA/RNA bases)		*	4
1495/6 cm⁻¹	C=C, deformation C–H	*		48
1499 cm⁻¹	C–C stretching in benzenoid ring		*	28
1500 cm⁻¹	In-plane CH bending vibration from the phenyl rings	*		6
	CH in-plane bending	*		6
1504 cm⁻¹	$CH_{2',5',6'}$ in-plane band	*		6
	In-plane CH bending vibration from the phenyl rings	*		6
1506 cm⁻¹	N–H bending		*	28
1506/08/10 cm⁻¹	Cytosine		*	21
1510 cm⁻¹	In-plane CH bending vibration from the phenyl rings	*		6
	$CH_{2,6}$ in-plane bend	*		6
	(A ring breathing modes in the DNA bases)		*	12
1513 cm⁻¹	Cytosine		*	21
1514 cm⁻¹	ν(C=C), diagnostic for the presence of a carotenoid structure; most likely a cellular pigment	*	*	22
	ν(C=C) carotenoid		*	33
1515 cm⁻¹	Cytosine		*	21
1517 cm⁻¹	Amide II	*		65
	β-carotene accumulation (C–C stretch mode)		*	49
1518 cm⁻¹	ν(C=C), porphyrin		*	20
	Carotenoid peaks due to C–C and conjugated C=C band stretch		*	20
1519 cm⁻¹	$CH_{2',5',6'}$ in-plane band	*		6
1520 cm⁻¹	Carotene		*	54
1520–38 cm⁻¹	–C=C– carotenoid		*	9
1521 cm⁻¹	$CH_{2,6}$ in-plane bend	*		6
1522 cm⁻¹	(–C=C–) stretching of carotenoid		*	60
1524 cm⁻¹	Carotenoid (absent in normal tissues)		*	45
	Stretching C=N, C=C	*		48
1525 cm⁻¹	In-plane vibrations of the conjugated –C=C–		*	81
1526 cm⁻¹	C=N guanine	*		48

(*Continued*)

TABLE 8.1 (*Continued*)

Major Characteristic Peak Frequencies Reported in the Literature

Peak	Assignment	FT-IR	Raman	Reference Number
1527 cm^{-1}	Stretching C=N, C=C	*		48
1528 cm^{-1}	C=N guanine	*		48
	Carotenoid (absent in normal tissues)		*	45
152829/30 cm^{-1}	C=N adenine, cytosine	*		23
1530 cm^{-1}	Stretching C=N, C=C	*		48
1531 cm^{-1}	Modified guanine?	*		48
1532 cm^{-1}	Stretching C=N, C=C	*		48
1534 cm^{-1}	Modified guanine?	*		48
	Amide II	*		77
1535/7 cm^{-1}	Stretching C=N, C=C	*		48
1540 cm^{-1}	Protein amide II absorption; predominately β-sheet of amide II	*		83
	Amide II	*		77
1540–650 cm^{-1}	Amide II	*		55
1540–680 cm^{-1}	Amide carbonyl group vibrations and aromatic hydrogens		*	84
1541 cm^{-1}	Amide II absorption (primarily an N–H bending coupled to a C–N stretching vibrational mode)	*		38, 52
	Amide II	*		50
1543 cm^{-1}	Amide II	*		76
	$v_{11}(C_\beta C_\beta)$, observed in the spectra of single human RBC		*	22
1544 cm^{-1}	Amide II		*	24
	Amide II bands (arises from C–N stretching and CHN bending vibrations)	*		52, 62
1545 cm^{-1}	Protein band	*		71
	Amide II (δN–H, νC–N)	*		65
	Peptide amide II	*		53, 100
	C_6–H deformation mode		*	5
1546 cm^{-1}	Bound and free NADH		*	54
1548 cm^{-1}	Tryptophan		*	9, 10
1549 cm^{-1}	Amide II	*		77
	Amide II of proteins	*		51
1550 cm^{-1}	Amide II	*		74
	Amide II of proteins	*		78
	N–H bending and C–N stretching	*		101

TABLE 8.1 (*Continued*)

Major Characteristic Peak Frequencies Reported in the Literature

Peak	Assignment	FT-IR	Raman	Reference Number
	Due to the stretching of –C–N and bending of –N–H group of R– CONR2		*	30
1550–650 cm⁻¹	Ring stretching vibrations with little interaction with CH in-plane bending	*		13
1550–700 cm⁻¹	C=C		*	1
1550–800 cm⁻¹	Region of the base vibrations	*		48
1552 cm⁻¹	Tryptophan		*	20
	ν(C=C), tryptophan (protein assignment)		*	20
	ν(C=C), porphyrin		*	20
	Ring base	*		48
1553 cm⁻¹	CO stretching	*		64
	Predominately α-sheet of amide II (Amide II band mainly stems from the C–N stretching and C–N–H bending vibrations weakly coupled to the C=O stretching mode.)	*		83
1554 cm⁻¹	Amide II		*	15
1555 cm⁻¹	Ring base	*		48
1558 cm⁻¹	Tryptophan		*	3
	ν(CN) and δ (NH) amide II (protein assignment)		*	33
	ν(C=C) porphyrin		*	33
	Tyrosine, amide II, COO⁻		*	43
1559 cm⁻¹	Ring base	*		48
1560 cm⁻¹	Tryptophan		*	42
1560–600 cm⁻¹	COO⁻		*	15
1567 cm⁻¹	Ring base	*		48
1570 cm⁻¹	Amide II	*		42
1571/3 cm⁻¹	C=N adenine	*		48
1573 cm⁻¹	Guanine, adenine, TRP (protein)		*	9
1574 cm⁻¹	C=N adenine	*		48
	Tryptophan, nucleic acids (guanine, adenine), protein in tumours		*	17
1575 cm⁻¹	Ring breathing modes in the DNA bases		*	12
	G, A (ring breathing modes of the DNA/RNA bases)			

(*Continued*)

TABLE 8.1 (*Continued*)

Major Characteristic Peak Frequencies Reported in the Literature

Peak	Assignment	FT-IR	Raman	Reference Number
	Histidine (ring breathing modes of the DNA/RNA bases)		*	4
	C=N adenine	*		48
1576 cm⁻¹	Nucleic acid mode		*	45
	C=N adenine	*		48
	Nucleic acid modes indicating the nucleic acid content in tissues		*	93
	Nucleic acid vibrational mode		*	4
	Protein peak		*	17
	Tryptophan, nucleic acids (guanine, adenine), protein in tumours		*	17
1576/7 cm⁻¹	Guanine (N_3)		*	21
1577 cm⁻¹	Bound and free NADH		*	54
	Ring C–C stretch of phenyl (1)	*		6
	IgG?		*	43
1578 cm⁻¹	Guanine (N_3)		*	21
1579 cm⁻¹	Pyrimidine ring (nucleic acids) and heme protein		*	9, 10
1580 cm⁻¹	C–C stretching		*	28
1581 cm⁻¹	Ring C–C stretch of phenyl (2)	*		6
1582 cm⁻¹	δ(C=C), phenylalanine		*	20
	Phenylalanine		*	20
	Adenine, guanine		*	16
1583 cm⁻¹	C=C bending mode of phenylalanine		*	42
1585 cm⁻¹	C=C olefinic stretch		*	15
	C=C olefinic stretch (protein assignment)		*	33
	Urea/triglycerides, which are abundant in the plasma	*		30
1586 cm⁻¹	Phenylalanine, hydroxyproline		*	3
1587 cm⁻¹	C=C olefinic stretching of lipids, retinols		*	60
1588 cm⁻¹	Phenylalanine, hydroxyproline		*	3
1589 cm⁻¹	Ring C–C stretch of phenyl (1)	*		6
1590 cm⁻¹	Carbon particles		*	94
1592 cm⁻¹	C=N, NH_2 adenine	*		48
1593 cm⁻¹	C=N and C=C stretching in quinoid ring		*	28
1594 cm⁻¹	Ring C–C stretch of phenyl (2)	*		48
1596 cm⁻¹	Methylated nucleotides	*		48

TABLE 8.1 (*Continued*)

Major Characteristic Peak Frequencies Reported in the Literature

Peak	Assignment	FT-IR	Raman	Reference Number
1597 cm^{-1}	C=N, NH$_2$ adenine	*		48
1600 cm^{-1}	Ring stretch with CO conjugation		*	14
1600–50 cm^{-1}	C=C stretching vibrations from the aromatic rings		*	1
1600–700 cm^{-1}	Amide I		*	4
	Amide I (due mostly to the C=O stretching vibrations of the peptide backbone)		*	30
1600–715 cm^{-1}	In-plane double bond vibrations of the nucleotide bases	*		47
1600–720 cm^{-1}	The region of the amide I band of tissue proteins (highly sensitive to the conformational changes in the secondary structure) (Amide I band is due to in-plane stretching of the C=O bond, weakly coupled to stretching of the C–N and in-plane bending of the N–H bond.)	*		83
1600–800 cm^{-1}	Amide I band of proteins; due to C=O stretching		*	12, 89
	Amide I (which is due mostly to the C=O stretching vibrations of the peptide backbone; has been used the most for structural studies due to its high sensitivity to small changes in molecular geometry and hydrogen bonding of peptide group)		*	24
	Amide I of proteins		*	35
1601/2 cm^{-1}	C=N cytosine, N–H adenine	*		48
1602 cm^{-1}	Phenylalanine		*	20
	δ(C=C), phenylalanine (protein assignment)		*	3
1603 cm^{-1}	C=C in-plane bending mode of phenylalanine and tyrosine		*	9, 10
	Ring C–C stretch of phenyl (1)		*	6
1603/4 cm^{-1}	C=N, NH$_2$ adenine	*		48
1604 cm^{-1}	Adenine vibration in DNA	*		23
1605 cm^{-1}	Cytosine (NH$_2$)		*	21
	Ring C–C stretch of phenyl (1)		*	6
	Phenylalanine, tyrosine, C=C (protein)		*	12

(*Continued*)

TABLE 8.1 (*Continued*)

Major Characteristic Peak Frequencies Reported in the Literature

Peak	Assignment	FT-IR	Raman	Reference Number
	$v_{as}(COO^-)$ (polysaccharides, pectin)	*		7
1606 cm^{-1}	Adenine vibration in DNA	*		23
1607 cm^{-1}	Tyrosine, phenylalanine ring vibration		*	43
1608/9 cm^{-1}	Cytosine (NH$_2$)		*	21
1609 cm^{-1}	Adenine vibration in DNA	*		23
	Haemoglobin		*	60
1610 cm^{-1}	Cytosine (NH$_2$)		*	21
	C=C bending in phenylalanine and tyrosine		*	16
1613 cm^{-1}	Ring stretch with CO conjugation		*	14
1614 cm^{-1}	Tyrosine		*	3
1615 cm^{-1}	Tyrosine, Tryptophan, C=C (protein)		*	12
1616 cm^{-1}	C=C stretching mode of tyrosine and tryptophan		*	9, 10
1617 cm^{-1}	$v(C_aC_b)$, observed in the spectra of single human RBC		*	22
1618 cm^{-1}	v(C=C), tryptophan (protein assignment)		*	20
	v(C=C), porphyrin		*	20
	Tryptophan		*	20
	Bound and free NADH		*	54
	Ring C–C stretch of phenyl(2)	*		6
1620 cm^{-1}	Peak of nucleic acids due to the base carbonyl stretching and ring breathing mode	*		50
	v(C=C), porphyrin		*	33
	Haemoglobin		*	60
1620–40 cm^{-1}	Bending vibration of R-CONHR group		*	30
1620–750 cm^{-1}	In-plane double end vibrations of bases. The spectra in this region are very sensitive to base-pairing interactions and base-stacking effects, i.e., effects of hydrogen bond formation.		*	24, 30
1622 cm^{-1}	Tryptophan		*	3
	Tryptophan and/or β-sheet		*	3
	Tryptophan (IgG?)		*	43
1628 cm^{-1}	C_α=C_α stretch		*	6

TABLE 8.1 (*Continued*)

Major Characteristic Peak Frequencies Reported in the Literature

Peak	Assignment	FT-IR	Raman	Reference Number
	Amide C=O stretching absorption for the β-form polypeptide films		*	102
1630–80 cm^{-1}	Due to the carbonyl stretching of an amide group		*	30
1630–700 cm^{-1}	Amide I region	*		38
1632 cm^{-1}	Ring C–C stretch of phenyl (2)	*		6
1632/4 cm^{-1}	C=C uracyl, C=O	*		48
1634 cm^{-1}	Amide I		*	24
1635 cm^{-1}	Differences in collagen content		*	24
	β-sheet structure of amide I	*		83
1637 cm^{-1}	Amide I band		*	15
	C=C uracyl, C=O	*		48
	Amide I band (both α-helix and β-structure)		*	43
1638 cm^{-1}	Intermolecular bending mode of water		*	9
	Very weak and broad v_2 mode of water		*	43
1638/9 cm^{-1}	C=C thymine, adenine, N–H guanine	*		48
1639 cm^{-1}	Amide I	*		77
1640 cm^{-1}	Amide I band of protein and H–O–H deformation of water	*		101
1640–80 cm^{-1}	Amide I band (protein band)		*	30, 95
1640–740 cm^{-1}	C=O stretching modes		*	1
1642 cm^{-1}	C$_5$ methylated cytosine?	*		48
1643 cm^{-1}	Amide I band (arises from C=O stretching vibrations)	*		62
1644 cm^{-1}	Amide I	*		77
1645 cm^{-1}	Amide I (α-helix)		*	9
1646 cm^{-1}	Amide I	*		74
	C$_5$ methylated cytosine?	*		48
	C=O, stretching C=C uracyl, NH$_2$ guanine	*		48
1647 cm^{-1}	Random coils		*	43
1647/8 cm^{-1}	Amide I in normal tissues; in lower frequencies in cancer	*		74
1649 cm^{-1}	Unordered random coils and turns of amide I	*		83
	C=O, C=N, N–H of adenine, thymine, guanine, cytosine	*		23

(Continued)

TABLE 8.1 (*Continued*)

Major Characteristic Peak Frequencies Reported in the Literature

Peak	Assignment	FT-IR	Raman	Reference Number
1650 cm^{-1}	Amide I absorption (predominantly the C=O stretching vibration of the amide C=O)	*		38, 52
	(C=C) amide I		*	54
	Protein amide I absorption			
	C=O, stretching C=C uracyl, NH$_2$ guanine	*		48
	Peptide amide I	*		72, 100
	Amide I		*	103
	Carbonyl stretching		*	85
1650–60 cm^{-1}	C=C stretch band for the nonconjugated *cis* configuration		*	58, 104, 105
1650–60 cm^{-1}	C–C stretch		*	58
1652 cm^{-1}	Amide I	*		69
1652/3 cm^{-1}	C$_2$=O cytosine	*		48
	Lipid (C=C stretch)		*	9
1653 cm^{-1}	Carbonyl stretch (C=O)		*	5
	vC=O		*	1
1653/4 cm^{-1}	C=O, C=N, N–H of adenine, thymine, guanine, cytosine	*		23
1654 cm^{-1}	Due to a combination of C=C stretch and the amide I bands + Amide I		*	24, 96
	Amide I (collagen assignment)			
	C=C stretch		*	19
	Amide I (C=O stretching mode of proteins, α-helix conformation)/ C=C lipid stretch		*	19
			*	9
	Collagen			
	ν(C=C) of lipids		*	26
			*	17
1655 cm^{-1}	Amide I (of collagen)		*	3
	C=C (of lipids in normal tissues; not that of amide I)		*	54
	ν(C=O) amide I, α-helix, collagen		*	20, 33, 98
	Amide I (C=O stretching mode of proteins, α-helix conformation)/ C=C lipid stretch		*	9, 10
	In normal tissues : C=C of lipids (and not amide I)		*	20
	C=O stretching of collagen and elastin (protein assignment)		*	20

TABLE 8.1 (*Continued*)

Major Characteristic Peak Frequencies Reported in the Literature

Peak	Assignment	FT-IR	Raman	Reference Number
	Amide I (of proteins in α-helix conformation)	*		56, 65
	Amide I (νC=O, δC–N, δN–H)	*		48
	C=O cytosine	*		23
	C=O, C=N, N–H of adenine, thymine, guanine, cytosine	*		48
	Peak of nucleic acids due to the base carbonyl stretching and ring breathing mode	*		50
	Amide I has some overlapping with the carbonyl stretching modes of nucleic acid	*		51
	Amide I of proteins		*	33, 35, 41
	Amide I (collagen assignment)		*	19
	Amide I (typically associated with collagen)		*	91
	Amide I (α-helix)	*		7
	Protein amide I band (C=O stretching mode of proteins, indicating mainly α-helix conformation)		*	35
	C=O stretching of collagen and elastin		*	35
1655–80 cm⁻¹	T, G, C (ring breathing modes of the DNA/RNA bases); amide I (protein)		*	12
1656 cm⁻¹	Amide I	*		19, 76
	C₂=O cytosine	*		48
	ν(C=C) *cis* (phospholipids)		*	43
	C=C (lipids)		*	43
	Amide I (proteins)		*	43
	Amide I, α-helix, ν(C=O) of proteins collagen and elastin		*	17
1657 cm⁻¹	Fatty acids		*	24, 25
	α-helical structure of amide I	*		83
	Amide I (collagen assignment)		*	19
	Triglycerides (fatty acids)		*	49
1658 cm⁻¹	C=O, stretching C=C uracyl, NH₂ guanine	*		49
	Amide I	*		50
	Amide I (α-helix)		*	43
	νC=O		*	1

(*Continued*)

TABLE 8.1 (*Continued*)

Major Characteristic Peak Frequencies Reported in the Literature

Peak	Assignment	FT-IR	Raman	Reference Number
	Proteins, amide I (α helix) in normal tissues		*	46
	Amide I, α-helix, ν(C=O) of proteins collagen and elastin		*	17
1659 cm⁻¹	Amide I	*		77
	Amide I vibration (collagen like proteins)		*	50, 94
	Amide C=O stretching absorption for the α-folded polypeptide films		*	102
	Cholesterol band (associated to atherosclerotic spectrum)		*	49
1660 cm⁻¹	Amide I band	*	*	22
	Amide I		*	22
	Amide I vibration mode of structural proteins		*	49
	ν(C=C) *cis*, lipids, fatty acids	*	*	7
	C=C groups in unsaturated fatty acids		*	2
	Ceramide backbone		*	2
	Amide I		*	60
	Amide I, β-pleated sheet, and/or random coil conformation		*	35
	Proteins, amide I (α helix) in cancerous tissues		*	46
1661 cm⁻¹	Amide I, α-helix			60
1662 cm⁻¹	Nucleic acid modes		*	45
	Nucleic acid modes indicating the nucleic acid content in tissues		*	68
	Nucleic acid vibrational mode		*	4
1663 cm⁻¹	DNA		*	8
	Proteins, including collagen I		*	8
1664/5 cm⁻¹	Amide I		*	3
	C=O cytosine, uracyl	*		48
1665 cm⁻¹	Amide I (of collagen)		*	3
	Amide I		*	3
	Amide I (collagen assignment)		*	19
	Amide I (disordered structure, solvated)	*	*	7
	ν_s(C=O)		*	106
	C=C stretch band (*cis* form)		*	58

TABLE 8.1 (*Continued*)

Major Characteristic Peak Frequencies Reported in the Literature

Peak	Assignment	FT-IR	Raman	Reference Number
1665–80 cm⁻¹	C=C stretch band for the nonconjugated *trans* configuration		*	34
1666 cm⁻¹	Collagen		*	97
1667 cm⁻¹	Protein band		*	24, 25
	C=C stretching band		*	45
	α-helical structure of amide I		*	20
	Structural protein modes of tumours		*	27
	Carbonyl stretch (C=O)		*	5
1669 cm⁻¹	Carbonyl stretch (C=O)		*	5
	Cholesterol ester		*	2
	C=C stretch band (*trans* form)		*	58
	νC=O		*	1
1670 cm⁻¹	Amide I		*	99
	C=C stretching vibrations		*	25
	Cholesterol and its esters		*	49
	C=C stretching vibration mode of steroid ring		*	49
	Amide I (anti-parallel β-sheet)	*	*	7
	ν(C=C) *trans*, lipids, fatty acids	*	*	7
1671 cm⁻¹	νC=O		*	1
1672 cm⁻¹	C=C stretch		*	6
	Amide I band (C=O stretch coupled to a N–H bending)		*	2
	Ceramide	*	*	2
1673 cm⁻¹	Amide I		*	15
	C=O		*	1
1674 cm⁻¹	C=C stretch vibration		*	2
	Cholesterol		*	2
1676 cm⁻¹	Amide I (β-sheet)		*	43
1678 cm⁻¹	Bound and free NADH		*	54
1679 cm⁻¹	Stretching C=O vibrations that are H-bonded. (Changes in the C=O stretching vibrations could be connected with destruction of old H-bonds and creation of the new ones.)	*		48
	C=O guanine deformation N–H in plane	*		48
	C=O		*	1

(*Continued*)

TABLE 8.1 (*Continued*)

Major Characteristic Peak Frequencies Reported in the Literature

Peak	Assignment	FT-IR	Raman	Reference Number
1680 cm⁻¹	Unordered random coils and turns of amide I	*		83
	vC=O		*	1
1681 cm⁻¹	C=O guanine deformation N–H in plane	*		48
	vC=O		*	1
1682 cm⁻¹	One of the absorption positions for the C=O stretching vibrations of cortisone		*	102
	C=O		*	1
1684 cm⁻¹	C=O guanine deformation N–H in plane	*		48
1685 cm⁻¹	Amide I (disordered structure; nonhydrogen bonded)	*	*	7
1690 cm⁻¹	Peak of nucleic acids due to the base carbonyl stretching and ring breathing mode	*		50
1694 cm⁻¹	A high-frequency vibration of an antiparallel β-sheet of amide I	*		83
	(The amide I band is due to in-plane stretching of the C=O band weakly coupled with stretching of the C–N and in-plane bending of the N–H bond.)			
1695 cm⁻¹	vC=O		*	1
1697 cm⁻¹	Amide I (turns and bands)		*	43
1698/9 cm⁻¹	C₂=O guanine	*		48
	N–H thymine	*		48
1700–15 cm⁻¹	The region of the bases	*		23
1700–50 cm⁻¹	v(C=O)OH (amino acids aspartic and glutamic acid)		*	7
1700–800 cm⁻¹	Fatty acid esters	*		56
	Lipids observed during cell apoptosis (more specifically at 1740 cm⁻¹)	*		39
1700/2 cm⁻¹	C=O guanine	*		48
1702 cm⁻¹	C=O thymine	*		48
	Stretching C=O vibrations that are H-bonded. (Changes in the C=O stretching vibrations could be connected with destruction of old H-bonds and creation of the new ones.)	*		48

TABLE 8.1 (*Continued*)

Major Characteristic Peak Frequencies Reported in the Literature

Peak	Assignment	FT-IR	Raman	Reference Number
	C=O		*	1
1703 cm⁻¹	νC=O		*	1
1706/7 cm⁻¹	C=O thymine	*		48
1707 cm⁻¹	C=O Guanine	*		48
1708 cm⁻¹	C=O thymine	*		48
1710 cm⁻¹	One of the absorption positions for the C=O stretching vibrations of cortisone		*	102
1712/9 cm⁻¹	C=O	*		48
1713/4/6 cm⁻¹	C=O thymine	*		48
1716–41 cm⁻¹	C=O		*	2
1717 cm⁻¹	C=O thymine	*		48
1719 cm⁻¹	C=O	*		48
1725–45 cm⁻¹	C=O stretching band mode of the fatty acid ester	*		56
1728/9 cm⁻¹	C=O band	*		56
1729 cm⁻¹	Ester group		*	2
1730 cm⁻¹	Absorption band of fatty acid ester	*		56
	Fatty acid ester band	*		56
	Out-of-phase C=O stretch		*	14
1732 cm⁻¹	One of the absorption positions for the C=O stretching vibrations of cortisone		*	102
1733 cm⁻¹	Out-of-phase C=O stretch		*	14
1734 cm⁻¹	νC=O		*	1
1736 cm⁻¹	Phospholipid C=O stretching band	*		74
1738 cm⁻¹	Lipids		*	24
1739 cm⁻¹	Ester group		*	2
	ν(C=O) (polysaccharides, hemicellulose)	*		7
1740 cm⁻¹	C=O	*		74
	C=O	*		40
	Ester group		*	2
	A small band that is predicted to be an indication of cell		*	30
1742 cm⁻¹	νC=O		*	1
1743 cm⁻¹	C=O stretching mode of lipids	*		100
1744 cm⁻¹	Carbonyl feature of lipid spectra		*	27
	Ester group		*	2

(Continued)

TABLE 8.1 (*Continued*)

Major Characteristic Peak Frequencies Reported in the Literature

Peak	Assignment	FT-IR	Raman	Reference Number
1745 cm^{-1}	Phospholipids		*	20
	Ester group (C=O) vibration of triglycerides	*		74
	v(C=O), phospholipids (lipid assignment)		*	20, 33
	v(C=O), phospholipids		*	50
	Triglycerides (fatty acids)		*	49
	v(C=O) (polysaccharides, pectin)	*	*	7
	The ester group (C=O) vibration of triglycerides in normal tissues (can reliably identify fat content in the tissue C=O stretching mode of phospholipids)	*	*	74
			*	35
	Phospholipids		*	35
1746 cm^{-1}	C=O stretch (lipid)		*	9
	C=O stretch of phospholipids		*	17
	v(C=O) of phospholipids		*	17
1747 cm^{-1}	C=O, lipids		*	43
1750 cm^{-1}	C=O (lipid in normal tissues)		*	54
	v(C=C) lipids, fatty acids	*	*	7
	The ester group (C=O) vibration in malignant tissues	*		74
1750/1 cm^{-1}	Out-of-phase C=O stretch		*	14
1754 cm^{-1}	C=O (lipid)		*	43
1755 cm^{-1}	vC=O		*	14
	In-phase carbonyl C=O stretch		*	14
1756 cm^{-1}	One of the absorption positions for the C=O stretching vibrations of cortisone		*	102
1762 cm^{-1}	In-phase carbonyl C=O stretch		*	14
1776/8 cm^{-1}	In-phase carbonyl C=O stretch		*	14
1997/2040/53/ 58 cm^{-1}	The band of second order	*		48
2095 cm^{-1}	C–N stretch (one of three thiocyanate peaks, with 445 and 735 cm^{-1})		*	5
2225 cm^{-1}	C≡N		*	107
2300–3800 cm^{-1}	Region of the OH–NH–CH stretching vibrations		*	48
2343 cm^{-1}	Asymmetric stretching band of CO_2^- hydrates		*	48
2550–80 cm^{-1}	v(S–H) (amino acid methionine)		*	7

TABLE 8.1 (*Continued*)

Major Characteristic Peak Frequencies Reported in the Literature

Peak	Assignment	FT-IR	Raman	Reference Number
2600 cm^{-1}	H-bonded NH vibration band	*		23
2633/678 cm^{-1}	Stretching N–H (NH$_3$$^+$)	*		48
2700–3300 cm^{-1}	C–H stretches		*	107
2700–3500 cm^{-1}	Stretching vibrations of CH, NH, and OH groups		*	2
2727/731 cm^{-1}	Stretching N–H (NH$_3$$^+$)	*		48
2761 cm^{-1}	CH$_3$ modes	*		57
2765/66/69/99 cm^{-1}	Stretching N–H (NH$_3$$^+$)	*		48
2800 cm^{-1}	Stretching N–H (NH$_3$$^+$)	*		48
2800–3000 cm^{-1}	C–H	*		65
	Lipid region	*		40
	CH$_3$, CH$_2$ lipid and protein	*		56
2800–3020 cm^{-1}	C–H stretching vibrations		*	85
2800–3050 cm^{-1}	Contributions from acyl chains		*	2
2800–3100 cm^{-1}	C–H stretching vibrations of methyl (CH$_3$) and methylene (CH$_2$) groups and olefins	*		74
	CH, CH$_2$, and CH$_3$ symmetric and antisymmetric stretching		*	108
	C–H stretching vibrations of methyl (CH$_3$) and methylene (CH$_2$) groups and olefins	*		17
2800–3500 cm^{-1}	Cholesterol, phospholipids, and creatine (higher in normal tissues)	*		62
	Stretching vibrations of CH$_2$ and CH$_3$ of phospholipids, cholesterol, and creatine	*		62
2802/12/20/21/4/ 34 cm^{-1}	Stretching N–H (NH$_3$$^+$)	*		48
2834 cm^{-1}	Symmetric stretching of methoxy (1)	*		6
2836 cm^{-1}	Stretching N–H (NH$_3$$^+$)	*		48
2838 cm^{-1}	Stretching C–H	*		48
	Symmetric stretching of methoxy (3)	*	*	6
2846 cm^{-1}	Symmetric stretching of methoxy (4)	*		6
2848 cm^{-1}	Cholesterol, phospholipids, and creatine (higher in normal tissues)	*		62
	Stretching vibrations of CH$_2$ and CH$_3$ of phospholipids, cholesterol, and creatine	*		62

(Continued)

TABLE 8.1 (*Continued*)

Major Characteristic Peak Frequencies Reported in the Literature

Peak	Assignment	FT-IR	Raman	Reference Number
2849 cm^{-1}	Stretching C–H	*		48
2850 cm^{-1}	C–H stretching bands	*		74
	Stretching C–H	*		48
	v_sCH$_2$, lipids, fatty acids		*	7
	CH$_2$ symmetric			
2850–3100 cm^{-1}	CH stretching modes, both symmetric and asymmetric		*	1
2851 cm^{-1}	Symmetric CH$_2$ stretch	*		48
2852 cm^{-1}	v_sCH$_2$			65
2853 cm^{-1}	v_sCH$_2$ of lipids	*		51
	Asymmetric CH$_2$ stretching mode of the methylene chains in membrane lipids	*		51
	Symmetric stretching of methoxy (2)	*		6
2854 cm^{-1}	v_sCH$_2$		*	1
2860 cm^{-1}	Stretching C–H	*		48
2864 cm^{-1}	Symmetric stretching of methoxy (2)		*	6
	Symmetric stretching of methoxy (1)	*		6
2865 cm^{-1}	C=O		*	1
2867 cm^{-1}	Symmetric stretching of methoxy (3)	*		6
2870 cm^{-1}	Symmetric stretching of methoxy (4)	*		6
2871 cm^{-1}	Symmetric stretching of methoxy (2)	*		6
2874 cm^{-1}	v_sCH$_3$	*		65
	Stretching C–H, N–H	*		6
2884/85 cm^{-1}	Stretching C–H	*		6
2885 cm^{-1}	v_sCH$_3$, lipids, fatty acids		*	7
2886 cm^{-1}	Fermi resonance CH$_2$ stretch		*	2
2886/7/8/9/90 cm^{-1}	Stretching C–H	*		6
2893/4/5 cm^{-1}	CH$_3$ symmetric stretch		*	95
2893/4/6 cm^{-1}	CH$_3$ symmetric stretch	*		95
2900 cm^{-1}	CH stretch		*	109
2900–3040 cm^{-1}	CH$_3$ stretching modes (both symmetric and asymmetric)		*	1
2910–65 cm^{-1}	CH$_3$ stretching vibrations		*	103

TABLE 8.1 (*Continued*)

Major Characteristic Peak Frequencies Reported in the Literature

Peak	Assignment	FT-IR	Raman	Reference Number
2915 cm^{-1}	CH band of lipids and proteins (The lipid band has a tendency toward higher levels of energy, such as 2855 cm^{-1}, while the protein band's tendency is toward lower energy levels like 2920 cm^{-1}.)		*	27
2916 cm^{-1}	Cholesterol, phospholipids, and creatine (higher in normal tissues)	*		62
	Stretching vibrations of CH$_2$ and CH$_3$ of phospholipids, cholesterol, and creatine	*		62
2917/8/9 cm^{-1}	Stretching C–H	*		48
2923–33 cm^{-1}	The strongest C–H stretching bands in malignant tissues	*		74
2923/5 cm^{-1}	Stretching C–H	*		48
2925 cm^{-1}	The strongest C–H stretching bands in normal tissues	*		74
	$\nu_{as}CH_2$ lipids	*		51
2928 cm^{-1}	Stretching C–H	*		48
	Symmetric CH$_3$ stretch		*	27
	Due primarily to protein	*	*	27
2930 cm^{-1}	C–H stretching bands	*		74
	$\nu_{as}CH_2$	*		65
2934 cm^{-1}	Asymmetric stretching of methoxy (2)	*		6
2934/9 cm^{-1}	Asymmetric stretching of methoxy (3)	*	*	6
2935 cm^{-1}	Chain end CH$_3$ symmetric band		*	2
2940 cm^{-1}	C–H vibrations in lipids and proteins		*	89
	$\nu_{as}CH_2$, lipids, fatty acids		*	7
2944 cm^{-1}	Asymmetric stretching of methoxy (1)	*		6
2947/8 cm^{-1}	Stretching C–H	*		48
2950 cm^{-1}	Asymmetric stretching of methoxy (4)	*		6
2951 cm^{-1}	Stretching C–H	*		48
2952 cm^{-1}	CH$_3$ asymmetric stretch	*		95
2951/3/5/6 cm^{-1}	Stretching C–H	*		48
2956/7 cm^{-1}	CH$_3$ asymmetric stretch		*	95

(*Continued*)

TABLE 8.1 (*Continued*)

Major Characteristic Peak Frequencies Reported in the Literature

Peak	Assignment	FT-IR	Raman	Reference Number
2959 cm⁻¹	C–H stretching	*		74
	$v_{as}CH_3$ of lipids, DNA, and proteins	*		51
	Asymmetric stretching mode of the methyl groups from cellular proteins, nucleic acids, and lipids	*		51
	Asymmetric stretching of methoxy (2)	*		6
2960 cm⁻¹	$v_{as}CH_3$	*		65
	Out-of-plane chain-end antisymmetric CH_3 stretch band		*	2
2963 cm⁻¹	CH_3 modes	*		57
2965 cm⁻¹	Stretching C–H	*		48
2969 cm⁻¹	Asymmetric stretching of methoxy (3)	*		6
2970 cm⁻¹	Asymmetric stretching of methoxy (1)	*		6
	$v_{as}CH_3$, lipids, fatty acids		*	7
	Cholesterol and cholesterol ester		*	2
2971/3 cm⁻¹	Asymmetric stretching of methoxy (4)		*	6
2975 cm⁻¹	Stretching N–H, stretching C–H	*		48
2984 cm⁻¹	$CH_{\alpha,\alpha'}$ stretch	*		6
2993/4 cm⁻¹	C–H ring	*		48
2994 cm⁻¹	$CH_{\alpha,\alpha'}$ stretch	*		6
2998/9 cm⁻¹	C–H ring	*		48
3000 cm⁻¹	C–H ring	*		48
	CH stretching vibrations (remain unaltered by the methoxy and hydroxyl substitution)	*		6
>3000 cm⁻¹	CH stretching		*	90
3000–600 cm⁻¹	N–H stretching	*		83
3007 cm⁻¹	C–H	*		74
3007–10 cm⁻¹	=C–H groups that are related to olefins bands or unsaturated fatty acids (absent in cancer samples)	*		74
3008 cm⁻¹	C–H ring	*		48
	$v_{as}(=C–H)$, lipids, fatty acids		*	7
3010 cm⁻¹	Unsaturated =CH stretch		*	2
3015 cm⁻¹	$v=CH$ of lipids	*		51
	CH stretching band of the phenyl rings		*	6

TABLE 8.1 (*Continued*)

Major Characteristic Peak Frequencies Reported in the Literature

Peak	Assignment	FT-IR	Raman	Reference Number
3015/17/20 cm^{-1}	CH$_2$, aromatic stretch	*	*	6
3021/2 cm^{-1}	C–H ring	*		48
3050 cm^{-1}	Amide B (N–H stretching)	*		65
3064 cm^{-1}	C$_2$^{C}–H$_2$ aromatic stretching	*		6
3070 cm^{-1}	Fermi-enhanced overtone of the amide II band (at 1550 cm^{-1})	*		83
3072/4 cm^{-1}	C–H ring	*		48
3074 cm^{-1}	CH stretching band of the phenyl rings	*		6
	C$_2$^{C}–H$_2$ aromatic stretching	*		6
3075 cm^{-1}	Amide B bands stemming from N–H stretching modes in proteins and nucleic acids	*		48
3078 cm^{-1}	C–H ring	*		48
3082 cm^{-1}	C$_2$^{C}–H$_2$ aromatic stretching		*	6
3111/4/6 cm^{-1}	C–H ring	*		48
3163/82 cm^{-1}	Stretching N–H symmetric	*		48
3190 cm^{-1}	N–H stretching bands of mainly *cis*-ordered substructures	*		83
3194/5/7/9/200 cm^{-1}	Stretching N–H symmetric	*		48
3200–550 cm^{-1}	Symmetric and asymmetric vibrations attributed to water, so it would be better not to consider this region for detailed analysis.	*		62
	Ring OH stretching modes		*	1
3201 cm^{-1}	Stretching N–H symmetric	*		48
3216/17/26 cm^{-1}	Stretching O–H symmetric	*		48
3232 cm^{-1}	O–H & N–H stretching vibrations (hydrogen bonding network may vary in the malignant tissues)		*	83
3269 cm^{-1}	OH		*	1
3273/87/89 cm^{-1}	Stretching O–H symmetric	*		48
3293 cm^{-1}	OH stretching (associated)	*		6
3295 cm^{-1}	Amide A (N–H stretching)	*		65
3300 cm^{-1}	Amide A bands stemming from N–H stretching modes in proteins and nucleic acids	*		83
	Attributed to OH stretch		*	2
3301 cm^{-1}	Amide A band	*		77

(Continued)

TABLE 8.1 (*Continued*)

Major Characteristic Peak Frequencies Reported in the Literature

Peak	Assignment	FT-IR	Raman	Reference Number
3309 cm^{-1}	OH		*	1
3313 cm^{-1}	Amide A band	*		77
3320 cm^{-1}	NH band	*		23
3327 cm^{-1}	Stretching N–H asymmetric	*		48
3328 cm^{-1}	Amide A band	*		77
3329 cm^{-1}	N–H vibration of proteins		*	103
3330/5/7/9/43 cm^{-1}	Stretching N–H asymmetric	*		48
33349 cm^{-1}	OH		*	1
33343 cm^{-1}	Stretching N–H asymmetric	*		48
3350 cm^{-1}	O–H, N–H, C–H	*		23
3350–550 cm^{-1}	OH stretching		*	103
3353 cm^{-1}	Stretching N–H asymmetric	*		48
3354 cm^{-1}	O–H, N–H, C–H	*		23
3359 cm^{-1}	Stretching N–H asymmetric	*		48
	O–H, N–H, C–H	*		23
3362 cm^{-1}	O–H, N–H, C–H	*		23
3394 cm^{-1}	OH		*	1
3396 cm^{-1}	Stretching O–H asymmetric	*		48
3400–550 cm^{-1}	OH vibrations, stretching modes		*	1
3401 cm^{-1}	O–H & N–H stretching vibrations (hydrogen bonding network may vary in the malignant tissues)	*		74
3410/16/20/22 cm^{-1}	Stretching O–H asymmetric	*		48
3430 cm^{-1}	N–H stretching bands of mainly *trans*-ordered substructures	*		83
3435/442 cm^{-1}	Stretching O–H asymmetric	*		48
3463 cm^{-1}	OH		*	1
3484 cm^{-1}	OH		*	1
3498 cm^{-1}	OH		8	1
3500–600 cm^{-1}	OH bonds	*		23
3506 cm^{-1}	OH stretching (free)	*		6
3524/28/42 cm^{-1}	Stretching O–H	*		48
3550 cm^{-1}	O–H stretching vibration		*	110
3561 cm^{-1}	OH stretching (free)	*		6
3570/77/78/82/90/9 cm^{-1}	Stretching O–H	*		48
3611 cm^{-1}	O–H and N–H stretching vibrations (hydrogen bonding network may vary in the malignant tissues)	*		74

Characteristic Peak Frequencies

It is believed that accurate peak definitions can have a great influence on reliability of the results provided through different spectroscopic techniques. As a result, and with the aim of putting this shortcoming aside, a wide range of the most frequently seen peaks are tabulated in this chapter. This will enhance understanding of the major peak assignments to specific functional groups that are present in complex biological molecules. For example, peaks in the area of 900 cm^{-1} are attributed to phosphodiester groups in both Raman and FTIR studies. The same finding can be seen in the area of 990–940 cm^{-1} (peaks related to C–C and C–O). More important peaks, such as those on the area of 1060–1030 cm^{-1} (C–C and C–O, mainly due to proteins and carbohydrates), 2000–1080 cm^{-1} (C–C, C–O, and phosphate of nucleic acids and membrane phospholipids, and partially proteins), 1350–1200 cm^{-1} (amide III region), 1600–1480 cm^{-1} (amide II region), 1800–1600 cm^{-1} (amide I region), and 3100–2800 cm^{-1} (CH region) also show considerable similarities in both techniques.

Summary

This chapter attempts to illustrate the significant attention being given to spectroscopy as a newly emerging method for biological studies, particularly cancer detection. In addition, it leads to the conclusion that a considerable amount of information can be obtained from spectroscopic methods. It also proves the similarities, which can be seen in peak definitions in both FTIR and Raman spectroscopy techniques. In other words, regardless of the method used, the peak intensities could be defined similarly.

This chapter can be used as a unique and reliable database of peak definitions for both Raman and FTIR studies, and it can surely provide significant assistance to those who concentrate on these methods. However, it should be mentioned that this is an ongoing project. Newly conducted researches along with their introduced peak frequencies will gradually be used in order to make this table more comprehensive.

Because the IR and Raman spectroscopic techniques are complementary, similarities of peaks attributed to functional groups are also observed in biological molecules. In closing, this chapter can be used as a unique and reliable database of peak definitions for IR and Raman studies of biological molecules, specifically cancer studies.

References

1. Calheiros, R., Machado, N. F. L., Fiuza, S. M., Gaspar, A., Garrido, J., Milhazes, N., Borges, F., and Marques, M. P. 2008. Antioxidant phenolic esters with potential anticancer activity: A Raman spectroscopy study. *J. Raman Spectroscopy* 39, 95–107.

2. Krafft, C., Neudert, L., Simat, T., and Salzer, R. 2005. Near-infrared Raman spectra of human brain lipids. *Spectrochimia Acta* 61, 1529–1535.

3. Cheng, W. T., Liu, M. T., Liu, H. N., and Lin, S. Y. 2005. Micro-Raman spectroscopy used to identify and grade human skin pilomatrixoma. *Micrsc. Res. Tech.* 68, 75–79.

4. Pinzaru S. C., Andronie L. M., Domsa I., Cozar, O., and Astilean, S. 2008. Bridging biomolecules with nanoparticles: Surface-enhanced Raman scattering from colon carcinoma and normal tissue. *J. Raman Spectroscopy* 39, 331–334.

5. Farguharson, S., Shende, C., Inscore, F. E., Maksymiuk, P., and Gift, A. 2005. Analysis of 5-fluorouracil in saliva using surface-enhanced Raman spectroscopy. *J. Raman Spectroscopy* 36, 208–212.

6. Schultz, H., and Baranska, M. 2007. Identification and qualification of valuable plant substances by IR and Raman spectroscopy. *Vibrational Spectroscopy* 43, 13–25.

7. Shetty, G., Kedall, C., Shepherd, N., Stone, N., and Barr, H. 2006. Raman spectroscopy: Evaluation of biochemical changes in carcinogenesis of oesophagus. *Br. J. Cancer* 94, 1460–1464.

8. Binoy, J., Abraham, J. P., Joe, I. H., Jayakumar, V. S., Petit, G. R., and Nielsen, O. F. 2004. NIR-FT Raman and FT-IR spectral studies and *ab initio* calculations of the anti-cancer drug combretastatin-A4. *J. Raman Spectroscopy* 35, 939–946.

9. Stone, N., Kendall, C., Smith, J., Crow, P., and Barr, H. 2004. Raman spectroscopy for identification of epithelial cancers. *Faraday Discussion* 126, 141–157.

10. Stone, N., Kendell, C., Shepherd, N., Crow, P., and Barr, H. 2002. Near-infrared Raman spectroscopy for the classification of epithelial pre-cancers and cancers. *J. Raman Spectroscopy* 33, 564–573.

11. Marzullo, A. C. D., Neto, O. P., Bitar, R. A., Martinho, H. D. S., and Abrahao, A. 2007. FT-Raman spectra of the border of infiltrating ductal carcinoma lesions. *Photomed. Laser Surg.* 25, 455–460.

12. Chan, J. W., Taylor, D. S., Zwerdling, T., Lane S. T., Ihara, K., and Huser, T. 2006. Micro-Raman spectroscopy detects individual neoplastic and normal hematopoietic cells. *Biophys. J.* 90, 648–656.

13. Chiang, H. P., Song, R., Mou, B., Li, K. P., Chiang, P., Wang, D., Tse, W. S., and Ho, L. T. 1999. Fourier transform Raman spectroscopy of carcinogenic polycyclic aromatic hydrocarbons in biological systems: Binding to heme proteins. *J. Raman Spectroscopy* 30, 551–555.

14. Cipriani, P., and Smith, C.Y. 2008. Characterization of thalidomide using Raman spectroscopy. *Spectrochem. Acta. A Mol. Biomol. Spectroscopy* 69, 333–337.

15. Faoláin, E. O., Hunter, M. B., Byrne, J. M., Kelehan, P., McNamer, M., Byrne, H. J., and Lyng, F. M. 2005. A study examining the effects of tissue processing on human tissue sections using vibrational spectroscopy, *Vibrational Spectroscopy* 38, 121–127.

16. Jess, P. R. T., Smith, D. D. W., Mazilu, M., Dholakia, K., Riches, A. C., and Herrington, C. S. 2007. Early detection of cervical neoplasia by Raman spectroscopy by Raman spectroscopy. *Int. J. Cancer* 121, 2723–2728.
17. Kast R. E., Serhatkulu G. K., Cao A., Pandya, A. K., Dai, H., Thakur, J. S., Naik, V. M., Naik, R., Klein, M. D., Auner, G. W., and Rabah, R. 2008. Raman spectroscopy can differentiate malignant tumors from normal breast tissue and detect early neoplastic changes in a mouse model. *Biopolymers* 89, 134–241.
18. Deng, J. L., Wei, Q., Zhang, M. H., Wang, Y. Z., and Li, Y. Q. 2005. Study of the effect of alcohol on single human red blood cells using near-infrared laser tweezers Raman spectroscopy. *J. Raman Spectroscopy* 36, 257–261.
19. Frank, C. J., McCreecy, R. L., and Redd, D. C. B. 1995. Raman spectroscopy of normal and diseased human breast tissues. *Anal. Chem.* 67, 777–783.
20. Huang, Z., McWilliams, A., Lui, M., McLean, D. I., Lam, S., and Zeng, H. 2003. Near-infrared Raman spectroscopy for optical diagnosis of lung cancer. *Int. J. Cancer* 107, 1047–1052.
21. Ruiz-Chica, A. J., Medina, M. A., Sanchez-Jimenez, F., and Ramirez, F. J. 2004. Characterization by Raman spectroscopy of conformational changes on guanine-cytosine and adenine-thymine oligonucleotides induced by aminooxy analogues of spermidine. *J. Raman Spectroscopy* 35, 93–100.
22. Naumann, D. 1998. Infrared and NIR Raman spectroscopy in medical microbiology. *Proc. S.P.I.E.* 3257, 245–257.
23. Dovbeshka, G. I., Chegel, V. I., Gridina, N. Y., Repnytska, O. P., Shirshov, Y. M., Tryndiak, V. P., Todor, I. M., and Solyanik, G. I. 2002. Surface enhanced IR absorption of nucleic acids from tumor cells: FTIR reflectance study. *Biopolymer* 67, 470–486.
24. Dukor, R. K. 2002. Vibrational spectroscopy in the detection of cancer. *Biomed. Appl.* 3335–3359.
25. Hanlon, E. B., Manoharan, R., Koo, T. W., Shafer, K. E., Motz, J. T., Fitzmaurice, M., Kramer, J. R., Itzkan, I., Dasari, R. R., and Feld, M. S. 2000. Prospects for in vivo Raman spectroscopy. *Phys. Med. Biol.* 45, 1–59.
26. Utzinger, U. R. S., Heintzelman, D. L., Mahadevan-Jansen, A., Malpica, A., Follen, M., and Richards-Kortum, R. 2001. Near-infrared Raman spectroscopy for in vivo detection of cervical precancers. *Appl. Spectroscopy* 55, 955–959.
27. Kline, N. J., and Treado, P. J. 1997. Raman chemical imaging of breast tissue. *J. Raman Spectroscopy* 28, 119–124.
28. Laska, J., and Widlarz, J. 2005. Spectroscopic and structural characterization of low molecular weight fractions of polyaniline. *Polymer* 46, 1485–1495.
29. Gniadecka, M., Wulf, H. C., Mortensen, N. N., Nielsen, O. F., and Christensen, D. H. 1997. Diagnosis of basal cell carcinoma by Raman spectroscopy. *J. Raman Spectroscopy* 28, 125–129.
30. Khanmohammadi, M., Nasiri, R., Ghasemi, K., Samani, S., and Garmarudi, A. B. 2007. Diagnosis of basal cell carcinoma by infrared spectroscopy of whole blood samples applying soft independent modeling class analogy. *J. Cancer Res. Clin. Oncol.* 133, 1001–1010.
31. Kateinen, E., Elomaa, M., Laakkonen, U. M., Sippola, E., Niemela, P., Suhonen, J,. and Jarninen, K. 2007. Qualification of the amphetamine content in seized street samples by Raman spectroscopy. *J. Forensic Sci.* 52, 88–92.

32. Krishna, C. M., Prathima, N. B., Malini, R., Vadhiraja B. M., Bhatt, R. A., Fernandes, D. J., Kushtagi, P., Vidyasagar, M. S., and Kartha, V. B. 2006. Raman spectroscopy studies for diagnosis of cancers in human uterine cervix. *Vibrational Spectroscopy* 41, 136–141.

33. Huang, Z., Lui, H., McLean, D. I., Korbelik, M., and Zeng, H. 2005. Raman spectroscopy in combination with background near-infrared autofluorescence enhances the in vivo assessment of malignant tissues. *Photochem. Photobiol.* 81, 1219–1226.

34. Vidyasagar, M. S., Maheedhar, K., Vadhiraja, B. M., Fernendes, D. J., Kartha, V. B., and Murali Krishna, C. 2008. Prediction of radiotherapy response in cervix cancer by Raman spectroscopy: A pilot study. *Biopolymers* 89, 530–537.

35. Teh S. K., Zheng W., Ho K. Y., Teh, M., Yeoh, K. G., and Huang, Z. 2008. Diagnostic potential of near-infrared Raman spectroscopy in the stomach: Differentiating dysplasia from normal tissue. *Br. J. Cancer* 98, 457–465.

36. Li, X., Lin, J., Ding, J., Wang, S., Liu, Q., and Qing, S. 2004. Raman spectroscopy and fluorescence for the detection of liver cancer and abnormal liver tissue. Annual International Conference IEEE EMBS, San Francisco, CA, September 1–5.

37. Maheedhar, K., Bhat, R. A., Malini, R., Prathima, N. B., Keerthi, P., Kushtagi, P., and Murali Krishna, C. 2008. Diagnosis of ovarian cancer by Raman spectroscopy: A pilot study. *Photomed. Laser Surg.* 26, 83–90.

38. Wood, B. R., Quinn, M. A., Tait, B., Ashdown, M., Hislop, T., Romeo, M., and McNaughton, D. 1998. FTIR microspectroscopic study of cell types and potential confounding variables in screening for cervical malignancies. *Biospectroscopy* 4, 75–91.

39. Sahu, R. K., Mordechai, S., and Manor, E. 2008. Nucleic acids absorbance in mid-IR and its effect on diagnostic variates during cell division: A case study with lymphoblastic cells. *Biopolymers* 89, 993–1001.

40. Andrus, P. G. L., and Strickland, R. D. 1998. Cancer grading by Fourier transform infrared spectroscopy. *Biospectroscopy* 4, 37–46.

41. Huang, Z., McWilliams A., Lam, S., English, J., McLean, D., Lui, H., and Zeng, H. 2003. Effect of formalin fixation on the near-infrared Raman spectroscopy of normal and cancerous human bronchial tissues. *Int. J. Oncol.* 23, 649–655.

42. Lau, D. P., Huang, Z., Lui, H., Man, C. S., Berean, K., Morrison, M. D., and Zeng, H. 2003. Raman spectroscopy for optical diagnosis in normal and cancerous tissue of the nasopharynx: Preliminary findings. *Lasers Sur. Med.* 32, 210–214.

43. Lakshimi, R. J., Kartha, V. B., Krishna, C. M., Solomon, J. G. R., Ullas, G., and Uma Devi, P. 2002. Tissue Raman spectroscopy for the study of radiation damage: Brain irradiation of mice. *Radiat. Res.* 157, 175–182.

44. Sekhar, P. K., Ramgir, N. S., and Bhansali, S. 2008. Metal-decorated silica nanowires: An active surface-enhanced Raman substrate for cancer biomarker detection. *J. Phys. Chem.* 112, 1729–1734.

45. Mahadevan-Jansen, A., and Richards-Kortum, R. 1997. Raman spectroscopy for cancer detection. 19th International Conference IEEE EMBS, Chicago, IL, October 30–November 2.

46. Pichardo-Molina, J. L., Frausto-Reyes, C., Barbosa-Garcia, O., Huerta-Franco, R., Gonzalez-Trujillo, Ramirez-Alvarado, A., Gutierrez-Jurarez, G., and Medina-Gutierrez, C. 2007. Raman spectroscopy and multivariate analysis of serum samples from breast cancer patients. *Lasers Med. Sci.* 22, 229–236.

47. Kondepati, V. R., Heise, H. M., Oszinda T., Mueller, R., Keese, M., and Backhaus, J. 2008. Detection of structural disorders in colorectal cancer DNA with Fourier-transform infrared spectroscopy. *Vibrational Spectroscopy* 46, 150–157.

48. Dovbeshko, G. I., Gridina, N. Y., Kruglova, E. B., and Pashchuk, O. P. 2000. FTIR spectroscopy studies of nucleic acid damage. *Talana* 53, 233–246.

49. Silveira, L., Sathaiah, S., and Zangaro, R. A. 2002. Correlation between near-infrared Raman spectroscopy and the histopathological analysis of atherosclerosis in human coronary arteries. *Lasers Surg. Med.* 30, 290–297.

50. Chiriboga, L., Xie, P., Yee, H., Vigorita, V., Zarou, D., Zakim, D., and Diem, M. 1998. Infrared spectroscopy of human tissue. I. Differentiation and maturation of epithelial cells in the human cervix. *Biospectroscopy* 4, 47–53.

51. Fung, M. F. K., Senterman, M. K., Mikhael, N. Z., Lacelle, S., and Wong, P. T. T. 1996. Pressure-tuning Fourier transform infrared spectroscopic study of carcinogenesis in human endometrium. *Biospectroscopy* 2, 155–165.

52. Wood, B. R., Quinn, M. A., Burden, F. R., and McNaughton, D. 1996. An investigation into FT-IR spectroscopy as a bio-diagnostic tool for cervical cancer. *Biospectroscopy* 2, 143–153.

53. Argov, S., Sahu, R. K., Bernshtain, E., Salam, A., Shohat, G., Zelig, U., and Mordechai, S. 2004. Inflammatory bowel diseases as an intermediate stage between normal and cancer: A FTIR-microspectroscopy approach. *Biopolymers* 75, 384–392.

54. Malini, R., Venkatakrishma, K., Kurien, J., Pai, K. M., Rao, L., Kartha, V. B., and Murali Krishna, C. 2006. Discrimination of normal, inflammatory, premalignant, and malignant oral tissue: A Raman spectroscopy study. *Biopolymers* 81, 179–193.

55. Gazi, E., Dwyer, J., Gardner, P., Ghanbari Siakhani, A., Wde, A. P., Lockyer, N. P., Vickerman, J. C., Clarke, N. W., Shanks, J. H., Scott, L. J., Hart, C. A., and Brown, M. 2003. Applications of Fourier transform infrared microspectroscopy in studies of benign prostate and prostate cancer. A pilot study. *J. Pathol.* 201, 99–108.

56. Yoshida, S., Miyazaki, M., and Sakai, K. 1997. Fourier transform infrared spectroscopic analysis of rat brain microsomal membranes modified by dietary fatty acids: Possible correlation with altered learning behavior. *Biospectroscopy* 3, 281–290.

57. Alfano, R. R., Tang, G. C., Pradhan, A., Lam, W., Choy, D. S. J., and Opher, E. 1987. Optical spectroscopic diagnosis of cancer and normal breast tissues. *IEEE J. Quantum Electronics*, QE 23, 1806–1811.

58. Onogi, C., Motoyama, M., and Hamaguchi, H. 2008. High concentration *trans* form unsaturated lipids detected in a HeLa cell by Raman microspectroscopy. *J. Raman Spectroscopy* 39, 555–556.

59. Lord, R. C., and Yu, N. T. 1970. Laser-excited Raman spectroscopy of biomolecules. II. Native ribonuclease and alpha-chymotrypsin. *J. Mol. Biol.* 51, 203–213.

60. Hu Y. G., Shen, A. G., and Jiang, T. 2008. Classification of normal and malignant human gastric mucosa tissue with confocal Raman microspectroscopy and wavelet analysis. *Pectrochemica Acta Part A Mol. Biomol. Spectroscopy* 69, 378–382.

61. Bogomolny E., Argov, S., Mordechai, S., and Huleihel, M. 2008. Monitoring of viral cancer progression using FTIR microscopy: A comparative study of intact cells and tissues. *Biochimica et Biophysica Acta* 1780, 1038–1046.

62. Huleihel, M., Salman, A., Erukhimovich, V., Ramesh, J., Hammody, Z., and Mordechai, S. 2002. Novel optical method for study of viral carcinogenesis in vitro. *J. Biochem. Biophys. Methods* 50, 111–121.

63. Mordechai, S., Mordechai, J., Ramesh, J., Levi, C., Huleihel, M., Erukhimovitch, V., Moser, A., and Kapelushnik, J. 2001. Application of FTIR microspectroscopy for the follow-up of childhood leukaemia chemotherapy. *Proc. SPIE Subsurface and Surface Sensing Technologies and Applications III* 4491, 243–250.

64. Wang, H. P., Wang, H. C., and Huang, Y. J. 1997. Microscopic FTIR studies of lung cancer cells in pleural fluid. *The Sci. Total Envir.* 204, 283–287.

65. Paluszkiewicz, C., and Kwiatek, W. M. 2001. Analysis of human cancer prostate tissues using FTIR microscopy and SXIXE techniques. *J. Mol. Structures* 565–566, 329–334.

66. Fujioka, N., Morimoto, Y., Arai, T., and Kikuchi, M. 2004. Discrimination between normal and malignant human gastric tissues by Fourier transform infrared spectroscopy. *Cancer Detect. Prev.* 28, 32–36.

67. Rigas, B., and Wong, P. T. T. 1992. Human colon adenocarcinoma cell lines display infrared spectroscopic features of malignant colon tissues. *Cancer Res.* 52, 84–88.

68. Lucassen, G. W., Van Veen, G. N., and Jansen, J. A. 1998. Band analysis of hydrated human skin stratum corneum attenuated total reflectance Fourier transform infrared spectra in vivo. *J. Biomed. Opt.* 3, 267–280.

69. Mordechai, S., Sahu, R.K., Hammody, Z., Mark, S., Kantarovich, K., Guterman, H., Podshyvalov, A., Goldstein, J., and Argov, S. 2004. Possible common biomarkers from FTIR microspectroscopy of cervical cancer and melanoma. *J. Microsc.* 215, 86–91.

70. Mordechai, S., Salman, A., Argov, S., and Goldstein, J. 2000. Fourier-transform infrared spectroscopy of human cancerous and normal intestine. *Proc SPIE.* 3918, 66–77.

71. Yano, K., Ohoshima, S., Grotou, Y., Kumaido, K., Moriguchi, T., and Katayama, H. 2000. Direct measurement of human lung cancerous and noncancerous tissues by Fourier transform infrared microscopy: Can an infrared microscope be used as a clinical tool? *Anal Biochem.* 287, 218–225.

72. Andrus, P. G. 2006. Cancer monitoring by FTIR spectroscopy. *Technol. Cancer. Res. Treat.* 5, 157–167.

73. Kondepati, V. R., Keese, M., Heise, H. M., and Backhaus, J. 2006. Detection of structural disorders in pancreatic tumour DNA with Fourier–transform infrared spectroscopy. *Vibrational Spectroscopy* 40, 33–39.

74. Wu, J. G., Xu, Y. Z., Sun, C. W., Soloway, R. D., Xu, D. F., Wu, Q. G., Sun, K. H., Weng, S. F., and Xu, G. X. 2001. Distinguishing malignant from normal oral tissues using FTIR fiber-optic techniques. *Biopolymer* 62, 185–192.

75. Ronen, S. M., Stier, A., and Degani, H. 1990. NMR studies of the lipid metabolism of T47D human breast cancer spheroids. *FEBS Letter* 266, 147–149.

76. Richter, T., Steiner, G., Abu-Id, M. H., Salzer, R., Gergmann, R., Rodig, H., and Johannsen, B. 2002. Identification of tumor tissue by FTIR spectroscopy in combination with positron emission tomography. *Vibrational Spectroscopy* 28, 103–110.

77. Yang, Y., Sule-Suso, J., Sockalingum, G. D., Kegelaer, G., Manfait, M., and El Haj, A. J. 2005. Study of tumor cell invasion by Fourier transform infrared microspectroscopy. *Biopolymers*, 78, 311–317.

78. Fukuyama, Y., Yoshida, S., Yanagisawa, S., and Shimizu, M. 1999. A study on the differences between oral squamous cell carcinomas and normal oral mucosas measured by Fourier transform infrared spectroscopy. *Biospectroscopy* 5, 117–126.

79. Rigas, B., Morgello, S., Goldman, I. S., and Wong, P. T. T. 1999. Human colorectal cancers display abnormal Fourier-transform infrared spectra. *Proc. Natl. Acad. Sci. U S A.* 87, 8140–8144.

80. Wong, P. T. T., Papavassiliou, E. D., and Rigas, B. 1991. Phosphodiester stretching bands in the infrared spectra of human tissues and cultured cells. *Appl. Spectrosc.* 45, 1563–1567.

81. Puppeles, G. J., Garritsen, H. S. P., Kummer, J. A., and Greve, J. 1993. Carotenoids located in human lymphocyte subpopulations and natural killer cells by Raman microspectroscopy. *Cytometry* 14, 251–256.

82. Wong, P. T., Goldstein, S. M., Grekin, R. C., Godwin, T. A., Pivik, C., and Rigas, B. 1993. Distinct infrared spectroscopic patterns of human basal cell carcinoma. *Cancer Res.* 53, 762–765.

83. Eckel, R., Huo, H., Guan, H. W., Hu, X., Che, X., and Huang, W. D. 2001. Characteristic infrared spectroscopic patterns in the protein bands of human breast cancer tissue. *Vibrational Spectroscopy* 27, 165–173.

84. Kachrimanis, K., Braun, D. B., and Griesser, U. J. 2007. Quantitative analysis of paracetamol polymorphs in powder mixtures by FT-Raman spectroscopy and PLS regression. *J. Pharm. Biomed. Anal.* 43, 407–412.

85. Matthaeus, C., Kale, A., Chernenko, T., Torchilin, V., and Diem, M. 2008. New ways of imaging uptake and intracellular fate of liposomal drug carrier systems inside individual cells, based on Raman microscopy. *Mol. Pharm.* 5, 287–293.

86. McIntosh, I. M., Jackson, M., Mantsch, H. H., Stranc, M. F., Pilavdzic, D., and Crowson, A. N. 1999. Infrared spectra of basal cell carcinomas are distinct from non-tumor-bearing skin components. *J. Invest. Dermatol.* 112, 951–956.

87. Lau, D. P., Huang, Z., Lui, H., Anderson, D. W., Beren, K., Morrison, M. D., Shen, L., and Zeng, H. 2005. Raman spectroscopy for optical diagnosis in the larynx: Preliminary findings. *Lasers Surg. Med.* 37, 192–200.

88. Barry, B. W., Edwards, H. G. M., and Williams, A. C. 1992. Fourier transform Raman and infrared vibrational study of human skin: Assignment of spectral bands. *J. Raman Spectroscopy* 23, 641–645.

89. Sigurdsson, S., Philipsen, P. A., Hansen, L. K., Laesen, L., Gniadecka, M., and Wulf, H. C. 2004. Detection of skin cancer by classification of Raman spectra. *IEEE Transactions Biomed. Eng.* 51, 1784–1793.

90. Singha, A., Ghosh, A., Roy, A., and Ray, N. R. 2006. Quantitative analysis of hydrogenated diamond-like carbon films by visible Raman spectroscopy. *J. Appl. Phys.* 100, 1–8.

91. Viehoever, A. R., Anderson, D., Jansen, D., and Mahadevan Jansen, A. 2003. Organotypic raft cultures as an effective in vitro tool for understanding Raman spectral analysis of tissues. *Photochem. Photobiol.* 78, 517–524.

92. Takai, Y., Masuko, T., and Takeuchi, H. 1997. Lipid structure of cytotoxic granules in living human killer T lymphocytes studied by Raman microspectroscopy. *Biochimica et Biophysica Acta* 1335, 199–208.

93. Barr, H., Dix, T., and Stone, N. 1998. Optical spectroscopy for the early diagnosis of gastrointestinal malignancy. *Lasers Med. Sci.* 13, 3–13.

94. Min, Y. K., Yamamoto, T., Kohoda, E., Ito, T., and Hamaguchi, H. 2005. 1064-nm near-infrared multichannel Raman spectroscopy of fresh human lung tissues. *J. Raman Spectroscopy* 36, 73–76.

95. Agarwal, R., Tandon, P., and Gupta, V. D. 2006. Phonon dispersion in poly(dimethylsilane). *J. Organometal. Chemi.* 691, 2902–2908.

96. Shafer-Peltier, K. E., Haka, A. S., Fitzmaurice, M., Crowe, J., Dasari, R. R., and Feld, M. S. 2002. Raman microspectroscopic model of human breast tissue: Implications for breast cancer diagnosis in vivo. *J. Raman Spectroscopy* 33, 552–563.

97. Kaminaka, S., Yamazaki, H., Ito, T., Kohoda, E., and Hamaguchi, H. 2001. Near-infrared Raman spectroscopy of human lung tissues: Possibility of molecular-level cancer diagnosis. *J. Raman Spectroscopy* 32, 139–141.

98. Tan, Y. Y., Shen, A. G., Zhang, J. W., Wu, N., Feng, L., Wu, Q. F., Ye, Y., and Hu, J. M. 2003. Design of auto-classifying system and its application in Raman spectroscopy diagnosis of gastric carcinoma. 2nd International Conference Machine Learning & Cybernetics, November 2–5.

99. Kaminaka, S., Ito, T., Yamazaki, H., Kohoda, E., and Hamaguchi, H. 2002. Near-infrared multichannel Raman spectroscopy toward real-time in vivo cancer diagnosis. *J. Raman Spectroscopy* 33, 498–502.

100. Sukuta, S., and Bruch, R. 1999. Factor analysis of cancer Fourier transform infrared evanescent wave fiberoptical (FTIR-FEW) spectra. *Lasers Surg. Med*, 24, 382–388.

101. Li, Q. B., Sun, X. J,. Xu, Y. Z., Yang, L. M., Zhang, Y. F., Weng, S. F., Shi, J. S., and Wu, J. G. 2005. Diagnosis of gastric inflammation and malignancy in endoscopic biopsies based on Fourier transform infrared spectroscopy. *Clin Chem.* 51, 346–350.

102. Shaw, R. A., and Mantsch, H. H. 1999., Vibrational biospectroscopy: from plants to animals to humans: A historical perspective. *J. Mol. Struct.* 480–481, 1–13.

103. Caspers, P. J., Lucassen, G. W., Carter, E. A., Bruining, H. A., and Puppels, G. J. 2001. In vivo confocal Raman microspectroscopy of the skin: Non-invasive determination of molecular concentration profiles. *J. Invest. Dermatol.* 116, 434–442.

104. Tu, A. T. 1982. *Raman Spectroscopy in Biology: Principles and Applications.* New York: John Wiley & Sons.

105. Gunstone, F. D., Harwood, J. L., and Dijkstra, A. J. 2007. *The Lipid Handbook*, 3rd ed. Boca Raton, FL: CRC Press.

106. Mazurek, S., and Szostak, R. 2006. Quantitative determination of captopril and prednisolone in tablets by FT-Raman spectroscopy. *J. Pharm. Biomed. Anal.* 40, 1225–1230.

107. Conroy, J., Ryder, A. G., Leger, M. N., Hennessey, K., and Madden, M. G. 2005. Qualitative and quantitative analysis of chlorinated solvents using Raman Spectroscopy and machine learning. *Proc SPIE Int. Soc. Opt. Eng.* 5826, 131–142.

108. Quintas, G., Garrigues, S., Pastor, A., and de la Guardia, M. 2004. FT-Raman determination of Mepiquet chloride in agrochemical products. *Vibrational Spectroscopy* 36, 41–46.

109. Ortiz, C., Zhang, D., Xie, Y., Ribbe, A. E., and Ben Amotz, D. 2006. Validation of the drop coating deposition Raman method for protein analysis. *Anal. Biochem.* 353, 157–166.

110. Behrens, H., Roux, J., Neuville, D. R., and Siemann, M. 2006. Quantification of dissolved H_2O in silicate glasses using confocal micro-Raman spectroscopy. *Chem. Geol.* 229, 96–112.

9

Future Potential Clinical Applications of Spectroscopic Techniques in Human Diseases

Diagnosis

Spectroscopic techniques appear to have immense promising potential in helping with the diagnosis of human diseases including infections, inflammatory conditions, and various cancers. Currently, although imaging can aid in diagnosis of a lot of disease processes, histopathological tissue diagnosis remains the gold standard for diagnosis of most clinical conditions. In this respect, spectroscopic techniques can be of immense value to aid histopathological diagnosis. Below are a few of the many possible diagnostic applications of spectroscopy.

Breast Cancer

Breast cancer commonly presents as a palpable lump. However, screening mammography is the most common technique for detecting nonpalpable, highly curable breast cancer, but the technique has its limitations, as not all of mammographically detected lesions are found to be malignant upon needle biopsy. As a consequence of the limitations of current techniques, each year a large number of breast biopsies are performed on benign lesions. Improvement in this assessment may be possible with the help of vibrational spectroscopy, improving the accuracy of diagnosis.

The majority of breast cancers are invasive adenocarcinomas, also known as invasive ductal carcinomas (IDC), which are graded according to the Ellis and Elston modification of the Bloom and Richardson criteria, as grade I well-differentiated, grade II moderately differentiated, and grade III poorly differentiated carcinomas [1,2]. The preinvasive stage of the disease is considered to be ductal carcinoma in situ (DCIS), which is also graded into three different nuclear grades. The management of IDC and DCIS have some differences.

Rehman et al. reported on the use of both the Fourier transform infrared (FTIR) and Raman spectroscopic methods for analysis of breast cancer grades, both DCIS and IDC. This work has shown that not only it is possible to differentiate between normal breast, DCIS, and IDC, but the nuclear grades of breast cancer tissues can be successfully distinguished [3,4]. The initial results are sufficiently promising to warrant further investigation specifically looking at the differences between estrogen receptor positive (ER+) and estrogen receptor negative (ER−) breast tumours and between invasive and in situ components. Hence, spectroscopy could be employed to diagnose and type breast cancers.

Brain Tumours and Other Brain Disorders

Astrocytomas are the most common gliomas, accounting for about half of all primary brain and spinal cord tumours. Astrocytomas develop from star-shaped glial cells called astrocytes, part of the supportive tissue of the brain. They may occur in many parts of the brain, but most commonly in the cerebrum. They occur less commonly in the spinal cord. People of all ages can develop astrocytomas, but they are more prevalent in adults, particularly middle-aged men. In children, most are low-grade tumours, while in adults most are high-grade tumours. the World Health Organization (WHO) has established a four-tiered histologic grading guideline for astrocytomas that assigns a grade from 1 to 4, with 1 being the least aggressive and 4 being the most aggressive, also known as glioblastoma multiforme (GBM). A computed tomography (CT) or magnetic resonance imaging (MRI) scan is necessary to characterise the extent of these tumours (size, location, consistency). However, these techniques will not distinguish astrocytomas from lesions due to other causes and a neurosurgical biopsy is required for confirmation of diagnosis, which is not always feasible.

Studies have shown that proton MR spectroscopy readily distinguishes normal brain tissues from astrocytomas [5–7]. Some investigators have proposed that the presence of lactate correlates with a higher degree of malignancy and that it is commonly observed in glioblastoma multiforme [8,9]. Proton MR spectroscopy may also be used to distinguish infection from a tumour [10]. However, proton MR spectroscopy may not be able to distinguish between different histologic grades of malignancy in astrocytomas [11]. Vibrational spectroscopy may provide additional information by differentiating between the histologic grades of astrocytomas, and hence could be more precise in diagnosis. Vibrational spectroscopy, like proton MR spectroscopy, may also play a role in monitoring the response of astrocytomas to treatment [8].

As with proton MR spectroscopy, vibrational spectroscopy may also be useful in the diagnosis of difficult cases of meningiomas as lactate and alanine may be elevated in some meningiomas, which can be detected with this technique [12]. Meningiomas are tumours that grow from the meninges, the layers of tissue covering the brain and spinal cord.

Alzheimer's Disease

Biochemically, Alzheimer's disease is characterised by decreased cortical acetylcholine caused by a loss of cholinergic cells that are critical for normal memory and cognition. The clinical manifestations of the disease may be subtle and insidious, and often the diagnosis is difficult. A noninvasive tool with which to diagnose early Alzheimer's disease is therefore highly desirable. In two studies of Alzheimer's disease employing proton MR spectroscopy, *N*-acetyl aspartate (NAA) was decreased and myoinositol was increased [13,14]. These findings were present even in cases of mild to moderate dementia. It should be possible to detect these chemical changes with vibrational spectroscopy with even greater precision.

Colon Cancer

British researchers looked at changes affecting 18 genes that play a key role in the very early stages of colon cancer and found a pattern of chemical changes in people who had precancerous polyps likely to develop into a tumour. This could help us to predict people more likely to get colon cancer in the future according to research at Britain's Institute of Food Research. The researchers focused on a chemical process called *methylation*, which turns genes on and off [15]. We believe that spectroscopic analysis of colonic adenomas versus carcinomas would allow detection of these changes.

Gynaecological Cancers

Spectroscopic techniques have the potential to differentiate between normal and malignant ovarian, endometrial, and cervical tissues, with good sensitivity and specificity demonstrated in some studies, as detailed in Chapter 5 of this text. Changes associated with preinvasive cervical cancer have also been studied and described.

Lung Cancer

Lung cancers are largely divided into non-small cell lung cancer (NSCLC) accounting for 80–85% of lung cancers, and small cell lung cancer (SCLC) accounting for 15–20% of lung cancers. NSCLC is a heterogeneous aggregate of histologies, including epidermoid or squamous cell carcinoma, adenocarcinoma, and large cell carcinoma. These histologies are classified together as NSCLC because approaches to diagnosis, staging, prognosis, and treatment are similar.

Lung cancer diagnosis and staging's based on Chest X-ray, CT scan, PET-CT (Positron Emission Tomography-CT) scan and biopsy (commonly via bronchoscopy or CT-guided biopsy). Chest x-ray has its limitations and can miss some small lung cancers, and does not always reliably distinguish between lung cancer and other lung diseases [16]. CT of the chest is an important informative tool that helps in detailed imaging of the primary tumour and its anatomic relationship to other structures, and it provides information with respect to the size of mediastinal lymph nodes and the status of the pleural space, and hence staging. However, CT criteria for adenopathy are based on size alone and do not always accurately reflect the presence or absence of tumour metastases. In addition, CT cannot reliably distinguish between benign and malignant lung nodules, hence the need for biopsy and histological confirmation [17].

Positron emission tomography (PET)-CT scanning appears to be valuable in deciding whether a nodule is benign or malignant, as well as in staging loco-regional and distant metastatic disease. However, it cannot distinguish between inflammatory conditions and malignancy, and is of limited value in lesions smaller than one centimetre [18,19].

Biopsy, of course, provides the precise type of lung cancer, but it is possible to have a negative biopsy due to sampling error, necessitating the need for repeat biopsy. It is not always easy to repeat procedures for obtaining a biopsy, as the procedures that allow biopsy, including bronchoscopy and mediastinoscopy, are not without significant risks.

It should be possible in the future to diagnose lung cancer with spectroscopy and to distinguish between various subtypes of lung cancer, as well as reliably detect cancer in mediastinal lymph nodes, with the development of a probe for use with spectroscopy. It should be possible to use the probe while obtaining a biopsy because if the first biopsy happens to miss tissue, a second biopsy could be avoided, due to live confirmation of diagnosis via probe at the time of the first biopsy. In addition, it should be possible to differentiate between benign and malignant lung nodules, and to determine the origin of metastases in the lung from various primary cancers.

In addition, as discussed in Chapter 5 in great detail, spectroscopy could potentially be employed in many other cancer diagnostics, including head and neck carcinomas, skin cancers, pancreatic cancers, gastric and oesophageal cancers, hepatocellular carcinomas, and others.

Determination of Prognostic Factors

Amongst prognostic factors, tumour grade is an excellent indicator of biological potential. However, reliable assessment of tumour grades has been hindered in the past by difficulties in determining reproducible criteria and the problem of interobserver variability [20,21].

In recent years, it has been learned that breast cancer is not a single entity, but rather a heterogeneous disease comprised of distinct biological subtypes with diverse natural history, and presents a varied spectrum of clinical, pathologic, and molecular features with different prognostic and therapeutic implications. The four breast cancer subtypes are currently defined by immunohistochemistry (IHC) expression of oestrogen receptor (ER) or progesterone receptor (PR) and human epidermal growth factor receptor 2 (HER2) and include: ER/PR$^+$, HER2$^+$; ER/PR$^+$, HER2$^-$, ER/PR$^-$, HER2$^+$, and ER/PR, HER2$^-$. The ER/PR$^-$, HER2$^-$ (also called *triple negative*) subtype of breast cancer is clinically characterised as more aggressive and less responsive to standard treatment and associated with poorer overall patient prognosis [22]. It should be possible to differentiate between these subtypes of breast cancers with the help of spectroscopy, as they have different prognostic implications.

For most cancers, the stage of cancer, or the extent of disease process, carries prognostic implications, where spectroscopy can also play a role as discussed in the following section.

Determination of Extent of Disease Process

The stage of a cancer is a description (usually numbers I to IV depending on degree of progression) of the extent to which the cancer has spread. The stage often takes into account the size of a tumour (T), how deeply it has penetrated, whether it has invaded adjacent organs, the number of lymph nodes to which it has metastasized (N, if any), and whether it has spread to distant organs (M). Staging of cancer is depicted as TNM and is the most important predictor of survival, and cancer treatment is primarily determined by staging.

Cancer staging can be divided into a clinical stage and a pathologic stage. In the TNM (tumour, node, and metastasis) system, clinical stage and pathologic stage are denoted by a lowercase "c" or "p" before the stage (e.g., cT3N1M0 or pT2N0).

Clinical stage is based on all of the available information obtained before surgery to remove the tumour. Thus, it may include information about the tumour obtained by physical examination, radiologic examination (x-rays, CT scans, MRI scans, PET-CT scans, etc.), and endoscopy. Pathologic stage adds additional information gained by examination of the tumour microscopically by a pathologist.

Because they use different criteria, the clinical stage and pathologic stage often differ. Pathologic staging is usually considered the "better" stage because it allows direct examination of the tumour and its spread, contrasted with clinical staging, which is limited by the fact that the information is obtained by making indirect observations of a tumour that is still in the body. However, clinical staging and pathologic staging should complement each other. Not every tumour is treated surgically; therefore, pathologic staging is not always available.

Spectroscopy can be of immense value in providing additional information to help with staging, in addition to radiological imaging, particularly in cases where obtaining tissue biopsy is difficult, as the management is different depending on the stage. Possible examples in which spectroscopy could help include determination of mediastinal lymph node involvement in cases of lung cancers, and differentiation of nonspecific lung nodules, liver lesions, brain lesions, and pelvic masses, to determine if these are benign or malignant (primary or metastatic) to help guide management.

Designing Therapy and Guiding Tissue Biopsy

Tissue biopsies are usually obtained from lesions after localisation by radio-logical imaging. However, as we have already pointed out, imaging cannot reliably distinguish among normal, benign, and malignant tissue areas, and hence, some tissue biopsies are not diagnostic due to sampling error, requiring a second attempt at biopsy. This obviously results in patient discomfort, distress, and anxiety, as well as a delay in diagnosis and initiation of treatment. The diagnostic pathway can be shortened by precise localisation of the disease process including malignancy and better yield of tissue biopsy, where spectroscopy can help.

Evaluating Efficacy and Response to Treatment

In a significant number of conditions, including cancers, it is desirable to assess response to treatment, which can be a prognostic factor in itself. In some cancers, for example, oesophageal, breast, lung, rectal, ovarian, and bladder cancers, chemotherapy and/or other treatments are administered prior to surgery for various reasons. The response is determined radio-logically, which may not be entirely accurate as it is difficult to distinguish between viable residual cancer and tissue necrosis as a result of treatment response. This differentiation can help in accurate planning of subsequent surgical treatment.

Earlier Detection of Recurrence and Metastases

In adults with multiple brain lesions, the primary differential diagnosis is that of metastases. In the presence of a single lesion, differentiating between primary and secondary brain tumours is important but not often possible. Unfortunately, proton MR spectroscopic findings are also nonspecific in this situation [11,23].

It should be possible for vibrational spectroscopy to detect not only the difference between the primary and secondary brain tumours, due to its precise nature, but also the tumour recurrence before MR images become abnormal.

Spectroscopy should be similarly able to distinguish among lung, liver, bone, peritoneal, adrenal, and other metastases from various primary cancers with good precision, which would be particularly useful in cases where these lesions are deep-seated and not easily amenable to biopsy.

Detection of Tissue Injury

Similar to proton MR spectroscopy, vibrational spectroscopy may also be promising as a tool for the detection of radiation injury before it becomes evident by MR imaging. Radiation necrosis may be indistinguishable from residual and recurrent tumours by CT, MRI, SPECT single-photon emission computed tomography (SPECT), and even PET-CT imaging [19,24]. Tissue necrosis due to radiation is likely to result in cellular breakdown, probably consisting of free fatty acids, lactate, and amino acids.

Detection of Therapeutic Drug Levels

Due to precise determination of chemical structural changes by spectroscopy, it should be possible to detect the concentration of various drugs in blood or serum. This could, in turn, help with studies of drug pharmacokinetics and pharmacodynamics.

In summary, we believe that vibrational spectroscopy has immense potential to help at various stages of different disease processes, in particular, with cancer diagnosis, staging, and treatment design. This will result in improved patient care pathways. A wealth of data now exists, describing chemical structural changes associated with various human disease conditions, including cancers. There still is the unmet need for creation of a single comprehensive and standardised database of these chemical change spectra for future applications and reference.

References

1. Bloom, H. J., and Richardson, W. W. 1957. Histological grading and prognosis in breast cancer: A study of 1409 cases of which 359 have been followed for 15 years. *Br. J. Cancer.* 11, 359–377.
2. Elston, C. W., and Ellis, I. O. 1991. Pathologic prognostic factors in breast cancer. I. The value of histological grades in breast cancer: Experience from a large study with long-term follow-up. *Histopathology* 19, 403–410.
3. Rehman, S., Movasaghi, Z., Tucker, A. T., and Rehman, I. U. 2007. Raman spectroscopic analysis of breast cancer tissues: Identifying differences between normal, invasive ductal carcinoma and ductal carcinoma in situ of the breast tissue. *J. Raman Spectroscopy* 38, 1345–1351.

4. Rehman, S., Movasaghi, Z., Darr, J. A., and Rehman, I. U. 2010. Fourier transform infrared spectroscopic analysis of breast cancer tissues: Identifying differences between normal breast, invasive ductal carcinoma and ductal carcinoma in situ of the breast. *Appl. Spectrosc. Rev.* 45, 355–368.

5. Luyten, P. R., Marien, A. J. H., Heindel, W., Gerwen, P. H. V., Herholz, K., Hollander, J. A. D., Friedmann, G., and Heiss, W. D. 1990. Metabolic imaging of patients with intracranial tumors: H-1 MR spectroscopic imaging and PET. *Radiology* 176, 791–799.

6. Bruhn, H., Frahm, J., Gyngell, M. L., Merboldt, K. D., Hanicke, W., Sauter, R. and Hamburger, C. 1989. Noninvasive differentiation of tumors with use of localized H-1 MR spectroscopy in vivo: Initial experience in patients with cerebral tumors. *Radiology* 172, 541–548.

7. Gill, S. S., Thomas, D. G., Van Bruggen, N., Gadian, D. G., Peden, C. J., Bell J. D., Cox, I. J., Menon, D. K., Iles, R. A., and Bryant, D. J. 1990. Proton MR spectroscopy of intracranial tumors: In vivo and in vitro studies. *J. Comput. Assist. Tomogr.* 14, 497–504.

8. Fulham, M. J., Bizzi, A., Dietz, M. J., Shih, H. H., Raman, R., Sobering, G. S., Frank, J. A., Dwyer, A. J., Alger J. R., and Di Chiro, G. 1992. Mapping of brain tumor metabolites with proton MR spectroscopic imaging: Clinical relevance. *Radiology* 185, 675–686.

9. Demaerel, P., Johannik, K., Van Hecke, P., Van Ongeval, C., Verellen, S., Marchal, G., Wilms, G., Plets, C., Goffin, J., and Van Calenbergh, F. 1991. Localized H-1 NMR spectroscopy in fifty cases of newly diagnosed intracranial tumors. *J. Comput. Assist. Tomogr.* 15, 67–76.

10. Miller, B. L. 1991. A review of chemical issues in 1H NMR spectroscopy: N-acetyl-L-aspartate, creatine, and choline. *NMR Biomed.* 4, 47–52.

11. Kugel, H., Heindel, W., Ernestus, R. I., Bunke, J., du Mesnil, R. and Friedmann, G. 1992. Human brain tumors: Spectral patterns detected with localized H-1 MR spectroscopy. *Radiology* 183, 701–709.

12. Bruhn, H., Frahm, J., Gyngell, M. L., Merboldt, K. D., Hanicke, W., Sauter, R., and Hamburger, C. 1989. Noninvasive differentiation of tumors with use of localized H-1 MR spectroscopy in vivo: Initial experience in patients with cerebral tumors. *Radiology* 172, 541–548.

13. Shonk, T. K., Moats, R. A., Gifford, P., Michaelis, T., Mandigo, J. C., Izumi, J., and Ross, B. D. 1995. Probable Alzheimer disease: Diagnosis with proton MR spectroscopy. *Radiology* 195, 65–72.

14. Miller, B. L., Moats, R. A., Shonk, T., Ernst, T., Woolley, S., and Ross, B. D. 1993. Alzheimer disease: Depiction of increased cerebral myo-inositol with proton MR spectroscopy. *Radiology* 187, 433–437.

15. Belshaw, N. J., Elliott, G. O., Foxall, R. J., Dainty, J. R., Pal, N., Coupe, A., Garg, D., Bradburn, D. M., Mathers, J. C., and Johnson, I. T. 2008. Profiling CpG island field methylation in both morphologically normal and neoplastic human colonic mucosa. *Br. J. Cancer*. 99, 136–142.

16. Miller, W. T. 1990. Value of clinical history. *AJR Am. J. Roentgenol.* 155, 653–654.

17. Swensen, S. J., Brown, L. R., and Colby, T. V. 1996. Lung nodule enhancement at CT: Prospective findings. *Radiology* 201, 447–455.

18. Hellwig, D., Baum, R. P., and Kirsch, C. 2009. FDG-PET, PET/CT and conventional nuclear medicine procedures in the evaluation of lung cancer: A systematic review. *Nuklearmedizin* 48, 59–69.

19. Griffeth, L. K. 2005. Use of PET/CT scanning in cancer patients: Technical and practical considerations. *Proc. Bayl. Univ. Med. Cent.* 18, 321–330.

20. Robbins, P., Pinder, S., de Klerk, N., Dawkins, H., Harvey, J., Sterrett, G., Ellis, I., and Elston, C. 1995. Histological grading of breast carcinomas: A study of interobserver agreement, *Hum. Pathol.* 26, 873–879.

21. Ellis, I. O., Coleman, D., Wells, C., Kodikara, S., Paish, E. M., Moss, S., Al-Sam, S., Anderson, N., Bobrow, L., Buley, I., Connolly, C. E., Dallimore, N. S., Hales, S., Hanby, A., Humphreys, S., Knox, F., Lowe, J., Macartney, J., Nash, R., Parham, D., Patnick, J., Pinder, S. E., Quinn, C. M., Robertson, A. J., Shrimankar, J., Walker, R. A., and Winder, R. 2006. Impact of a national external quality assessment scheme for breast pathology in the UK. *J. Clin. Pathol.* 59, 138–145.

22. Onitilo, A. A., Engel, J. M., Greenlee, R. T., and Mukesh B. N. 2009. Breast cancer subtypes based on ER/PR and HER2 expression: Comparison of clinicopathologic features and survival. *Clin. Med. Res.* 7, 4–13.

23. Ott, D., Hennig, J., and Ernst, T. 1993. Human brain tumors: Assessment with in vivo proton MR spectroscopy. *Radiology* 186, 745–752.

24. Gober, J. R. 1993. Noninvasive tissue characterization of brain tumor and radiation therapy using magnetic resonance spectroscopy. *Neuroimaging Clin. North Am.* 3, 779–802.

Index

Printed and bound by CPI Group (UK) Ltd, Croydon, CR0 4YY

01/11/2024

01782637-0004